SUPPLY CHAIN STRUCTURES

INTERNATIONAL SERIES IN
OPERATIONS RESEARCH & MANAGEMENT SCIENCE
Frederick S. Hillier, Series Editor
Stanford University

Saigal, R. / *LINEAR PROGRAMMING: A Modern Integrated Analysis*

Nagurney, A. & Zhang, D. / *PROJECTED DYNAMICAL SYSTEMS AND VARIATIONAL INEQUALITIES WITH APPLICATIONS*

Padberg, M. & Rijal, M. / *LOCATION, SCHEDULING, DESIGN AND INTEGER PROGRAMMING*

Vanderbei, R. / *LINEAR PROGRAMMING: Foundations and Extensions*

Jaiswal, N. K. / *MILITARY OPERATIONS RESEARCH: Quantitative Decision Making*

Gal, T. & Greenberg, H. / *ADVANCES IN SENSITIVITY ANALYSIS AND PARAMETRIC PROGRAMMING*

Prabhu, N. U. / *FOUNDATIONS OF QUEUEING THEORY*

Fang, S.-C., Rajasekera, J. R. & Tsao, H.-S. J. / *ENTROPY OPTIMIZATION AND MATHEMATICAL PROGRAMMING*

Yu, G. / *OPERATIONS RESEARCH IN THE AIRLINE INDUSTRY*

Ho, T.-H. & Tang, C. S. / *PRODUCT VARIETY MANAGEMENT*

El-Taha, M. & Stidham, S. / *SAMPLE-PATH ANALYSIS OF QUEUEING SYSTEMS*

Miettinen, K. M. / *NONLINEAR MULTIOBJECTIVE OPTIMIZATION*

Chao, H. & Huntington, H. G. / *DESIGNING COMPETITIVE ELECTRICITY MARKETS*

Weglarz, J. / *PROJECT SCHEDULING: Recent Models, Algorithms & Applications*

Sahin, I. & Polatoglu, H. / *QUALITY, WARRANTY AND PREVENTIVE MAINTENANCE*

Tavares, L. V. / *ADVANCED MODELS FOR PROJECT MANAGEMENT*

Tayur, S., Ganeshan, R. & Magazine, M. / *QUANTITATIVE MODELING FOR SUPPLY CHAIN MANAGEMENT*

Weyant, J. / *ENERGY AND ENVIRONMENTAL POLICY MODELING*

Shanthikumar, J. G. & Sumita, U. / *APPLIED PROBABILITY AND STOCHASTIC PROCESSES*

Liu, B. & Esogbue, A. O. / *DECISION CRITERIA AND OPTIMAL INVENTORY PROCESSES*

Gal, T., Stewart, T. J., Hanne, T. / *MULTICRITERIA DECISION MAKING: Advances in MCDM Models, Algorithms, Theory, and Applications*

Fox, B. L. / *STRATEGIES FOR QUASI-MONTE CARLO*

Hall, R. W. / *HANDBOOK OF TRANSPORTATION SCIENCE*

Grassman, W. K. / *COMPUTATIONAL PROBABILITY*

Pomerol, J.-C. & Barba-Romero, S. / *MULTICRITERION DECISION IN MANAGEMENT*

Axsäter, S. / *INVENTORY CONTROL*

Wolkowicz, H., Saigal, R., Vandenberghe, L. / *HANDBOOK OF SEMI-DEFINITE PROGRAMMING: Theory, Algorithms, and Applications*

Hobbs, B. F. & Meier, P. / *ENERGY DECISIONS AND THE ENVIRONMENT: A Guide to the Use of Multicriteria Methods*

Dar-El, E. / *HUMAN LEARNING: From Learning Curves to Learning Organizations*

Armstrong, J. S. / *PRINCIPLES OF FORECASTING: A Handbook for Researchers and Practitioners*

Balsamo, S., Personé, V., Onvural, R. / *ANALYSIS OF QUEUEING NETWORKS WITH BLOCKING*

Bouyssou, D. et al / *EVALUATION AND DECISION MODELS: A Critical Perspective*

Hanne, T. / *INTELLIGENT STRATEGIES FOR META MULTIPLE CRITERIA DECISION MAKING*

Saaty, T. & Vargas, L. / *MODELS, METHODS, CONCEPTS & APPLICATIONS OF THE ANALYTIC HIERARCHY PROCESS*

Chatterjee, K. & Samuelson, W. / *GAME THEORY AND BUSINESS APPLICATIONS*

Hobbs, B. et al / *THE NEXT GENERATION OF ELECTRIC POWER UNIT COMMITMENT MODELS*

Vanderbei, R. J. / *LINEAR PROGRAMMING: Foundations and Extensions, 2nd Ed.*

Kimms, A. / *MATHEMATICAL PROGRAMMING AND FINANCIAL OBJECTIVES FOR SCHEDULING PROJECTS*

Baptiste, P., Le Pape, C. & Nuijten, W. / *CONSTRAINT-BASED SCHEDULING*

Feinberg, E. & Shwartz, A. / *HANDBOOK OF MARKOV DECISION PROCESSES: Methods and Applications*

Ramík, J. & Vlach, M. / *GENERALIZED CONCAVITY IN FUZZY OPTIMIZATION AND DECISION ANALYSIS*

SUPPLY CHAIN STRUCTURES
Coordination, Information and Optimization

Edited by
JING-SHENG SONG
University of California, Irvine

DAVID D. YAO
Columbia University

Kluwer Academic Publishers
Boston/Dordrecht/London

Distributors for North, Central and South America:
Kluwer Academic Publishers
101 Philip Drive
Assinippi Park
Norwell, Massachusetts 02061 USA
Telephone (781) 871-6600
Fax (781) 871-6528
E-Mail <kluwer@wkap.com>

Distributors for all other countries:
Kluwer Academic Publishers Group
Distribution Centre
Post Office Box 322
3300 AH Dordrecht, THE NETHERLANDS
Telephone 31 78 6392 392
Fax 31 78 6546 474
E-Mail <orderdept@wkap.nl>

 Electronic Services <http://www.wkap.nl>

Library of Congress Cataloging-in-Publication Data

Supply chain structures : coordination, information and optimization / edited by
Jing-Sheng Song, David D. Yao
    p. cm. -- (International series in operations research & management science ; 42)
  Includes bibliographical references and index.
  ISBN 0-7923-7534-3 (alk. paper)
    1. Business logistics. 2. Materials management. I. Song, Jing-Sheng. II. Yao,
David D., 1950- III. Series.

HD38.5 .S8962 2001
658.5--dc21

                                                          2001038773

Printed on acid-free paper.

Printed in the United States of America

Foreword

For managers and students of supply chains, the last decade has been exhilarating and exhausting. For several decades before that, the field was calm, orderly, perhaps a bit dull. The structures, methods, and scope of business transactions were fairly stable. But now we are in the midst of three distinct but related revolutions:

- in communication technologies, highlighted by the explosive growth of the internet;

- in political structures, including the fall of the iron curtain and the rise of stable economies in Latin America;

- and in the emergence of a global managerial culture, with a more-or-less common understanding of such basic elements as financial performance, quality and customer service.

These changes have already enabled vast shifts in the micro-level methods of transactions and the macro-level structures of supply chains in many industries, and it is a fair guess that even greater shifts will come soon. Companies that understand and exploit these developments will prosper, while those that don't will suffer.

In this book, editors Jeannette Song and David Yao have collected a spectrum of approaches to these challenges from some of the leading scholars of supply chains, from both the academic and commercial worlds. Each of the articles offers an interesting and illuminating way to think about the key issues in supply-chain management. Some also offer practical techniques to solve important problems. Together they provide an excellent survey of the current state of the art in research and practice.

Paul Zipkin
Duke University, Durham, NC, USA

Contents

Chapter 1

INTRODUCTION AND OVERVIEW

Jing-Sheng Song
Graduate School of Management
University of California, Irvine, CA 92697
jssong@uci.edu

David D. Yao
IEOR Department, 302 Mudd Building
Columbia University, New York, NY 10027
yao@ieor.columbia.edu

The structure of a supply chain is fundamentally a reflection of a firm's business model. For example, in the personal computer (PC) industry, the supply chain often takes a hybrid form: the components — processor, memory, hard disk, monitors, and other peripherals — are built to stock, whereas the end product, the "box", is assembled to order. In such a system, the time it takes to assemble the end product is quite negligible (provided all the components are available), while the production/procurement leadtime for each component is more substantial. Hence, by keeping inventory at the component level, customer orders can be filled quickly. On the other hand, postponing the final assembly until order arrival provides a high level of flexibility, in terms of both product variety and risk pooling. Indeed, this hybrid of make-to-stock and assemble-to-order appears to be an ideal business model in providing both mass customization and quick response to order fulfillment.

To deploy this business model, however, a careful design is needed to coordinate the make-to-stock and the assemble-to-order segments of the supply chain. Furthermore, managing the coordination between the PC producer and the suppliers of the components that go into the PC is another important issue. Should the PC producer go for dual/multiple sourcing, or to seek strategic partnership with one particular supplier?

What is the right kind of incentive schemes that will result in a win-win collaboration between partners? What is the role of information sharing in constructing supply contracts, so that an efficient flow of information will enhance the distribution of materials?

The usual inventory-service tradeoff, a key factor in supply chain management, becomes even more prominent in the above example, as each customer order typically involves a number of components, and the stockout of any component will cause a delay in supplying the order. Hence, it is critically important to characterize the order fill rate and the necessary inventory investment. The analysis, however, is made difficult as the inventory dynamics among the components are highly correlated, driven by a common demand stream. Another challenge is that for firms to move from a traditional make-to-stock operation to a hybrid make-to-stock and assemble-to-order model, it is essential to decide the right level of component commonality and to resolve related design issues such as modularization.

The objective of this volume is to bring forth an overview of some of the recent research on supply chain structures, focusing on modeling, performance analysis and optimization, and issues in information dynamics, incentives, and supply contracting.

The remaining chapters in the volume are organized roughly into four topical sections, overviewed below. These chapters, we believe, can serve as supplementary readings or references for graduate courses in production planning, operations management, logistics, operations research, and industrial engineering. Other researchers and practitioners may also find the volume a useful technical reference in supply chain management.

1.   Structures, Flexibility, and Coordination

In the personal computer (PC) industry, original equipment manufacturers (OEMs) and their partners in the distribution channel have been actively working on removing redundant inventory in supply chains while providing a high level of customer service. Whereas successes and failures of different channel structures have been frequently reported in trade journals, formal models are needed so as to quantify the impact of different channel structures on operational performance. In Chapter 2, Lingxiu Dong and Hau Lee compare four distribution-channel structures currently used in the PC industry. They investigate the cost drivers of each structure, the condition under which a particular structure leads to the minimum cost in the supply chain, and the differential impacts on the manufacturer and the distributor. They also explore the postponement boundary in channel assembly, and demonstrate the operational complexity associated with such industry initiative.

Enabled through new technologies and distribution channels, the automotive industry is rapidly changing in response to heightened customer expectations. These changes in the marketplace are forcing automobile manufacturers, whose production has traditionally been driven by demand forecasts, to transform their production into make-to-order systems. This transformation necessitates a new manufacturing strategy since demand variability can no longer be hedged against with finished goods inventory. One strategy to address demand variability in this make-to-order environment is to invest in flexible manufacturing capacity. These investments have to be made long before production starts and have a large impact on the performance of the supply chain. In Chapter 3, Stephan Biller, Ebru Bish and Ana Muriel examine the magnitude of this impact when designing a multi-plant multi-product manufacturing system. They illustrate how the cost and benefit of manufacturing flexibility can be traded off at the strategic planning (investment) stage.

It is widely recognized that conflicting interests among the various parties in a supply chain can cause inefficiency. In the literature, substantial efforts have been devoted to analyzing coordination mechanisms in supply chains in which multiple decision makers pursue their individual objectives. One common assumption is a monolithic preference structure within each organization. Chapter 4, by Narendra Agrawal and Andy Tsay, departs from this assumption by considering a manufacturer-retailer supply chain in which the operational-level decisions at the retailer are made by a subordinate, the manager. While the manufacturer and the owner of the retail firm seek to maximize their individual expected profits, the retail manager seeks to maximize the likelihood of achieving a profit target set by the owner. The aim of this research is to study how the incongruence of goals impacts the behavior of the decision makers in both organizations, and the welfare of end customers. The analysis sheds light to the preference of all parties, in terms of wholesale and retail prices, and inventory levels, and suggests certain key supply chain strategies and their implications.

2. Value of Information

The advancement of information technology, especially the Internet, has made it much cheaper and faster for companies to obtain and exchange information. The availability of various kinds of information is expected to have a positive impact on the effectiveness of supply chain management. This is understood conceptually and qualitatively. However, questions still remain: a) What kind of information is most valuable to acquire or to share? b) How should the information be used to improve management? c) What is the value of information sharing?

Chapter 5, by Guillermo Gallego and Ozalp Ozer, surveys some of the models recently developed by the authors addressing these issues. The focus is on the optimal use of demand information on inventory replenishment decisions. Both the current and advance demand information are considered.

In Chapter 6, Lode Li and Hongtao Zhang examine the incentives for firms to share information vertically in the presence of horizontal competition. They consider a two-level supply chain consisting of one upstream firm, the manufacturer, and multiple downstream firms, the retailers. The retailers are endowed with some private information. Issues studied include how vertical information sharing will impact the competition among retailers, whether the confidentiality of shared information should be a concern, and when information sharing can be achieved on a voluntary basis. Key insights from the authors' recent research are overviewed.

3. Optimization, with Industrial Applications

Supply chain optimization in industrial applications presents great challenges. Each specific supply chain has its own unique and complex features, which require creative modeling and solution techniques. The three articles in this part report some recent successes.

In Chapter 7, Nico Vandaele and Marc Lambrecht study a problem faced by a division of Dana Corporation, which produces key parts for off-highway vehicles. Final products are assembled to order, but the components are manufactured in house. Due to the large variety of components and the complexity of the production process, the production leadtimes are long, resulting in mismatches at the time of the actual customer commitment. The authors present an integrated capacity planning and scheduling system that leads to substantial operational improvement, such as reducing leadtimes by a factor of 2 to 3. A queueing network model is used to analyze layout changes, product-mix and lot-sizing decisions, and to evaluate response times. The output of the model is then used in an optimization scheme to determine optimal sequencing and scheduling decisions.

Chapter 8, by Dirk Beyer and Julie Ward, describes an inventory management project conducted at Hewlett-Packard Laboratories for HP's Network Server Division. The division manufactures a major subassembly of network servers in Singapore and ships it to four distribution centers (DCs) worldwide, where the assembly is completed to customer specifications. Two modes of transportation, air and ocean shipments, are available between the factory and DCs, and customer demand at the

DCs is stochastic and non-stationary. The project goal is to compute inventory policies that reduce inventory- and shipment-related costs for this subassembly by enabling more cost-effective usage of ocean shipments, all without compromising order fulfillment. The authors discuss the challenges associated with this project in the context of recent trends in supply chain management, along with the implementation and related qualitative results.

In Chapter 9, Alexander Brown, Markus Ettl, Grace Lin, Raja Petrakian and David Yao present a multi-echelon, multi-item inventory allocation model that aims at lowering inventory and increasing serviceability at Xilinx, a firm that produces programmable logic integrated circuits (ICs) for sale to original equipment manufacturers. The paper focuses on a postponement strategy, by which inventory is held at an intermediate point in a generic, non-differentiated form and is only differentiated when demand is better known. To take full advantage of this strategy, one must determine the inventory levels in the intermediate (generic) and finished goods stocking points that allow for the best service at the lowest cost. The problem is formulated as a constrained nonlinear program, so as to find the safety stock targets that optimize a system-wide service measure subject to constraints on allowable inventory holding costs by echelons. An efficient algorithm is developed to find the optimal solution, based on an analytical approximation of the objective function. Insights are discussed on how inventory should be allocated between echelons and stocking points. Implementation results are also presented, based on a Xilinx data set.

4. Inventory-Service Tradeoff under Demand Correlation

One key issue in supply chain optimization is how to deal with complications caused by demand correlation. For instance, in the assemble-to-order (ATO) system mentioned earlier, demand for a certain set of components originates from a common order for the end-product that uses this set of components. This makes it very difficult to evaluate any product-based performance measures, which depend on the joint distribution of the component inventories. The three articles in this part present some recently developed tools to analyze ATO systems, focusing on the inventory-service tradeoff.

In Chapter 10, Yashan Wang applies some asymptotic techniques to analyze the relationship among capacity, inventory, delivery lead-time, and service level in an assemble-to-order system. The production capacity for each component is finite. It is shown that at high fill rates there is an approximately linear relationship between inventory and delivery leadtime, with parameters depending on the production capacity. Based on this relationship, a closed-form formula is developed for near-optimal component base-stock levels that minimize inventory-holding cost while satisfying a fill rate requirement. The effectiveness of this closed-form solution is illustrated through several numerical examples.

Using stochastic comparison techniques, in Chapter 11 Susan Xu presents a general framework to study dependence properties of several ATO systems. These properties reveal how and in what sense product-based and system-based performance measures in an ATO system are affected by demand correlation. It is shown that a performance measure can be affected, favorably or adversely, by different types of dependencies. The analysis also suggest effective means to compute bounds for several key performance measures. Numerical examples show that these bounds can be used as approximations with good accuracy in a wide range of parameter settings.

Chapter 12, by John Mamer and Stephen Smith, studies a problem related to the ATO system. It focuses on decisions on the starting inventory for multi-item systems that face sequences of demands. The problem arises in the application of repair-kit planning and mission provisioning for space and ocean voyages. There are constraints on the aggregate value, weight, and volume of items placed in inventory, the objective is to maximize either the expected number of demands that can be met or the probability that a sequence of demands of a given length can be filled. Here, too, the difficulty is that each demand requires a different (and possibly overlapping) set of items, and the precise contents of future demands are not known at the time stock levels are to be set. Stocking decisions based on the marginal demand rates for each item can be quite misleading. The authors suggest some performance bounds, which have simple forms, are easy to use in optimization models.

Acknowledgments

Our research and professional work in supply chain modeling and optimization have been supported by our universities, UC Irvine and Columbia, by NSF grants DMI-0084922 and DMI-0085124, and by IBM Research Division (DDY). We are very grateful to Makus Ettl for his generous assistance in processing the latex files and figures. We thank Yindong Lu, Seungjin Whang, Kevin Zhu, and Paul Zipkin, as well as many contributors to this volume who also served as reviewers, for their very helpful comments and critique on the chapters.

Chapter 2

EFFICIENT SUPPLY CHAIN STRUCTURES FOR PERSONAL COMPUTERS

Lingxiu Dong
John M. Olin School of Business
Washington University
St. Louis, MO 63130-4899
dong@olin.wustl.edu

Hau L. Lee
Graduate School of Business
Stanford University
Stanford, CA 94305
haulee@stanford.edu

1.   Introduction

As computing technologies are revolutionizing business efficiency and productivity and improving quality of life, the personal computer (PC) industry that supports such changes has been struggling through the difficult process of improving its supply chain efficiency.

Traditional PC distribution is similar to that in many other industries: original equipment manufacturers (OEMs) or their subcontractors acquire components from suppliers and assemble PCs to demand forecasts. These completely assembled PCs are stocked at the OEMs' distribution centers, and then shipped to distributors' warehouses. Distributors, who mainly make their profit on quantity discounts for huge volumes, usually place large orders to OEMs. Resellers/Value added resellers (VARs), who have direct contact with end customers, usually order small numbers of PCs from distributors and configure them to customer requirements. In such a distribution channel, inventory must be stocked at every tier of the supply chain to guarantee high service levels to customers.

As computing and telecommunication technologies advanced and started to play a more important role in business operations and people's daily lives, the PC market evolved and more sophisticated and efficient distribution is needed. Today's PCs have two important characteristics: every household can use a computer for communication, web access, education, and entertainment; and different individuals, businesses, schools, and organizations need different hardware, software, and network configurations. These characteristics have led to the rapidly increasing market and the fragmentation of the marketplace with respect to customer requirements, preferences, and knowledge. Traditional distribution channels have not kept up with the rapid changes in the PC marketplace. The huge pipeline inventory in channels makes it difficult for OEMs to offer enough product variety to meet diversified customer needs. This provides opportunities to many resellers/VARs to build "white boxes" (no-brand computers) which can be configured to customer orders. Today, white boxes comprise 35% of the PC market. The heavy channel inventory burden can be a major liability to OEMs due to the shorter PC product life cycles. New generations of components are being rolled out at an accelerating speed, forcing OEMs to drop PC prices along the product life cycles. Prices for memory and chips are dropping at roughly one to two percent a week, so that an assembled computer loses six percent of its value every one to two weeks (Keck et al. [11]). OEMs must either offer generous price protection to push channel inventory towards the market or warrant returns to take back obsolete products and move new products to the market. Both strategies lead to a reduced profit margin.

Faced with these challenges, it is difficult for channel-based OEMs to rely on the traditional make-to-stock manufacturing and the order-to-stock distribution strategies to compete with the direct sales rivals who build to customer orders. Founded in the mid-1980's, Dell and Gateway recognized and seized the opportunity to sell computers more efficiently and cost effectively by using a completely different supply chain structure. Dell's supply chain mainly consists of Dell's assembly plants with component suppliers' supply hubs located next to them. This structure enables Dell to assemble to meet customer orders in a just-in-time fashion. On the distribution side, Dell takes orders directly from customers and ships products directly to them. Using build-to-order manufacturing and direct-sales strategies, Dell is able to eliminate component and finished goods inventory and bypass the channel markup, and thus offer competitive prices to customers. Such low pricing and fast delivery has empowered direct vendors to take market share quickly from indirect OEMs. Today, Dell has become a top 3 PC vendor worldwide.

The success of the direct sales model has threatened the survival of PC distributors and also sent a clear message to the PC industry: the value of a product lies not only in its technology and quality, but also in the supply chain through which the product is delivered. OEMs have become aware of the benefit of the direct-sales model and have also set up on-line stores to sell some products directly to customers. However, they have also been sensitive to the potential conflict between the new on-line direct and the existing channel distribution. Indeed, indirect channels can still play an important role in today's diversified customer base. Only a subset of customers, especially those self-sufficient with technology, prefer to have direct relationships with vendors. Others still need the value provided by channels: product selection, evaluation, networking installation, training, and after-sales service, etc. Hence, indirect-sales models will co-exist with the direct-sales model, but they will have to be much more efficient. Since the mid-1990's, indirect vendors and their channel partners have experimented with different supply chain structures so as to bring the logistics cost down to a level comparable with the direct model. In today's PC industry, four distribution strategies predominate:

Traditional manufacturer make-to-stock (MTS): The manufacturer procures components to stock, assembles them into different end products to stock, and then ships end products to the distributor according to orders. The distributor stocks the end-product inventory in warehouses to fill customer orders.

Manufacturer assemble-to-distributor-order (ATO): To eliminate the possibility of stocking unwanted end products, the manufacturer does not stock end-product inventory, but only assembles to distributor orders, and then ships end products to the distributor. The distributor still stocks end-product inventory for customer orders.

Channel assembly (CA): The manufacturer stocks only core components needed for each end product and builds them into "vanilla boxes," leaving final assembly to the distributor. The distributor procures vanilla boxes from the manufacturer and other components from manufacturer-authorized suppliers, and *assembles* to customer orders.

Sales agent (SA): The distributor does not stock end products, but rather, passes orders through to the manufacturer, who assembles to order and ships directly ("drop ships") to customers. In this model, the distributor acts like the manufacturer's sales agent who helps customers decide what type of product best suits their needs in terms of price and

technical efficiency, and offers after-sales service such as network setup
and asset management.

In the process of pursuing such structural change, both successes and
failures have been observed. No single strategy has been proven to
be dominant and critical issues remain unanswered. Are these channel structures remedies for all types of supply chains? What makes a
structure suitable for a given supply chain? How do members of a supply chain share the risk and benefit associated with such channel structures? Aiming at these questions, we start by comparing three existing
alternative channel structures — manufacturer assemble-to-distributor-
order, channel assembly, sales agent — with the traditional manufacturer
make-to-stock structure. This comparison, which is a first attempt to
quantitatively study the trade-offs in channel structural change, will be
based on two key measures: inventory cost and response time to customers. Our study shows that changing channel structures often times
leads to differential impacts on participating parties: a structure that
benefits the system might not lead to Pareto improvement for all participating parties. Moreover, channel structure alone does not guarantee a
reduction of the system cost: system parameters such as leadtimes and
product characteristics, and operational decisions could all be decisive
factors in its success. These findings partially explain the on-going de-
bates and confusion in choosing the right channel structures in the PC
industry, and also shed light on how to design new channel structures.

Austin and Lee's [1] survey of the PC industry provides an excellent
review and analysis of the industry turmoil. They point out that, although synchronized planning and execution of the coordinated tasks of
OEMs, distributors, and resellers/VARs are important, they have not
been well exploited by industry. For example, the controversial channel assembly structure reassigns the roles of the OEM and the channel,
bringing challenges such as assembly responsibility sharing and component (subassembly) inventory management to distributors. Recent
distributors' ratings of channel assembly has declined due to conflicts
arisen from the vendors' quick steps in slashing inventory and price protection, and the relatively slow development of the channel assembly
infrastructure. Debate among channel members on channel assembly
focuses on three aspects: the content of vanilla boxes, the component inventory management at the distributor's side, and the facility investment
in channel assembly. Ignoring these problems or implementing unsound
solutions could create channel conflicts and backfire on the newly formed
structure. In the second part of this chapter, we address the postponement boundary/vanilla-box configuration in channel assembly. Through
a simple model we show that setting postponement boundary, although

an operational decision, has significant impact on the manufacturer's, distributors', and component suppliers' investment decisions, as well as coordination across supply chain.

The organization of this chapter is as follows. In section 2 we give a brief literature review on research in production-distribution systems and postponement. In section 3, we compare four channel structures: the traditional manufacturer-make-to-stock, manufacturer-build-to-distributor-order, channel assembly, and sales agent. We explore the cost drivers of each strategy, compare their cost efficiencies and customer response times, and study conditions under which a structure best suits a supply chain. In section 4, we focus on the postponement boundary issue in channel assembly and study how assembly process sequence could affect the cost efficiency of supply chain. We conclude our findings in section 5.

2.   Literature Review

Traditional studies of manufacturing systems attempt to optimize the manufacturing process by minimizing the manufacturer's component and finished goods inventory costs. Traditional multi-echelon inventory research focuses on finding the optimal inventory control policy for one particularly structured supply system. Little work on improving supply chain performance has considered the manufacturing strategy and distribution channel structure in one framework. Graves and Willems [9] study a multi-stage network in which service times are decision variables and each stage provides 100% service (demands are bounded) within the service time. They develop an optimization algorithm to minimize the inventory holding cost for a spanning-tree structured supply chain. Ettl et al. [6] build an inventory-queue model to estimate the network performance such as end-customer service level and fill rates. Cohen and Lee [4] develop an analytical procedure that captures the stochastic, dynamic interactions of a multi-stage production/distribution system. The linkages among the material control, production, finished goods inventory, and distribution sub-models are the optimal control and fill rates determined by each sub-model. By adjusting the local control of each of the sub-models, the integrated model provides a means to optimize the performance of the entire supply chain. Lee and Billington [12] include more supply uncertainty factors and develop a model of material flow control in a more complex decentralized supply network. The application of their model to industry practice not only uncover opportunities for improvement, but also suggest that structural changes rather than control parameter adjustment are sometimes needed to improve supply chain efficiency. The latter finding is confirmed by the revolutionary process the PC industry is currently going through: local optimization of existing manufacturing processes or distribution processes is limited in its ability to achieve cost efficiency; therefore more PC vendors and their channel partners are cooperating to re-engineer the entire supply chain so as to remove the redundant processes.

Channel assembly is a postponement strategy used across enterprises in supply chain. Postponement strategies have been extensively studied within one assembly facility. Lee and Tang [13] develop a general framework modeling the costs and benefits of delayed product differentiation. Based on product structures, analytical models on postponement strategies can be categorized in three types. First, one differentiation operation with fixed assembly sequence (Aviv and Federgruen [2]). Second, multiple differentiation operations with flexible assembly sequence (Garg and Tang [7]; Lee and Tang [14]; Kapuscinski and Tayur [10]). Third, for modular designed products, using vanilla boxes to achieve postponement (Swaminathan and Tayur [18]&[19]). We refer readers to Aviv and Federgruen [2] for a comprehensive review on the costs and benefits of design for postponement. The above literature focuses on postponement strategies used within *one facility* where the postponement boundary only affects the product differentiation point. The postponement boundary in channel assembly, however is different in that it also determines the components that have to be decentralize stocked at and the assembly operations that have to performed by the *multiple distributors' sites*. Motivated by a different industry initiative, Lee and Dong [15] study setting optimal postponement boundary in a similar to channel assembly environment. They assume the assembly sequence is pre-specified and show that various cost factors affect the optimal postponement boundary.

3.   Distribution Channel Structures and Supply Chain Performance

In this section, we compare the four distribution channel structures described in the previous section. Table 2.1 characterizes the impact of transition from one channel structure to another on the manufacturer's and distributor's inventory, and shows that the inventory removed at a certain place in the supply chain may be transferred to another place in the supply chain. A transition from the traditional manufacturer make-to-stock (MTS) structure to the manufacturer assemble-to-order (ATO) structure removes the manufacturer's end-product inventory buffer but increases the distributor's end-product inventory buffer

Table 2.1. Inventory Changes Associated with Transition between Channel Structures

Transition	Effect to Manufacturer	Effect to Distributor
MTS→ATO	End-product Inventory Removed	End-product Inventory Increased
ATO→CA	Component Inventory Removed	End-product Inventory Removed
CA→SA	Component Inventory Increased	Component Inventory Removed

because the distributor must wait longer for the manufacturer to fulfill the order. A transition from the manufacturer assemble-to-order (ATO) structure to the channel assembly (CA) structure removes the manufacturer's component inventory and replaces the distributor's end-product inventory with the component inventory. A transition from the channel assembly (CA) structure to the sales agent (SA) structure increases the manufacturer's component inventory but removes the distributor's component inventory.

3.1. Model

In this subsection, we describe general assumptions for all four channel structures, including the description of product structure, demand process, and other operational assumptions. We formulate cost functions for each of the four structures, assuming a single manufacturer and distributor. All notation is shown in the Appendix.

We assume that each end product is composed of m component each of which belongs to a distinct component category. Within each component category, components are classified by functional characteristics. In component category j, there are $N_j (\geq 1)$ types of configurations. For the PC example, the CPU, hard drive, and memory card are component categories and CPU's can be further classified by speed or frequency. We assume that the selection of one component configuration from each of the m component categories yields a legitimate bill of material for an end product, and hence the size of the end-product family is $n = \prod_{j=1}^{m} N_j$. We assume that the purchase cost is c_j for any component in category j; thus the costs of all end products are the same, which is $\sum_{j=1}^{m} c_j$.

We assume that demand is independent across time periods and the joint demand for end products follows a multi-Normal distribution $\mathbf{N}(\mu, \Sigma)$, where $\mu = (\mu_i)_{i=1}^{n}$ and $\Sigma = [\sigma_{i_1 i_2}]$ are the respective mean vector and covariance matrix. We assume that marginal distributions of end-product demands are identical, i.e., $\xi_i \sim N(\mu, \sigma^2)$, $i = 1, 2, \ldots, n$; demand correlation between any two end products is uniform, i.e.,

$$\sigma_{i_1 i_2} = \rho \sigma^2, i_1, i_2 = 1, \ldots, n, i_1 \neq i_2, \quad \text{and} \quad \rho \in \left(-\left(\prod_{j=1}^{m} N_j - 1 \right)^{-1}, 1 \right)$$

to ensure a positive variance for total demand. It follows immediately that the marginal distribution of demand for component configuration k in component category j also follows a Normal distribution with mean and variance:

$$\xi_{j,k} \sim N\left(\mu_j, \sigma_c^2(j)\right), \quad \mu_j = \mu \prod_{j' \neq j} N_{j'}, \sigma_c^2(j)$$

$$= \sigma^2 \left[1 + \left(\prod_{j' \neq j} N_{j'} - 1 \right) \rho \right] \prod_{j' \neq j} N_{j'},$$

$$j = 1, \ldots, m, k = 1, \ldots, N_j.$$

We begin by considering only the inventory cost associated with components and end products. Inventory contributes the major cost in each of the four channel structures. However, a complete business model should include other cost factors such as transportation cost, processing cost, and investment cost.

Given the complex product structure, we limit our study to the use of separate periodic base-stock policies for components and end products. Schmidt and Nahmias [17] study a system in which an end product is assembled from two components supplied by an external supplier, and show that the optimal assembly and order policies are approximate base-stock policies. However, the optimal system cost is too complex to be expressed explicitly. Rosling [16] shows that an arborescent assembly structure is equivalent to a serial system with certain initial inventory stocking levels, and shows that a base-stock policy is optimal using the Clark and Scarf's [3] procedure. We use the base-stock policy as a first-cut approximation of the optimal inventory control policy for the manufacturer and the distributor. It also mimics the common practice observed in industry. Since the manufacturer and the distributor are typically different entities who make replenishment decisions separately, we separate the manufacturer model from the distributor model.

For the manufacturer, we assume that the unit inventory holding cost for components in category j is $H_j^c = \beta_m^c c_j$, where β_m^c represents the percentage of cost charged to opportunity and shelf space cost for component per period. We assume an end product with m components costs $(1 + v) \sum_{j=1}^{m} c_j$ to the manufacturer where v is the percentage value added by the assembly process. The unit inventory holding cost of the end product is $H^e = \beta_m^e (1+v) \sum_{j=1}^m c_j$, where β_m^e represents the corresponding percentage of cost charged to opportunity and shelf space cost per period for the end product. Similarly, we define the component unit backorder cost as $G_j^c = \alpha_m^c c_j$, where α_m^c represents the percentage component backorder cost per period; we define the end-product unit penalty cost as $G^e = \alpha_m^e (1+v) \sum_{j=1}^m c_j$, where α_m^e represents the percentage of end-product cost charged to backorder per period. For the distributor, a similar definition applies except that the distributor purchases end product from the manufacturer at price $(1+\Delta)(1+v) \sum_{j=1}^m c_j$, where Δ is the percentage profit margin the manufacturer charges for each end product. And we use the lower case letters, h_j^c, h^e, g_j^c, and g^e to denote component and end-product inventory holding and backorder costs, respectively.

3.2.   Cost structures

3.2.1   Traditional manufacturer make-to-stock structure.

We start with the distributor's activity. (1) At the beginning of a review period, end products on-order that are due in this period arrive. (2) The distributor reviews the inventory position (on-hand stock + on-order − the backlog) for each end product and places orders to the manufacturer so that inventory positions are raised to the target levels. (3) Customer orders arrive some time later in this period, and the distributor fills these orders. (4) At the end of the period, unused inventory incurs inventory holding costs and the excess demand is backlogged and incurs backorder costs to the distributor. The manufacturer's activity is as follows. (1) At the beginning of the review period, components on-order that are due in this period arrive and products in the assembly process that are due to be completed in this period are finished and join the manufacturer's end-product stockpiles. (2) The manufacturer reviews end-product inventory positions and component inventory positions, and then decides the assembly quantity and places orders to component suppliers to raise end-product inventory positions and component inventory positions to target levels, respectively. (3) Later in the period, when the distributor's orders arrive, the manufacturer issues a shipment to the distributor; unused component and end-product inventory incur inventory holding costs and excess demand for component and end products is backlogged and incurs backorder costs to the manufacturer. To simplify the analysis, we assume that the manufacturer can always assemble the entire internal order for end products and fill the entire external order for the distributor. The validity of this assumption requires either (a) that the manufacturer has very high component and end-product service levels or (b) that the manufacturer can always instantaneously "borrow" components and end products from other sources at some cost, and the borrowed quantity is counted as the backlog to the manufacturer and will be returned as soon as the next replenishment arrives. In this chapter we assume that the manufacturer can borrow at an extra cost, which is reflected in the model as the backorder cost.

We use l to denote distributor related leadtime and use L to denote manufacturer related leadtime. Assuming a stationary base-stock system, in every period the distributor places orders to raise end-product inventory positions to target levels so as to cover the demand in $l^e + 1$ periods (current period + the following transportation leadtime, l^e periods). Similarly, the manufacturer's target levels for end-product inventory positions are set to cover the demand in $L^a + 1$ periods (current period + assembly leadtime, L^a periods) and target levels for the component inventory positions are set to cover the demand in $L^c + 1$ periods (current period + component procurement leadtime, L^c periods). Let y_i^e be the distributor's target inventory level for end product i. Throughout the chapter we use y and Y to denote the distributor target inventory level and the manufacturer's respectively; use superscript to indicate the inventory type (e for end product, c for components, and V for vanilla box) of inventory; and use subscripts i, j, and k to index end product, component category, and component configuration type, respectively. We use $\xi^{(l)}$ to denote the lth convolution of random variable ξ. The distributor's cost minimization problem as follows:

$$C_d^{MTS} = \min_{\mathbf{y}^{MTS}} \left\{ \sum_{i=1}^{n} E\left[h^e \left(y_i^e - \xi_i^{(l^e+1)} \right)^+ + g^e \left(\xi_i^{(l^e+1)} - y_i^e \right)^+ \right] \right\},$$

where

$$\mathbf{y}^{MTS} = \left(y_i^e \right)_{i=1}^{n}.$$

The assumption that the manufacturer can also borrow components from other sources enables us to decompose the manufacturer's inventory management problem into an end-product inventory management problem and a component inventory management problem. Hence the manufacturer's inventory cost per period is:

$$C_m^{MTS} = \min_{\mathbf{Y}^{MTS}} \left\{ \sum_{i=1}^{n} E\left[H^e \left(Y_i^e - \xi_i^{(L^a+1)} \right)^+ + G^e \left(\xi_i^{(L^a+1)} - Y_i^e \right)^+ \right] \right.$$

$$\left. + \sum_{j=1}^{m} \sum_{k=1}^{N_j} E\left[H_j^c \left(Y_{j,k}^c - \xi_{j,k}^{(L^c+1)} \right)^+ + G_j^c \left(\xi_{j,k}^{(L^c+1)} - Y_{j,k}^c \right)^+ \right] \right\},$$

where

$$\mathbf{Y}^{MTS} = \left(Y_i^e ; Y_{j,k}^c \right)_{i=1,\dots,n; \ j=1,\dots,m; \ k=1,\dots,N_j}.$$

3.2.2 **Manufacturer** **assemble-to-distributor-order structure.** In the manufacturer assemble-to-distributor-order structure, the manufacturer does not stock end products. Hence, the event sequence in the manufacturer assemble-to-distributor-order structure differs from that of the traditional manufacturer make-to-stock structure in two ways. First, the lead time for the distributor is longer, and equals $L^a + l^e$ (the manufacturer assembly leadtime + transportation leadtime). The cost formulation for the distributor is:

$$C_d^{ATO} = \min_{\mathbf{y}^{ATO}} \left\{ \sum_{i=1}^{n} E\left[h^e \left(y_i^e - \xi_i^{(L^a + l^e + 1)} \right)^+ + g^e \left(\xi_i^{(L^a + l^e + 1)} - y_i^e \right)^+ \right] \right\},$$

where

$$\mathbf{y}^{ATO} = \left(y_i^e \right)_{i=1}^{n}.$$

Second, the manufacturer only makes stocking decisions for components:

$$C_m^{ATO} = \min_{\mathbf{Y}^{ATO}} \left\{ \sum_{j=1}^{m} \sum_{k=1}^{N_j} E\left[H_j^c \left(Y_{j,k}^c - \xi_{j,k}^{(L^c+1)} \right)^+ + G_j^c \left(\xi_{j,k}^{(L^c+1)} - Y_{j,k}^c \right)^+ \right] \right\},$$

where

$$\mathbf{Y}^{ATO} = \left(Y_{j,k}^c \right)_{j=1,\dots,m;\ k=1,\dots,N_j}.$$

3.2.3 **Channel assembly structure.** In the channel assembly structure, the manufacturer shifts partial assembly responsibility to the distributor. Assume that the manufacturer assembles the first J components in the vanilla box and builds vanilla boxes to stock; and the distributor stocks both vanilla boxes and the other $m - J$ categories of components and builds to customer orders. We denote the demand for vanilla boxes incorporating J components as ξ_i^J. Straightforwardly, ξ_i^J follows the Normal distribution $N(\mu_V(J), \sigma_V^2(J))$, where

$$\mu_V(J) = \mu \prod_{j=J+1}^{m} N_j \quad \text{and}$$

$$\sigma_V^2(J) = \sigma^2 \left[1 + \left(\prod_{j=J+1}^{m} N_j - 1 \right) \rho \right] \prod_{j=J+1}^{m} N_j, \quad i = 1, \dots, V_J,$$

where

$$V_J = \prod_{j=1}^{J} N_j.$$

In the channel assembly structure, the distributor follows an assemble-to-order manufacturing strategy. Following the rationale in the previous section we have the distributor's cost:

$$
C_d^{CA} = \min_{\mathbf{y}^{CA}} \left\{ \sum_{i=1}^{V_J} E\left[h_J^V \left(y_i^V - \xi_i^{J(l^e+1)} \right)^+ + g_J^V \left(\xi_i^{J(l^e+1)} - y_i^V \right)^+ \right] \right.
$$

$$
\left. + \sum_{j=J+1}^{m} \sum_{k=1}^{N_j} E\left[h_j^c \left(y_{j,k}^c - \xi_{j,k}^{(l^c+1)} \right)^+ + g_j^c \left(\xi_{j,k}^{(l^c+1)} - y_{j,k}^c \right)^+ \right] \right\},
$$

where

$$
\mathbf{y}^{CA} = \left(y_i^V ; y_{j,k}^c \right)_{i=1,\dots V_J;\ j=J+1,\dots,\ m;\ k=1,\dots,N_j}.
$$

Notice that in the above formulation, the distributor's component procurement leadtime is l^c and the vanilla box leadtime is still l^e.

The manufacturer's cost formulation is similar to that of the traditional manufacturer make-to-stock structure but with a smaller set of component family and a new vanilla-box assembly leadtime L_J^V:

$$
C_m^{CA} = \min_{\mathbf{Y}^{CA}} \left\{ \sum_{i=1}^{V_J} E\left[H_J^V \left(Y_i^V - \xi_i^{J(L_J^V+1)} \right)^+ + G_J^V \left(\xi_i^{J(L_J^V+1)} - Y_i^V \right)^+ \right] \right.
$$

$$
\left. + \sum_{j=1}^{J} \sum_{k=1}^{N_j} E\left[H_j^c \left(Y_{j,k}^c - \xi_{j,k}^{(l^c+1)} \right)^+ + G_j^c \left(\xi_{j,k}^{(l^c+1)} - Y_{j,k}^c \right)^+ \right] \right\},
$$

where

$$
\mathbf{Y}^{CA} = \left(Y_i^V ; Y_{j,k}^c \right)_{i=1,\dots,V_J;\ j=1,\dots,J;\ k=1,\dots,N_j}.
$$

3.2.4   Sales agent structure.

The sales agent structure completely removes the inventory responsibility from the distributor and the manufacturer only manages component inventory:

$$
C_m^{SA} = \min_{\mathbf{Y}^{SA}} \left\{ \sum_{j=1}^{m} \sum_{k=1}^{N_j} E\left[H_j^c \left(Y_{j,k}^c - \xi_{j,k}^{(l^c+1)} \right)^+ + G_j^c \left(\xi_{j,k}^{(l^c+1)} - Y_{j,k}^c \right)^+ \right] \right\},
$$

where

$$
\mathbf{Y}^{SA} = \left(Y_{j,k}^c \right)_{j=1,\dots,m;\ k=1,\dots,N_j}.
$$

3.3.   Channel structure cost comparison

The cost functions for both the manufacturer and the distributor in all of the four channel structures share a similar newsvendor form: $C = \min_{y} E[h(y-\xi)^+ + g(\xi-y)^+]$. Given $\xi \sim N(\mu, \sigma^2)$ and $\phi(\cdot)$ and $\Phi(\cdot)$ are the PDF and the CDF of the standard Normal distribution, we have the optimal stocking level $y^* = \mu + q\sigma$, and the minimum cost $C^* = \sigma[qh + (g+h)L(q)]$, where $q = \Phi^{-1}[g/(g+h)]$, $L(q) = \int_q^\infty (\xi - q)\phi(\xi)\,d\xi$.

The value of the cost functions depends on parameters such as component and end-product inventory holding costs and backorder costs, for both the manufacturer and the distributor. For ease of comparison, we assume inventory holding and backorder costs follow the following relationships: $\beta_m^e = t\beta_m^c$, $\beta_d^e = t\beta_d^c$, $\alpha_m^e = t\alpha_m^c$, $\alpha_d^e = t\alpha_d^c$, $\beta_d^e = T\beta_m^e$, $\beta_d^c = T\beta_m^c$, $\alpha_d^e = T\alpha_m^e$, $\alpha_d^c = T\alpha_m^c$, $t \geq 1$ and $T \geq 1$. We interpret t as the ratio of opportunity costs of stocking end products to that of stocking components, and T as the ratio of opportunity costs of stocking at the distributor's site to that of stocking at the manufacturer's site. Next, we perform pairwise comparisons to get a comprehensive view of trade-offs in the channel structural change. Proofs of lemmas and theorems are given in Appendix.

3.3.1   Manufacturer make-to-stock vs. assemble-to-distributor-order.

Define the total system costs of the two structures as $C_t^{MTS} = C_d^{MTS} + C_m^{MTS}$ and $C_t^{ATO} = C_d^{ATO} + C_m^{ATO}$. We have the following results:

Theorem 2.1
(1) $C_m^{MTS} > C_m^{ATO}$ and $C_d^{MTS} < C_d^{ATO}$;
(2) $C_t^{MTS} > C_t^{ATO}$ if and only if $(\sqrt{L^a + l^e + 1} - \sqrt{l^e + 1})/\sqrt{L^a + 1} < 1/T(1+\Delta)$;
(3) $(\sqrt{L^a + l^e + 1} - \sqrt{l^e + 1})/\sqrt{L^a + 1}$ is an increasing function of L^a but a decreasing function of l^e.

Theorem 2.1(1) illustrates a typical manufacturer-distributor win-lose situation for which we have to turn to the system cost comparison for a complete investigation. Theorem 2.1(2) gives a necessary and sufficient condition that the system leadtimes and the cost parameters have to satisfy to guarantee that the total system cost is reduced under the assemble-to-order structure. The threshold, $1/T(1+\Delta)$, for the assembly and transportation leadtimes, is high if the difference of holding inventory at the distributor and the manufacturer is small or the margin that the manufacturer charges on the end product is low. Theorem 2.1(3) implies that if the assembly leadtime is long, the increase of the

finished goods inventory at the distributor cannot be offset by the reduction of the finished goods inventory at the manufacturer, when we
change from the manufacturer make-to-stock to the assemble-to-order
structure; similarly, if the transportation leadtime is long, then the net
result might be an inventory reduction to the system. Therefore, the
assemble-to-order structure might not suit a system with long assembly leadtime but might suit a system with long transportation leadtime.
The condition in Theorem 2.1 indicates the effect of reducing assembly
leadtime on the system cost. In practice, HP's guarantee to distributors
of two weeks turn-around time (delivery time) in its Economic Program
(assemble-to-distributor-order in this chapter), shows the manufacturer's
effort to reduce the total system cost besides its own costs.

Replacing the make-to-stock structure with the assemble-to
-distributor-order structure usually results in different cost savings/
increase to different supply chains. If we think of the inventory cost
difference, $C_t^{MTS} - C_t^{ATO}$, as the value that the assemble-to-distributor-
order structure brings to the system, then Theorem 2.2 shows this value
is affected by different factors.

Theorem 2.2
*(1) There exist two thresholds $\theta_1 = [T(1 + \Delta)]^2 - 1$ and $\theta_2 = \frac{l^e}{\theta_1} - 1$,
such that
if $l^e < \theta_1$ then $C_t^{MTS} - C_t^{ATO}$ decreases in L^a;
if $l^e > \theta_1$ then $C_t^{MTS} - C_t^{ATO}$ increases in $L^a \in [0, \theta_2)$ and decreases in
$L^a \in (\theta_2, \infty)$;
(2) $C_t^{MTS} - C_t^{ATO}$ increases in l^e.*

The results of Theorem 2.2 are shown in Figure 2.1 and can be interpreted in the following way. If l^e is small enough (i.e., $<\theta_1$), the
value of assemble-to-order structure is smaller when assembly leadtime,
L^a, is long. If l^e is substantive (i.e., $>\theta_1$), then the value increases as
the assembly leadtime increases within the range of $[0, \theta_2)$, but the value
starts to decrease as soon as the assembly leadtime exceeds θ_2. Also, the
longer the transportation leadtime, the greater is the value the assemble-
to-order structure brings to the system. Theorem 2.2 captures different
cost dominant situations led by the dynamic of two counteracting system factors. On the one hand, consolidating all finished goods inventory
at the distributor's site provides risk pooling to the system; on the other
hand it is more expensive to stock the inventory at the distributor's site.
The conflict between risk pooling and inventory holding cost is reflected
by the thresholds θ_1 and θ_2 (both are functions of the manufacturer's
profit margin and the difference in holding inventory at the manufacturer
and the distributor), which govern the behavior of $C_t^{MTS} - C_t^{ATO}$.

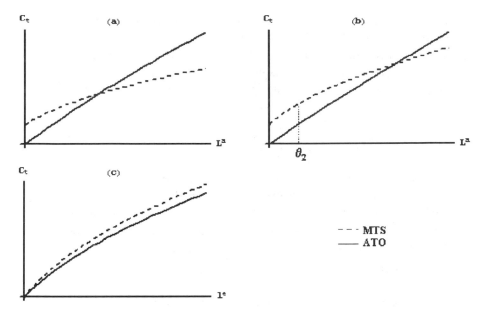

Figure 2.1. The Leadtime Effect on $C_t^{MTS} - C_t^{ATO}$: (a) $l^e < \theta_1$ (b) $l^e > \theta_1$ (c) effect of l^e

3.3.2 Manufacturer make-to-stock vs. channel assembly.

Since the manufacturer in channel assembly is only responsible for a subset of components and the vanilla boxes, he has a lower inventory cost than in the make-to-stock structure. The following Theorem 2.3 shows that, if the component transportation leadtime to the distributor is no longer than the vanilla box transportation leadtime, then channel assembly also incurs lower inventory cost to the distributor. Under that condition, channel assembly is a win-win structure for both the manufacturer and the distributor. Moreover, its value to systems is higher when assembly leadtime or transportation leadtimes are long. Theorem 2.3(4) shows that the demand uncertainty also plays a role in affecting the effectiveness of channel assembly. Since the vanilla boxes and the components are "common components" in the assembly process, whose demand uncertainties depend on the end-product demand correlation. Hence, channel assembly is more efficient when end-product demands are negatively correlated.

Theorem 2.3 *(1)* $C_m^{MTS} > C_m^{CA}$; *(2) If* $l^c \leq l^e$, *then* $C_d^{MTS} > C_d^{CA}$; *(3) if* $l^c = l^e$, *then* $C_t^{MTS} - C_t^{CA}$ *increases in* L^a *and* l^e; *(4) if* $l^c = l^e$ *and* $T \geq \sqrt{(L^a + 1)/(l^e + 1)}$, *then* $C_t^{MTS} - C_t^{CA}$ *decreases in* ρ.

3.3.3 Manufacturer assemble-to-distributor-order vs. channel assembly.

From the distributor's perspective, if the component procurement leadtime is shorter than the vanilla-box shipping leadtime, then channel assembly is always cheaper than the make-to-stock structure (Theorem 2.3(2)), which in turn is cheaper than the assemble-to-distributor-order structure.

Lemma 2.4 *If $l^c \leq l^e$, then $C_d^{ATO} > C_d^{CA}$.*

The comparison of manufacturer costs is rather complex and is closely related to the product structure, and hence depends on the content of vanilla boxes and the potential of improving vanilla-box assembly efficiency. Theorem 2.5 illustrates this point for the special case of uniform configuration options, i.e., $N_j = N, j = 1, \ldots, m$. For ease of exposition, we denote $n(J) = \sqrt{N^J + (N^m - N^J)\rho}$.

Theorem 2.5 *If $l^e = l^c = L^c$ and $N_j = N, j = 1, \ldots, m$, then*
(1) $C_m^{ATO} > C_m^{CA}$ if and only if

$$\frac{n(J)\left(\sum_{j=1}^{J} c_j\right)}{n(1)\left(\sum_{j=J+1}^{m} c_j\right)} < \frac{1}{t(1+v)} \frac{\sqrt{l^c+1}}{\sqrt{L_J^V+1}};$$

(2) $C_t^{ATO} > C_t^{CA}$ if and only if $\dfrac{n(J)\left(\sum_{j=1}^{J} c_j\right)}{n(m)\sum_{j=1}^{m} c_j} < \dfrac{\sqrt{L^a+l^e+1}(1+\Delta)}{\sqrt{l^e+1}(1+\Delta)+\sqrt{L_J^V+1}};$

(3) $\exists \theta_3$ such that $C_t^{ATO} - C_t^{CA}$ decreases in $L^a \in [0, \theta_3)$ and increases in $L^a \in (\theta_3, \infty)$, where

$$\theta_3 = \frac{(l^e+1)\left(\frac{J}{m}\iota\right)^2 - 1}{\frac{J}{m} - \left(\frac{J}{m}\iota\right)^2}, \quad \iota = \frac{1}{(1+\Delta)} \frac{n(J)\left(\sum_{j=1}^{J} c_j\right)}{n(m)\left(\sum_{j=1}^{m} c_j\right)};$$

(4) $\exists \theta_4$ such that $C_t^{ATO} - C_t^{CA}$ decreases in $l^e \in [0, \theta_4)$ and increases in $l^e \in (\theta_4, \infty)$, where $\theta_4 = \dfrac{L^a}{[(1+\Delta)\iota]^{-2}-1} - 1$;
(5) $C_t^{ATO} - C_t^{CA}$ decreases in ρ.

Theorem 2.5 shows how system leadtimes and the vanilla-box configuration interweave with each other to affect the cost efficiency of channel assembly. The value of the vanilla box should be smaller than a threshold so that the cost savings obtained at the distributor's site can offset the possible cost increase at the manufacturer's site. This insight supports the postponement of "processor plug-in" step to the channel (Swaminathan and Tayur [18] have a detailed study on vanilla-boxes in an assembly system). Dong and Lee [5] discusses the component inventory management issue in channel assembly. However, we will show some counter-intuitive findings in section 4. For systems with long assembly leadtime and transportation leadtime (exceeding corresponding thresholds), channel assembly achieves greater cost reduction. Similar to the previous argument, the value of channel assembly is higher when the end-product demands are negatively correlated.

3.3.4   Sales agent vs. other structures.   Here, $C_d^{SA} = 0$ and $C_m^{SA} = C_m^{ATO}$. From the distributor's and the system's perspectives, sales agent has the lowest inventory cost among the four structures. The manufacturer's cost in the sales agent structure is the same as in the assemble-to-distributor structure, which is lower than the make-to-stock structure; but compared with the channel assembly structure, the sales agent structure wins if the opposite of Theorem 2.5(1) is true, i.e., if the value of vanilla boxes is high.

3.3.5   A tentative ranking.   From the manufacturer's point of view, the make-to-stock structure incurs the highest inventory cost among the four channel structures. Although the channel assembly structure simplifies the manufacturer inventory management to the component and vanilla-box level, its cost efficiency to the manufacturer, however, depends on the product structure and the content of the vanilla boxes. Thus, the channel assembly structure may or may not be more favorable to the manufacturer compared to the sales agent structure or the manufacturer assemble-to-distributor-order structure, both of which incur the same component inventory cost to the manufacturer. Table 2.2

Table 2.2.   Channel Structure Ranking

Priority	Mfg.	Dist.	System
1	⋆ATO	SA	SA
2	⋆SA	CA	CA
3	⋆CA	MTS	⋆MTS
4	MTS	ATO	⋆ATO

shows the ranking of the four channel structures, where symbol $\star$ represents that the ranking is indefinite depending on the operational implementation or system parameters as previously discussed.

From the distributor's point of view, it is clear that the manufacturer assemble-to-distributor-order structure is most costly due to the longer time he has to wait for product delivery. The manufacturer make-to-stock structure reduces the distributor inventory by shortening the lead-time to the distributor. Channel assembly further reduces the inventory by replacing the end-product inventory with the component and vanilla-box inventory. The sales agent structure completely removes any form of inventory from the distributor and is therefore the first choice of the distributor if one considers only inventory.

From the system point of view, the sales agent structure which eliminates end-product inventory from the system and keeps all components inventory at the manufacturer's site is the most cost effective channel structure. The channel assembly structure, which removes the end-product inventory from the system but keeps the vanilla boxes, is more cost effective than the manufacturer make-to-stock or the manufacturer assemble-to-distributor-order structures, both of which stock end-product inventory in the system. Depending on parameters such as leadtimes and difference in stocking end products at the manufacturer's site and in the distribution channel, either of the last two channel structures could rank third in the system.

From the response time to customer perspective, the most cost effective channel structure may not offer the fastest service to customers. We define the *standard response time to a customer* as the leadtime from a customer placing an order to receiving the order under the non-stockout situation. Figure 2.2 illustrates the standard response times to customers associated with four distribution channel structures. The ranking of response times for the four channel structures is almost the reverse of the ranking of system inventory costs, reflecting the trade-offs between the cost and the response time. Taking measures of both inventory cost and response time, channel assembly has the greatest potential to offer the best of the combination. But the success of channel assembly relies on short supply leadtime, assembly responsibility sharing, and component inventory management scheme: inefficiency in any of these aspects could damage the interest of participating parties and result in supply chain conflicts.

3.3.6   Extension to multi-identical-distributor system.

The previous analysis, conducted on a single manufacturer and distributor system, can be easily extended to a multi-identical-distributor system

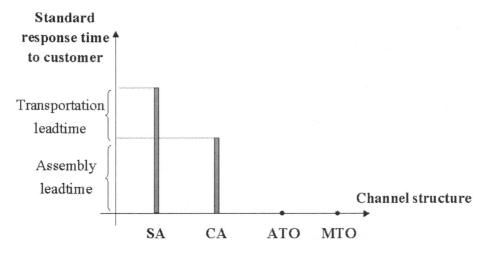

Figure 2.2. Standard Response Times to Customers of Channel Structures

in which each distributors faces an identical and independent demand process. For a particular structure, if we assume sufficient supply from the manufacturer, a distributor's inventory cost is the same regardless whether he is in a one-distributor system or a multi-distributor system. However, the distribution channel inventory cost increases linearly with the number of distributors in the structure; the manufacturer's inventory cost increases linearly with the square-root of the number of distributors. Therefore, if both the manufacturer and the distributor achieve cost reduction in a one-distributor system, then the cost reduction would increase as the number of distributors increases; if cost reduction only occurs at the distributor, then cost reduction would increase if more distributors participate; if cost reduction only occurs at the manufacturer in the single distributor system, then the more distributors participate, the higher cost this structure brings to the system.

4. Postponement in Channel Assembly

When only inventory cost is considered in channel assembly, intuitively we prefer to postpone the assembly steps with high value-adding and large product variety proliferation to the channel. The research presented in this section include both the inventory and assembly processing costs and show that the trade-off between the two costs could lead to some counter-intuitive decisions.

4.1.   Cost analysis

We assume a system of one manufacturer and M identical distributors participating in the channel assembly program. The general assumptions stated in the previous sections are valid in this section. In addition we make following modifications to the previous channel assembly model to simplify further analysis.

First, for each of the m component categories, we denote v_j as the incremental assembly/testing cost if the step of assembling component j is to be conducted by the distributors as opposed to the manufacturer, $j = 1, 2, \ldots, m$. We assume a positive value for v_j since it includes the amortized investment made by the distributors. Thus a component category j is described by three parameters, the component procurement cost, the incremental processing/testing cost, and the number of configuration options, hence is identifiable by vector (c_j, v_j, N_j). Second, we assume that the manufacturer builds vanilla boxes to distributors' order, and no vanilla-box inventory is stocked at the manufacturer's site. Third, we assume the manufacturer's assembly leadtime to build vanilla boxes is negligible, hence the supply leadtime from the manufacturer to the distributors is just the transportation leadtime l^e.

Finding the postponement boundary/vanilla-box configuration that minimizes the system cost requires the following two steps: 1. for a given number J, minimize the system cost for assembling J out of m components into vanilla boxes; 2. compare the vanilla-box configurations associated with $J = 1, 2, \ldots, m$, choose the one with the lowest cost as the optimal vanilla-box configuration. Such an approach poses a complicated combinatorial problem whose solution requires total enumeration or heuristic algorithms. Since the goal of this study is to provide insights for strategic planning rather than a tactical one, we decide to take the following approach. First we study the postponement boundary for a pre-specified assembly process. We then compare different assembly sequences to examine whether there exists one assembly sequence that incurs a lower total cost than any other assembly sequence wherever the postponement boundary is set. We refer to such a sequence as the lower-bound (LB) sequence. In particular, we intend to address the following questions: what characteristics does a LB sequence possess, i.e., can such a sequence be characterized by the monotonicity of c_j, v_j, or N_j? The answers not only provide a useful guideline for the product design for supply chain efficiency, also offer insights to the manufacturer and the distributors in designing a channel assembly program.

Assume a pre-specified assembly process and the manufacturer assembles the first J components into the vanilla boxes. Following the rationale in the previous section, we define an inventory cost function $I(J)$ as follows:

$$I(J) = I_m^c \sqrt{L^c + 1} \left(\sum_{j=1}^{J} N_j c_j \sigma_c(j) \right)$$

$$+ M I_m^c \sqrt{l^c + 1} \, T \left(\sum_{j=J+1}^{m} N_j c_j \sigma_c(j) \right)$$

$$+ M I_m^c t T \sqrt{l^e + 1} \left(\prod_{j=1}^{J} N_j \right) \sigma_V(J) \left(\sum_{j=1}^{J} c_j \right),$$

where the first term is the manufacturer's component inventory cost, the second term is the distributors' component inventory cost, the third term is the distributors' vanilla-box inventory cost. The distributors' incremental processing/testing cost is defined by $P(J)$

$$P(J) = M \mu \left(\prod_{j=1}^{m} N_j \right) \left(\sum_{j=J+1}^{m} v_j \right).$$

Therefore the system total cost is: $T(J) = I(J) + P(J)$.

Lemma 2.6 *(1) If $L^c = l^e = l^c = l$, $I(J)$ is increasing in J;*
(2) $P(J)$ is decreasing in J. $P(J)$ is convex in J if $v_j \geq v_{j+1}$, $j - 1, \ldots, m - 1$.

Lemmas 2.6 illustrates the fundamental cost trade-offs in channel assembly: less postponement (higher value of J) results in a higher inventory holding cost due to the higher value of vanilla boxes but a lower incremental processing/testing cost. Hence the optimal postponement boundary should reflect the cost balance between the inventory cost and the incremental processing cost. In general, the total system cost $T(J)$ does not bear the property of convexity. To simplify the analysis, we further assume $\rho = 0$, $L^c = l^e = l^c = l$.

Theorem 2.7 *$I(J)$ is convex if either of the following two conditions hold:*
(1) $c_j \leq \sqrt{N_{j+1}} c_{j+1}$ and $N_j \geq 4$;
(2) $c_j \leq \sqrt{N_{j+1}} c_{j+1}$ and $N_j \leq N_{j+1}$.

Lemma 2.6 and Theorem 2.7 provide sufficient conditions under which the total system cost function $T(J)$ is convex in the postponement boundary J. Moreover, we have $(\partial / \partial M)[I(J + 1) - I(J)] > 0$ and

$(\partial/\partial M)[P(J+1) - P(J)] < 0$. Hence, with more distributors in the channel assembly program, more postponement is favorable from the perspective of inventory cost saving, but less postponement is favorable from the perspective of investment in assembly line and testing process.

4.2.   Assembly process sequencing

If the product structure is flexible enough that the assembly process does not have to follow a fixed sequence, then the sequencing of the existing component set becomes an instrument for optimizing channel assembly. If the component set is sequenced as $S = \{s_j : s_j = (c_j, v_j, N_j), j = 1, \dots, m\}$ and the postponement boundary is set at J, then we define $I(J|S)$, $P(J|S)$, and $T(J|S)$ as the respective inventory cost, incremental processing/testing cost, and total cost. Let $c_j(S)$, $v_j(S)$, and $N_j(S)$ be the parameters associated with the jth component in sequence S.

Lemma 2.8 *Suppose the component set can be sequenced as follows:*

$$\underline{S} = \{\underline{s}_j : \underline{s}_j = (c_j, v_j, N_j), \quad \sqrt{N_j}c_j \leq \sqrt{N_{j+1}}c_{j+1}, \quad N_j \leq N_{j+1},$$
$$\text{and} \quad v_j \geq v_{j+1}\},$$

and

$$\overline{S} = \{\overline{s}_j : \overline{s}_j = (c_j, v_j, N_j), \quad \sqrt{N_j}c_j \geq \sqrt{N_{j+1}}c_{j+1}, \quad N_j \geq N_{j+1},$$
$$\text{and} \quad v_j \leq v_{j+1}\}.$$

Then, for any J, we have $T(J \mid \underline{S}) \leq T(J \mid S) \leq T(J \mid \overline{S})$ for any sequence S of the same component set.

From the postponement perspective, the ideal component set has the following property: the components are sequenced simultaneously in the increasing order of $\sqrt{N_j}c_j$ and N_j, but in decreasing order of v_j. Such a component set generates the assembly sequence that guarantees the lowest total cost for any postponement boundary. The LB sequence enables both the value-adding steps and the product variety proliferation steps to be postponed as late as possible. However, in reality, product design does not necessarily lead to a component set with such an ideal property. For instance, more expensive components might be more costly to assemble and test, i.e., if $c_j(S) \leq c_{j+1}(S)$ then $v_j(S) \leq v_{j+1}(S)$, postponing expensive components means to postpone the costly assembly process to channel as well. What is the LB sequence in this case? Are there circumstances that postponing expensive components or product proliferation is not necessarily cost effective? We will address those questions in the study of following special cases, which in fact, provide some interesting counter-intuitive results.

Case 2.9 Identical incremental processing/testing cost $(v_j = v)$

If the incremental processing/testing cost for each component is identical, then the different sequences result in the same incremental processing costs but different inventory costs. In particular, suppose $\sqrt{N_j}c_j = A$, which represents a special case where more expensive components have less number of options. Here, it is not possible to simultaneously postpone expensive components and components with more options. Such a component set can be sequenced exactly as those in Lemma 2.8:

$$\underline{S} = \{\underline{s}_j : \underline{s}_j = (c_j, v, N_j),\ N_j \leq N_{j+1}\} \quad \text{and}$$
$$\overline{S} = \{\overline{s}_j : \overline{s}_j = (c_j, v, N_j),\ N_j \geq N_{j+1}\}.$$

In the LB sequence, $\underline{S}$, we postpone the inexpensive components in order to get control over the vanilla-box proliferation.

Case 2.10 Identical configuration options $(N_j = N)$

If each component category has the same number of configuration options, then the different sequences result in the same product proliferation but different processing and inventory costs. If the more expensive components are less costly to assemble and test, then we have a LB sequence that postpones the expensive components to the later stage of the assembly process.

However, it is quite likely that the expensive components are also more costly to handle and assemble. For example, the processing cost might be proportional to the component cost, i.e., $v_j = rc_j$. We define the following two important sequences for the component set.

$$S_1 = \{s_j : s_j = (c_j, v_j, N),\ c_j \leq c_{j+1}\}$$

and the reverse sequence

$$S_2 = \{s_j : s_j = (c_j, v_j, N),\ c_j \geq c_{j+1}\}.$$

Theorem 2.11 *If $v_j = rc_j$, then*

$$\min_{i=1,2} (T(J \mid S_i)) \leq T(J \mid S) \leq \max_{i=1,2}(T(J \mid S_i)),$$

for any J and any sequence S.

Corollary 2.12 *If $v_j = rc_j$, then*
(1) for any given J, $\exists r_0$, such that $T(J \mid S_2) \leq T(J \mid S_1)$ for $r \geq r_0$;
(2) for any given J, $\exists \alpha_0$, such that $T(J \mid S_2) \leq T(J \mid S_1)$ for $\mu/\sigma \geq \alpha_0$;

(3) for any given J, $\exists m_0$, such that $T(J \mid S_2) \leq T(J \mid S_1)$ for $m \geq m_0$;
(4) for any given J, $\exists l_0$, such that $T(J \mid S_2) \leq T(J \mid S_1)$ for $l \leq l_0$.

Theorem 2.11 shows that any sequence of the component set can be bounded from below and above by two sequences, S_1 and S_2, that are monotonically ordered by the component cost and in reverse order to each other. However, it is not always true that the sequence in increasing component cost order, S_1, is the LB sequence. Corollary 2.12 provides conditions under which the sequence in decreasing component cost order, S_2, yields the lowest cost, for the simple reason that the incremental processing cost outweighs the inventory holding costs. Those conditions are: the incremental processing/testing cost is very high, product demand is stable, the bill of materials is large, or the component procurement lead-time is short.

Corollary 2.13 *If $v_j = v$, $L^c = l^c = l$, $l^e = 0$, and $t = T = 1$, then*
(1) for any given J, $\exists l_0$, such that $T(J \mid S_2) \leq T(J \mid S_1)$ for $l \geq l_0$;
(2) for any given J, if $l+1 > N^{J-1}$, then $\exists M_0$, such that $T(J \mid S_2) \leq T(J \mid S_1)$ for $M \geq M_0$.

Corollary 2.13 illustrates another situation in which postponing less expensive components is preferable. If the manufacturer's leadtime to distributors is negligible compared to the component procurement lead-time, then the system with long procurement leadtime and a large number of channel-assembly distributors would prefer to stock the more expensive components centrally at the manufacturer's site, hence postponing less expensive components to channel assembly.

Case 2.14 Identical component procurement cost $(c_j = c)$

If components with more options are more costly to assemble and test, e.g., $v_j = r\sqrt{N_j}$, then postponing such components might not be cost effective.

Theorem 2.15 *If $v_j = r\sqrt{N_j}$, we define sequences:*

$$S_1 = \{s_j : \underline{s}_j = (c, v_j, N_j), N_j \leq N_{j+1}\}$$

and the reverse sequence

$$S_2 = \{s_j : \bar{s}_j = (c, v_j, N_j), N_j \geq N_{j+1}\}.$$

For any J, we have:
(1) $\exists r_0$, such that $T(J \mid S_2) \leq T(J \mid S_1)$ for $r \geq r_0$;
(2) $\exists \alpha_0$, such that $T(J \mid S_2) \leq T(J \mid S_1)$ for $\mu/\sigma \geq \alpha_0$;
(3) $\exists m_0$, such that $T(J \mid S_2) \leq T(J \mid S_1)$ for $m \geq m_0$.

Theorem 2.15 gives conditions under which components with more options should be assembled first, again reflecting the trade-offs between the inventory and processing cost shown in Corollary 2.12.

4.3.   Managerial insights

Postponing the expensive components or components with more configuration options does not always result in a cost-efficient channel assembly practice. Through a simple model we show conditions under which postponing less expensive components or components with less options to channel will result in a lower total cost. Those conditions can be classified as two types. First, assembly cost (v_j) effect. When the incremental assembly cost v_j increases with the component procurement cost, inventory cost saving resulted from postponing expensive components is in direct conflict with the assembly cost saving from postponing the less expensive components. When incremental assembly cost at the channel is sufficiently high, demand is relatively stable, or component procurement leadtime is short, channel processing cost becomes the main cost driver in channel assembly, and expensive components should be assembled into the vanilla boxes. The same rationale applies to the case where the assembly cost increases with the number of component configuration options. The second type is the leadtime effect. When the component suppliers' leadtimes to the manufacturer and the distributors are long (either due to the geographical distance or other operational inefficiencies), it would be cost-effective to stock expensive components centrally at the manufacturer's site and postpone the inexpensive components to channel assembly.

5.   Conclusions and Discussions

In the war between the indirect distribution model and the direct sales model in the PC industry, strategies taken by indirect distribution channels can be categorized into two types: channel structural change and operation synchronization. The goal of structural change is to reassign responsibilities and eliminate operational redundancy so that the resulting supply chain can deliver products to customers at more competitive costs in a timely fashion. Channel assembly and sales agent are examples of structural change. On the other hand, a well structured supply chain should be supported by the seamless collaboration among all participating parties, each of whom must operate efficiently. In this chapter, starting from four different channel structures currently practiced in the PC industry, we compare the associated cost efficiencies and derive conditions under which a particular structure best suits a supply chain. We summarize our findings as follows:

1   The change of channel structure brings cost shifts among channel members and brings uneven differential impacts on the manufacturer and the distributor. Cost and risk sharing is needed among supply chain members to achieve Pareto improvements.

2   System cost efficiency is jointly determined by the channel structure and system characteristics: for the assemble-to-distributor-order structure, assembly and transportation leadtimes are the key factors to its cost efficiency; for the channel assembly structure, factors such as leadtimes, the product structure, and demand uncertainties all could affect its cost efficiency.

3   Overall, the sales agent structure incurs the lowest system inventory cost; while among the rest of the three structures, channel assembly provides the highest cost reduction for both the manufacturer and the distributor in a system with relatively long leadtimes.

4   A good channel structure to the supply chain is one that achieves the optimal balance between cost and response time, with a win-win cost sharing between the manufacturer and the distributor.

In this chapter, we have used the assumptions of 100% sufficient supply of components to the manufacturer, and the use of stationary base-stock inventory policies. In the PC industry, it not uncommon that the manufacturers experience short supply of key components such as processors and memory due to capacity limitation, unstable yields, or other unexpected events. Moreover, the product life cycles of many PC products could be as short as six months and demand is highly volatile; so that manufacturers and distributors often times need to adjust the target stocking levels along the product life cycle. These two assumptions may limit the significance of the results in this chapter.

We note also that we have focused primarily on inventory cost and response time performances. In reality, other costs such as investment in assembly, logistics, and information systems, and transportation costs are also important measures that could affect the desirability of a particular channel structure. We now briefly discuss some cost factors related to the sales agent and the channel assembly structures.

The sales agent structure was created by indirect vendors to imitate the direct sales model, involving channel partners. It changes the roles of members in the supply chain: distributors are no longer the traditional "box movers" for manufacturers but instead the service providers to end customers; manufacturers have to provide the hardware solutions to end customers in a timely fashion. Such a role change creates tremendous opportunities for third party logistic providers and various VARs in the new supply chain. But the drastic change requires significant investment. Product delivery will change from the fixed-route manufacturer-distributor shipment mode to the random-route order-driven manufacturer-end-customer mode. The transportation cost in the sales agent structure is thus higher than in the other three structures. Moreover, there is the added investment in information technology and new skill set for both manufacturers and distributors. Similarly, further channel assembly structure that changes the responsibilities and roles of channel members, investments in collaboration and new skill acquisition are needed.

Our model explicitly deals with one product family scenario in a single manufacturer and distributor setting. In practice, PC vendors have applied combinations of these structures to different customers. For example, HP's Extended Solutions Partnership Program (ESPP) comprises a channel assembly program, an economic program (manufacturer assemble-to-distributor-order), and a vendor express program, all of which serve customers in different segments (12 Technologies [8]). Hence we can extend the research to the following two directions: for a manufacturer who distributes products through multiple non-identical distributors (the difference could be in size, efficiency, etc.), the concern would be to apply the right distribution strategy to the right distributor; for a manufacturer who offers multiple product families to market, the concern would be to choose the right distribution strategy for the right product family.

While our research is taking place, the PC industry has been evolving at a fast pace: more new channel structures have been experimented with; even existing structures like channel assembly have developed alternatives like co-location. The model developed in this chapter could serve as a basic model, and some of the new developments can be evaluated by using the basic model with appropriate parameters and assumptions modifications. Ultimately, however, new models are needed to keep up with the industry's evolution.

6.   Appendix: Notation

i : index for the end product, $i = 1, 2, \ldots, n$;

j : index for the component category, $j = 1, 2, \ldots, m$;

k : index for the component within one component category, $k = 1, 2, \ldots N_j$, for $j = 1, 2, \ldots, m$;

n : number of end products;

m : number of component categories;

N_j : number of different component configurations in component category j;

Manufacturer related notation:

c_j : unit purchase cost of component j;

β_m^c : unit component holding cost per period in %;

H_j^c : unit inventory holding cost of component j per period, $\beta_m^c c_j$;

β_m^e : unit end-product holding cost per period in %;

v : value added to the end product through assembly process in %;

H^e : unit end-product holding cost per period, $H^e = \beta_m^e (1 + v) \sum_{j=1}^m c_j$;

H_J^V : unit inventory holding cost of a vanilla box with J components, $H_J^V = \beta_m^e (1 + v) \sum_{j=1}^J c_j$;

α_m^c : unit component backorder cost per period in %;

G_j^c : unit component backorder cost per period, $\alpha_m^c c_j$;

α_m^e : unit end-product backorder cost per period %;

G^e : unit end-product backorder cost per period, $G^e = \alpha_m^e (1 + v) \sum_{j=1}^m c_j$;

G_J^V : unit backorder cost of a vanilla box with J components, $G_J^V = \alpha_m^e (1 + v) \sum_{j=1}^J c_j$;

Δ : unit profit margin of the end product in %;

L^a : assembly leadtime;

L_J^V : assembly leadtime for the vanilla box with J components;

L^c : component procurement leadtime;

t : ratio of end-product and component holding/backorder cost, $\beta_m^e = t\beta_m^c$, $\alpha_m^e = t\alpha_m^c$;

I_m^c : expected unit component holding and penalty cost per period.

Distributor related notation:

β_d^c : unit component holding cost per period in %;

h_j^c : unit inventory holding cost of component j per period, $\beta_d^c c_j$;

β_d^e : unit end-product holding cost per period in %;

h^e : unit end-product inventory holding cost per period, $h^e = \beta_d^e (1 + \Delta)(1 + v) \sum_{j=1}^{m} c_j$;

h_J^V : unit inventory holding cost of a vanilla box with J components, $h_J^V = \beta^e (1 + \Delta)(1 + v) \sum_{j=1}^{J} c_j$;

α_d^c : unit component backorder cost per period in %;

g_j^c : unit backorder cost of component j per period, $g_j^c = \alpha_d^c c_j$;

α_d^e : unit end-product backorder cost per period in %;

g^e : unit end-product backorder cost per period, $g^e = \alpha_d^e (1 + \Delta)(1 + v) \sum_{j=1}^{m} c_j$;

g_J^V : unit backorder cost of a vanilla box with J components, $g_J^V = \alpha_d^e (1 + \Delta)(1 + v) \sum_{j=1}^{J} c_j$;

l^c : component transportation leadtime;

l^e : end produce transportation leadtime;

T : ratio of the distributor and the manufacturer holding/backorder cost, $\beta_d^e = T\beta_m^e$, $\beta_d^c = T\beta_m^c$, $\alpha_d^e = T\alpha_m^e$, $\alpha_d^c = T\alpha_m^c$.

Demand related notation:

ξ_i : per-period demand of end-product i;

$\xi_{j,k}$: per-period demand of component k of component type j;

ξ_i^J : per-period demand of vanilla box containing J components, $i = 1, \ldots, \prod_{j=1}^{J} N_j$.

7.   Appendix: Proofs

[Proof of Theorem 2.1]   (2) We first consider the manufacturer related cost:

$$C_m^{MTS} - C_m^{ATO}$$

$$= \sigma\sqrt{L^a + 1} \sum_{i=1}^{n} \left[q_m^e H^e + (G^e + H^e) L(q_m^e) \right]$$

$$= \sigma\sqrt{L^a + 1} \sum_{i=1}^{n} \left[q_m^e \beta^{cm} + (\alpha^{cm} + \beta^{cm}) L(q_m^e) \right] \left[t(1 + v) \right] \left(\sum_{j=1}^{m} c_j \right)$$

$$= \sigma\sqrt{L^a + 1} \sum_{i=1}^{n} \left[q_m^c \beta^{cm} + (\alpha^{cm} + \beta^{cm}) L(q_m^c) \right] \left[t(1 + v) \right] \left(\sum_{j=1}^{m} c_j \right)$$

$$= \sigma\sqrt{L^a + 1} I_m^c \left(\prod_{j=1}^{m} N_j \right) \left[t(1 + v) \right] \left(\sum_{j=1}^{m} c_j \right),$$

where

$$I_m^c = q_m^c \beta_m^c + \left(\alpha_m^c + \beta_m^c \right) L(q_m^c)$$

and

$$q_m^c = \Phi^{-1} \left(\frac{\alpha_m^c}{\alpha_m^c + \beta_m^c} \right) = \Phi^{-1} \left(\frac{t\alpha_m^c}{t(\alpha_m^c + \beta_m^c)} \right) = q_m^e.$$

We then consider the distributor related cost:

$$C_d^{MTS} - C_d^{ATO}$$

$$= \sigma\left(\sqrt{l^e + 1} - \sqrt{L^a + l^e + 1}\right) \sum_{i=1}^{n} \left[q_d^e h^e + (g^e + h^e) L(q_d^e) \right]$$

$$= \sigma\left(\sqrt{l^e + 1} - \sqrt{L^a + l^e + 1}\right)$$

$$\cdot \sum_{i=1}^{n} \left[q_m^c \beta_m^c + (\alpha_m^c + \beta_m^c) L(q_m^c) \right] Tt(1 + \Delta)(1 + v) \left(\sum_{j=1}^{m} c_j \right)$$

$$= \sigma\left(\sqrt{l^e + 1} - \sqrt{L^a + l^e + 1}\right)$$

$$\times I_m^c \left(\prod_{j=1}^{m} N_j \right) Tt(1 + \Delta)(1 + v) \left(\sum_{j=1}^{m} c_j \right),$$

where the second equality comes from

$$q_d^e = \Phi^{-1}\left(\frac{\alpha_d^e}{\alpha_d^e + \beta_d^e}\right) = \Phi^{-1}\left(\frac{Tt\alpha_m^c}{Tt(\alpha_m^c + \beta_m^c)}\right) = q_m^c.$$

Therefore,

$$C_t^{MTS} - C_t^{ATO} \tag{2.1}$$
$$= C_m^{MTS} - C_m^{ATO} + C_d^{MTS} - C_d^{ATO}$$
$$= \sigma I_m^c \left(\prod_{j=1}^{m} N_j\right)\left(\sum_{j=1}^{m} c_j\right)[t(1+v)]$$
$$\times \left\{\sqrt{L^a + 1} - (\sqrt{L^a + l^e + 1} - \sqrt{l^e + 1})T(1+\Delta)\right\}.$$

It is clear that $C_t^{MTS} - C_t^{ATO} > 0$ if and only if

$$\left(\sqrt{L^a + l^e + 1} - \sqrt{l^e + 1}\right)/\sqrt{L^a + 1} < 1/T(1+\Delta).$$

(3) Let $f(L^a, l^e) = \left(\sqrt{L^a + l^e + 1} - \sqrt{l^e + 1}\right)/\sqrt{L^a + 1}$. Then,

$$\frac{\partial f}{\partial L^a} = \frac{\sqrt{l^e + 1}\sqrt{L^a + l^e + 1} - l^e}{2\left(\sqrt{L^a + 1}\right)^3 \sqrt{L^a + l^e + 1}} > 0$$

and

$$\frac{\partial f}{\partial l^e} = \frac{1}{2\sqrt{L^a + 1}}\left(\frac{1}{\sqrt{L^a + l^e + 1}} - \frac{1}{\sqrt{l^e + 1}}\right) < 0.$$

[Proof of Theorem 2.2] (1) Taking the derivative of (2.1) w.r.t. L^a, we have

$$d\left(C_t^{MTS} - C_t^{ATO}\right)/dL^a = A\left(\sqrt{L^a + l^e + 1} - T(1+\Delta)\sqrt{L^a + 1}\right) \tag{2.2}$$

where

$$A = \frac{\sigma I_m^c \left(\prod_{j=1}^{m} N_j\right)\left(\sum_{j=1}^{m} c_j\right)[t(1+v)]}{2\sqrt{L^a + l^e + 1}\sqrt{L^a + 1}}.$$

It is straightforward to show that if $l^e < \theta_1$, where $\theta_1 = [T(1+\Delta)]^2 - 1$, then the second term of (2.2) is always negative for any positive value of L^a. If $l^e > \theta_1$, then the sign of the second term depends on the value of L^a, i.e., the second term is positive if and only if $L^a < \frac{l^e}{\theta_1} - 1$. (2) The desired result follows by taking the derivative of (2.1) w.r.t. l^e.

[Proof of Theorem 2.3] (1) The value of the intermediate product assembled by the manufacturer is smaller, therefore $C_m^{MTS} - C_m^{CA} > 0$.
(2) We have:

$$
C_d^{MTS} - C_d^{CA}
$$

$$
\geq \sigma\sqrt{l^e + 1}\, I_m^c T \left\{ \left(\prod_{j=1}^m N_j \right) t\,(1+\Delta)\,(1+v) \left(\sum_{j=1}^m c_j \right) \right.
$$

$$
- \sqrt{\left[1 + \left(\prod_{j=J+1}^m N_j - 1 \right)\rho \right] \prod_{j=J+1}^m N_j \left(\prod_{j=1}^J N_j \right)}
$$

$$
\times t\,(1+\Delta)\,(1+v) \left(\sum_{j=1}^J c_j \right)
$$

$$
\left. - \sum_{j=J+1}^m N_j c_j \sqrt{\left[1 + \left(\prod_{j'\neq j}^m N_{j'} - 1 \right)\rho \right] \prod_{j'\neq j}^m N_{j'}} \right\} ;
$$

and the right hand side above simplifies to:

$$
\sigma\sqrt{l^e + 1}\, I_m^c T \left\{ t\,(1+\Delta)\,(1+v) \left(\sum_{j=1}^J c_j \right) \right.
$$

$$
\times \left[\prod_{j=1}^m N_j - \sqrt{\left[1 + \left(\prod_{j=J+1}^m N_j - 1 \right)\rho \right] \prod_{j=J+1}^m N_j \left(\prod_{j=1}^J N_j \right)} \right]
$$

$$
+ \left[\left(\prod_{j=1}^m N_j \right) t\,(1+\Delta)\,(1+v) \left(\sum_{j=J+1}^m c_j \right) \right.
$$

$$
\left. \left. - \sum_{j=J+1}^m N_j c_j \sqrt{\left[1 + \left(\prod_{j'\neq j}^m N_{j'} - 1 \right)\rho \right] \prod_{j'\neq j}^m N_{j'}} \right] \right\} > 0,
$$

with the inequality following from

$$
\prod_{j=1}^m N_j > \sqrt{\left[1 + \left(\prod_{j=J+1}^m N_j - 1 \right)\rho \right] \prod_{j=J+1}^m N_j \left(\prod_{j=1}^J N_j \right)}
$$

and

$$\prod_{j=1}^{m} N_j > N_j \sqrt{\left[1 + \left(\prod_{j' \neq j}^{m} N_{j'} - 1\right)\rho\right] \prod_{j' \neq j}^{m} N_{j'}}$$

for $\rho < 1$.

(3) Taking derivative of $C_t^{MTS} - C_t^{CA}$ w.r.t. l^e, the desired result follows.

(4) If $l^e = l^c$, taking derivative of $C_t^{MTS} - C_t^{CA}$ w.r.t. ρ, the desired result follows.

[Proof of Theorem 2.5] (1)

$$C_m^{ATO} - C_m^{CA}$$
$$= \sigma I_m^c \sqrt{N^m} \left[\sqrt{L^c + 1} n(1) \sum_{j=J+1}^{m} c_j - \sqrt{L_J^V + 1} n(J) t(1+v) \left(\sum_{j=1}^{J} c_j\right)\right].$$

The desired result follows.

(2) If $L^c = l^c$, then

$$C_t^{ATO} - C_t^{CA} = C_m^{ATO} - C_m^{CA} + C_d^{ATO} - C_d^{CA}$$
$$= \sigma I_m^c \sqrt{N^m} t (1+v) \left\{ \sqrt{L^a + l^e + 1}\, n(m)\, (1+\Delta) \left(\sum_{j=1}^{m} c_j\right)\right.$$
$$\left. - n(J) \left(\sum_{j=1}^{J} c_j\right) \left[\sqrt{L_J^V + 1} + \sqrt{l^e + 1}\, (1+\Delta)\right]\right\}.$$

The desired result follows. (3), (4), and (5) are straightforward once the first derivative is taken.

[Proof of Lemma 2.6] If $L^c = l^e = l^c = l$, then

$$I(J) = I_m^c \sqrt{l+1} \left[\sum_{j=1}^{J} N_j c_j \sigma_c(j) + MTt \left(\prod_{j=1}^{J} N_j\right) \sigma_V(J) \left(\sum_{j=1}^{J} c_j\right)\right.$$
$$\left. + MT \sum_{j=J+1}^{m} N_j c_j \sigma_c(j)\right],$$

and

$$I(J+1) - I(J)$$

$$= I_m^c \sqrt{l+1} \Bigg\{ MTt \Bigg\{ \left(\sum_{j=1}^{J} c_j \right) \left[\left(\prod_{j=1}^{J+1} N_j \right) \sigma_V(J+1) - \left(\prod_{j=1}^{J} N_j \right) \sigma_V(J) \right]$$

$$+ c_{J+1} \left(\prod_{j=1}^{J+1} N_j \right) \sigma_V(J+1) \Bigg\} - (MT-1) N_{J+1} c_{J+1j} \sigma_c(J+1) \Bigg\}$$

$$\geq 0;$$

since

$$\left(\prod_{j=1}^{J} N_j \right) \sigma_V(J) = \sigma \sqrt{\prod_{j=1}^{m} N_j} \sqrt{\rho \prod_{j=1}^{m} N_j + \prod_{j=1}^{J} N_j (1-\rho)}$$

and

$$N_{J+1} \sigma_c(J+1) = \sigma \sqrt{\prod_{j=1}^{m} N_j} \sqrt{\rho \prod_{j=1}^{m} N_j + N_{J+1}(1-\rho)}.$$

[Proof of Lemma 2.7] Since

$$[I(J+2) - I(J+1)] - [I(J+1) - I(J)]$$

$$= \sigma \sqrt{l+1} I_m^c \sqrt{\prod_{j=1}^{m} N_j} \Bigg\{ (1-MT)\left(\sqrt{N_{J+2}} c_{J+2} - \sqrt{N_{J+1}} c_{J+1}\right)$$

$$+ MTt \sqrt{\prod_{j=1}^{J} N_j} \Bigg[\left(\sum_{j=1}^{J} c_j \right) \left(\sqrt{N_{J+2} N_{J+1}} - 2\sqrt{N_{J+1}} + 1 \right)$$

$$+ \sqrt{N_{J+1}} \left(\sqrt{N_{J+2}} c_{J+2} - c_{J+1} + \sqrt{N_{J+2}} c_{J+1} - c_{J+1} \right) \Bigg] \Bigg\},$$

the sufficient conditions for convexity follows.

[Proof of Lemma 2.8] We first calculate

$$P(J \mid S) - P(J \mid \underline{S}) = M\mu \left(\prod_{j=1}^{m} N_j \right) \left(\sum_{j=J+1}^{m} [v_j(S) - v_j(\underline{S})] \right) < 0.$$

We then calculate

$$I(J \mid S) - I(J \mid \underline{S})$$

$$= \sigma\sqrt{l+1} I_m^c \sqrt{\prod_{j=1}^{m} N_j} \left\{ (1-MT) \sum_{j=1}^{J} \left(\sqrt{N_j(S)} \, c_j(S) - \sqrt{N_j(\underline{S})} \, c_j(\underline{S}) \right) \right.$$

$$+ MTt \left[\left(\sum_{j=1}^{J} \sqrt{\prod_{\substack{j'=1, \\ j'\neq j}}^{J} N_{j'}(S)} \sqrt{N_j(S)} c_j(S) \right) \right.$$

$$\left. \left. - \left(\sum_{j=1}^{J} \sqrt{\prod_{\substack{j'=1, \\ j'\neq j}}^{J} N_{j'}(\underline{S})} \sqrt{N_j(\underline{S})} c_j(\underline{S}) \right) \right] \right\}$$

$$\geq 0.$$

The inequality is true because

$$\sum_{j=1}^{J} \sqrt{\prod_{\substack{j'=1, \\ j'\neq j}}^{J} N_{j'}(S)} \sqrt{N_j(S)} c_j(S) - \left(\sum_{j=1}^{J} \sqrt{\prod_{\substack{j'=1, \\ j'\neq j}}^{J} N_{j'}(\underline{S})} \sqrt{N_j(\underline{S})} c_j(\underline{S}) \right)$$

$$- \sum_{j=1}^{J} \left(\sqrt{N_j(S)} c_j(S) - \sqrt{N_j(\underline{S})} c_j(\underline{S}) \right)$$

$$= \sum_{j=1}^{J} \left[\left(\sqrt{\prod_{\substack{j'=1, \\ j'\neq j}}^{J} N_{j'}(S)} - 1 \right) \sqrt{N_j(S)} c_j(S) \right.$$

$$\left. - \left(\sqrt{\prod_{\substack{j'=1, \\ j'\neq j}}^{J} N_{j'}(\underline{S})} - 1 \right) \sqrt{N_j(\underline{S})} c_j(\underline{S}) \right]$$

$$\geq 0.$$

Similarly we can show that $P(J \mid \overline{S}) - P(J \mid S) \geq 0$ and $I(J \mid \overline{S}) - I(J \mid S) \geq 0$.

[Proof of Theorem 2.11]   If $v_j = rc_j$, then

$$T(J \mid S) - T(J \mid S_1) = \left(\sum_{j=1}^{J} c_j(S) - \sum_{j=1}^{J} c_j(S_1) \right) A(J)$$

and

$$T(J \mid S) - T(J \mid S_2) = \left(\sum_{j=1}^{J} c_j(S) - \sum_{j=1}^{J} c_j(S_2) \right) A(J)$$

where

$$A(J) = \left(\sigma\sqrt{l+1} I_m^c N^{\frac{m+1}{2}} \right) \left[N^{\frac{J-1}{2}} MTt - (MT-1) \right] - rM\mu N^m.$$

Since $T(J \mid S) - T(J \mid S_1)$ and $T(J \mid S) - T(J \mid S_2)$ have opposite signs, it follows that $\min\limits_{i=1,2}(T(J \mid S_i)) \leq T(J|S) \leq \max\limits_{i=1,2}(T(J \mid S_i))$.

[Proof of Corollary 2.12]   The sign of $T(J|S_2) - T(J|S_1)$ is determined by the sign of $A(J)$. Since $A(J)$ is a decreasing function of r, μ/σ, and m, and a increasing function of l, the desired result follows.

[Proof of Corollary 2.13]

$$T(J \mid S_1) - T(J \mid S_2)$$
$$= M\sigma I_m^c N^{\frac{m+1}{2}} \left[\sum_{j=1}^{J} c_j(S_1) - \sum_{j=1}^{J} c_j(S_2) \right] \left[\sqrt{l+1}\left(\frac{1}{M} - 1 \right) + N^{\frac{J-1}{2}} \right].$$

The desired result follows.

[Proof of Theorem 2.15]

$$T(J \mid S_2) - T(J \mid S_1)$$

$$= c\sigma\sqrt{l+1} I_m^c MTt \sqrt{\prod_{j=1}^{m} N_j} \left[J\sqrt{\prod_{j=1}^{J} N_j(S_2)} - J\sqrt{\prod_{j=1}^{J} N_j(S_1)} \right]$$

$$- \left[c\sigma\sqrt{l+1} I_m^c \sqrt{\prod_{j=1}^{m} N_j}(MT-1) + M\mu r \left(\prod_{j=1}^{m} N_j \right) \right]$$

$$\times \sum_{j=1}^{J} \left(\sqrt{N_j(S_2)} - \sqrt{N_j(S_1)} \right)$$

The above difference is a decreasing function of μ/σ, r, and m. The desired result follows.

References

[1] AUSTIN, T.A. AND H.L. LEE. (1999) Unlocking the Supply Chain's: A Case from the Personal Computer Industry. *Supply Chain Management Review*, Vol. 2, No. 2, 24–35.

[2] AVIV, Y. AND A. FEDERGRUEN. (1998) The Benefits of Design for Postponement. in *Quantitative Models for Supply Chain Management*. Edited by S. Tayur, M. Magazine, and R. Ganeshan. Kluwer Academic Publishers.

[3] CLARK, A. AND H. SCARF. (1960) Optimal Policies for a Multi-echelon Inventory Problem. *Management Science*, Vol. 6, 475–490.

[4] COHEN, M.A. AND H.L. LEE. (1988) Strategic Analysis of Integrated Production-Distribution Systems: Models and Methods. *Operations Research*. Vol. 36, No. 2, 216–228.

[5] DONG, L.X. AND H.L. LEE. (1999) Aligning Supply Chain Incentives in Channel Assembly with Component Commonality. Working Paper. Stanford University.

[6] ETTL, M., G.E. FEIGIN, G.Y. LIN, AND D.D. YAO. (2000) A Supply Network Model with Base-stock Control and Service Requirements. *Operations Research*. Vol. 48, No. 2, 216–232.

[7] GARG, A. AND C.S. TANG. (1997) On Postponement Strategies for Product Families with Multiple Points of differentiation. *IIE Transactions*, Vol. 29, 641–650.

[8] I2 TECHNOLOGIES. (1998) High Technology Industry Conference, Hewlett-Packard Company Presentation.

[9] GRAVES, S.C. AND S.P. WILLEMS. (2000) Optimizing Strategic Safety Stock Placement in SUpply Chains. *Manufacturing & Service Operations Management*, Vol. 2, No. 1, 68–83.

[10] KAPUSCINSKI, R. AND S. TAYUR. (1999) Variance vs. Standard Deviation — Variability Reduction through Operations Reversal. *Management Science*. Vol. 45, No. 5, 765–767.

[11] KECK, L., A. MARCHEVA, G. VALEN, AND S. YEARS. (1998) Channel Conflict: The Impact of Direct Internet Sales of Personal Computers on Traditional Retail Channels. Working Paper, Owen Graduate School of Management, Vanderbilt University.

[12] LEE, H.L. AND C. BILLINGTON. (1993) Material Management in Decentralized Supply Chains. *Operations Research*, Vol. 41, No. 5, 835–847.

[13] LEE, H.L. AND C.S. TANG. (1997) Modeling the Costs and Benefits of Delayed Product Differentiation. *Management Science.* Vol. 43, 40–53.

[14] LEE, H.L. AND C.S. TANG. (1998) Variability Reduction Through Operations Reversal. *Management Science.* Vol. 44, No. 2, 162–172.

[15] LEE, H.L. AND L.X. DONG. (2000) Postponement Boundary for Global Supply Chain Efficiency. Working Paper. Stanford University and Washington University.

[16] ROSLING, K. (1989) Optimal Inventory Policies for Assembly Systems Under Random Demands. *Operations Research,* Vol. 37, No. 4, 565–579.

[17] SCHMIDT, C.P. AND S. NAHMIAS. (1985) Optimal Policy for a Two-Stage Assembly System under Random Demand. *Operations Research.* Vol. 33, No. 5, 1130–1145.

[18] SWAMINATHAN, J.M. AND S.R. TAYUR. (1998A) Managing Broader Product Lines through Delayed Differentiation using Vanilla Boxes. *Management Science.* Vol. 44, No. 12, s161–72.

[19] SWAMINATHAN, J.M. AND S.R. TAYUR. (1998B) Stochastic Programming Models for Managing Product variety. in *Quantitative Modeling for Supply Chain Management.* Edited by S. Tayur, M. Magazine, and R. Ganeshan. Kluwer Academic Publishers.

Chapter 3

INTRAFIRM INCENTIVES AND SUPPLY CHAIN PERFORMANCE

Narendra Agrawal

Department of Operations and Management Information Systems
Leavey School of Business, Santa Clara University
Santa Clara, CA 95053
nagrawal@scu.edu

Andy A. Tsay

Department of Operations and Management Information Systems
Leavey School of Business, Santa Clara University
Santa Clara, CA 95053
atsay@scu.edu

1.   Introduction

Companies in many industries have begun to realize that conflicts of interest among the various parties in a supply chain can engender operationally inefficient behavior. Consequently, many researchers have become interested in identifying and evaluating methods of coordinating supply chains in which multiple decision makers pursue individual agendas (cf. [32]). The typical approach in the OM literature is to partition a traditional inventory model into a number of subproblems, each representing the decisions and objectives of a distinct organization. Most commonly, the supply chain is assumed to contain just two firms, e.g., a manufacturer and a retailer. The analysis then proceeds to pinpoint the root causes of inefficiency, and recommend mechanisms for appropriately adjusting individual incentives.

A shortcoming of the existing literature, however, is that it assumes a monolithic preference structure within each organization. In fact, evidence shows that operational-level decisions are often made by subordinates who may be motivated by the performance measurement structure to pursue objectives different from those of the firm. We believe that this has profound implications for the performance of each firm, and should be considered when structuring any contracts between firms in a supply chain. With this motivation, in this chapter we study how incongruence of goals within one firm impacts the behaviors of all parties involved in the supply contract, and the consequences for the welfare of end customers.

We consider a manufacturer-retailer supply chain serving a price-sensitive stochastic market demand. As in the extant supply chain literature, we assume that the manufacturer intends to maximize its individual profit. Our point of departure is to further partition the retail firm into two entities — an owner and a manager. The owner, as a long-term and deep-pocketed stakeholder, favors the maximization of expected retail profit. On the other hand, his store manager, charged with making operational decisions (inventory ordering and pricing to the market), is more interested in whether the profit outcome achieves some owner-specified threshold level[1], as this determines the manager's job security and prospects for advancement[2]. Such threshold-based performance measurement has been observed in a number of settings, as documented in the accounting literature ([6], [10], [11], [34])[3]. According to [19], "In many managerial situations, a budgeted profit is established, and the disutility resulting from not achieving this targeted profit level is much larger than the rewards for overachieving. A manager may then be interested primarily in maximizing the probability of meeting the budget, regardless of whether the target level is exceeded or barely attained." [8] refers to this objective as the "aspiration" criterion.

Several researchers suggest that this is a more realistic model of behavior in many settings, especially where risk preferences come into play. A number have examined its consequences in a single-node, single decision-maker, inventory context ([12], [14], [19], [20], [22], [23], [25], [28], [29]). It also appears in the financial literature on asset allocation (cf. [9]) and the decision theory literature ([3]).

In light of the potential for goal incongruence and therefore inefficiency, one might raise a number of questions of the described internal structure of the retail firm. First, why delegate? [1] argues that "since managers, by virtue of specialization, presumably develop better insights into the technology, process and external environment of the organization, it is natural for an owner to relinquish operational control of the organization to a manager." Moreover, the owner may simply be busy doing other things. We do not explicitly treat such motivations in our model, but assume this decision structure as a starting point. We therefore prohibit the owner from exerting direct control over the retail price and stocking level since this would defeat the purpose of delegation. This is analogous to the common modeling assumption that creates distinct manufacturers and retailers within supply chains, and grants each undisputed control over certain decisions, but without explicitly addressing why the parties exist independently in the first place.

Next, why does the owner use this type of incentive structure? Indeed, there may well exist alternatives that are theoretically more efficient (some possibilities are discussed in the concluding section). The empirical popularity of the above scheme and its variants may simply be due to ease of operationalization. For better or worse, real incentive systems tend to reward good *outcomes* (for instance, a large *actual* profit), rather than good *decisions* (such as ones that maximize *expected* profit). Our objective is certainly not to defend or advocate the described practices, but to explore the ramifications of their usage for supply chain behavior and performance. The specific, single-threshold form we study is the simplest such scheme and enjoys significant academic precedent, although it certainly has limitations. For instance, [9] correctly highlights the inability of this metric to distinguish among outcomes that fail to reach the threshold, while defending it as the most tractable of a useful class. In summary, our representation of the incentive structure should be considered a simple approximation that captures the flavor of an empirically well-documented phenomenon.

We extend the existing literature on supply contracts in a number of ways. First, by expanding upon the behavioral dynamics within the retailer organization, we provide a more realistic model of decision making within organizations in supply chains — both in terms of the *number of parties* making decisions as well as the *types of objective functions* that guide their individual behavior. Second, we discuss the effect of intrafirm goal incongruence not only on the firm itself, but also on those with which it does business. For instance, we demonstrate that the terms of the contract with the upstream supplier may depend upon who within the retail organization handles the negotiation. Finally, in order to quantify the consequences for the welfare of end customers, we develop a modification of the traditional measure of consumer surplus that is often used in the economics literature. We believe that we are unique in the supply chain literature in performing such analysis.

The remainder of the chapter is organized as follows. §2 details our key assumptions. §3 considers the control system outlined above, formulating and analyzing the behaviors of all three decision-makers (manufacturer, retail owner, and retail manager). Two benchmarks are then examined, one in which all retail decisions are made by the owner (§4) and the completely coordinated supply chain (§5). The various control regimes are compared in §6, followed by concluding remarks in §7. All proofs are deferred to an Appendix for clarity of exposition.

2.   Model Assumptions

We consider a single-period model of a supply chain in which a manufacturer provides product wholesale to a retailer, who in turn serves a price-sensitive stochastic demand. In the base model, denoted as control system M, the retail organization consists of the owner and a manager (see Figure 3.1). The manager makes all operational decisions for the retail firm, namely the market price p and the quantity Q to order from the manufacturer. For reasons described previously, the manager attempts to maximize the probability that retail profit will be at least T, a threshold set by the retail owner. The manufacturer, who builds exactly to the retailer's order, chooses a wholesale/transfer price c that maximizes his own profit in light of the order that will be induced. Customer demand D then occurs.

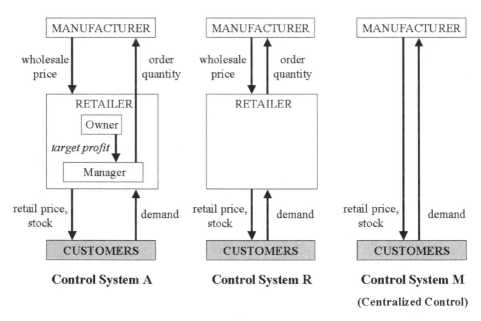

Figure 3.1.   The Supply Chain (arrows represent decision control)

2.1.   Cost structure

We modify the standard newsvendor model to represent this two-firm setting, hence the cost structure follows the newsvendor assumptions, augmented with a linear price for the transfer of product between the two firms:

m: manufacturer's unit production cost
c: unit wholesale/transfer price

p: retailer's unit selling price

s: unit salvage value

We assume that $0 < m \leq c \leq p$ and $s < m$. Goodwill loss is excluded to simplify the analysis and presentation.

2.2.   Demand model

D, the total demand per period, is a random variable with density and distribution of ϕ and Φ, respectively, and mean $\hat{\mu}$. We assume a multiplicative form of demand, with deterministic and stochastic components specified by $D = N \cdot g(p)$, where

N = the number of customer arrivals during the period, a random variable with density f, distribution F, and mean μ, and

$g(p)$ = demand per customer, a deterministic function of price that is defined on $[0, p_{max}]$ for some $p_{max} < \infty$, and satisfies the following structural assumptions:

(i) $g(p) \geq 0$, (ii) $g'(p) < 0$, (iii) $g(p_{max}) = 0$, (iv) $g''(p) \leq 0$.

Under these assumptions, the distribution of D may be written as $\Phi(d) = F(d/g(p))$. Also, by differentiation, $\phi(d) = f(d/g(p))/g(p)$ except when $p = p_{max}$, in which case $\phi(d)$ has unit mass precisely at $d = 0$. Also, $\hat{\mu} = \mu \cdot g(p)$.

This multiplicative form is a common way to model price-sensitive and stochastic demand (cf. [4], [7], [13]). Conditions (i) and (ii) are standard. (iii) implies that the market price cannot be increased indefinitely without eventually eliminating all demand. (iv) suggests that customers become more price-sensitive at higher prices. To focus attention on supply chain incentives, we make the fairly standard assumption of common beliefs about market demand (cf. [32]).

2.3.   Benchmark alternatives

Context for evaluating system M will be provided by comparison to two natural benchmarks described below:

Control System R: Here pricing and inventory management decisions are made directly by the retail owner, who wishes to maximize his expected profit. The manufacturer negotiates a wholesale price with the goal of maximizing his own expected profits.

Control System C: This is the first-best case of central control, in which all appropriate price-quantity tradeoffs are made from a global perspective.

For each regime, we will formulate the appropriate decision problem for each player, derive the rational behaviors, and characterize the equilibrium outcome (retail price, retail order quantity, target profit, wholesale price). We will then illuminate to the extent possible the structure of each system's equilibrium through comparative statics and cross-comparisons. M, R, and C will be used as subscripts on notation where appropriate.

3.   Analysis of Control System M

3.1.   The retail manager's problem

Retail profit is the following random variable:

$$Z_M(p, Q) = p \cdot \min(D, Q) - cQ + s \cdot \max(Q - D, 0).$$

Taking c (wholesale price) and T (profit target set by the owner) as given, the retail manager chooses p (retail selling price) and Q (quantity ordered from the manufacturer) to maximize the probability that this profit exceeds T. We denote this objective as $\theta(p, Q) = \Pr\{Z_M(p, Q) \geq T\}$. The p and Q that accomplish this are obtained in Theorem 3.1.

Theorem 3.1 *The retail manager's decisions in system M have the following properties:*

(i) For any fixed p, the order size that maximizes $\theta(p, Q)$ is

$$Q_M^*(T, p) = \frac{T}{p - c}, \text{ so that } \theta(p, Q_M^*(T, p)) = 1 - F\left(\frac{T}{(p - c)g(p)}\right)$$

$$(3.1)$$

(ii) For any profit target T, the optimal selling price is $p_M^*(T) = \bar{p}$, where $\bar{p} = argmax_{0 \leq p \leq p_{max}}(p - c)g(p)$, and may be obtained as the unique solution to $(\bar{p} - c)g'(\bar{p}) + g(\bar{p}) = 0$.

(iii) Hence, for a given T, the retail manager's decisions will be

$$\{p_M^*(T), Q_M^*(T, p_M^*(T))\} = \left\{\bar{p}, \frac{T}{\bar{p} - c}\right\}, \quad and, \qquad (3.2)$$

(iv) the corresponding optimal probability of achieving the profit target is

$$\theta(p_M^*(T), Q_M^*(T, p_M^*(T))) = 1 - F\left(\frac{T}{(\bar{p} - c) g(\bar{p})}\right). \qquad (3.3)$$

Interestingly, the probability of obtaining a profit of *at least* T is maximized by ordering a quantity such that T is the *largest possible attainable profit*[4]. So, the method of motivation rules out any pleasant surprises, since there is zero probability that this threshold will be exceeded. This formalizes the notion of "satisficing" behavior, and has implications for how the manager's actions might affect the organizational performance in the long run. Note also that both Q_M^* and p_M^* are independent of the demand distribution. That is, even though the manager's outcome (the resulting value of $\theta(p, Q_M^*(T, p))$) is related to the probability distribution of demand, his operational decisions are not. In fact, this invariance property provides the intuition for the form of p_M^*. Since the same p_M^* must result under any demand distribution, it is sufficient to consider the special case of deterministic demand, as in traditional Cost-Volume-Profit analysis (e.g., [34]). Because production can be matched perfectly to demand, avoiding both shortage and excess, $(p - c)g(p)$ is the retailer's profit-per-customer in this case. N is independent of p, so the manager's optimal price will be the one that maximizes this quantity, which is exactly how $\overline{p}$ is defined in part (ii) of Theorem 3.1.

3.2. The retail owner's problem

The owner's only control variable is the profit target T. Whether or not an immediate financial reward is involved, for reasons outlined in §1 the manager may believe that his job security, prospects for promotion, and generally favorable standing in the eyes of the owner will suffer if this threshold is not met. Hence, the manager will behave so as to maximize the probability that the profit will be at least T. Anticipating this, the owner will set T in hopes of maximizing expected retail profit, which we denote as $\pi_M(T) = E[Z_M(p_M^*(T), Q_M^*(T, p_M^*(T)))]$.

Theorem 3.2 *In system M, the retail owner will set a profit target of*

$$T^* = (\overline{p} - c) \cdot \Phi^{-1}\left(\frac{\overline{p} - c}{\overline{p} - s}\right) = (\overline{p} - c)g(\overline{p}) \cdot F^{-1}\left(\frac{\overline{p} - c}{\overline{p} - s}\right) \qquad (3.4)$$

Hence the retail manager orders

$$Q_M^*(T^*, p_M^*(T^*)) = g(\overline{p}) \cdot F^{-1}\left(\frac{\overline{p} - c}{\overline{p} - s}\right). \qquad (3.5)$$

The resulting expected retail profit is

$$\pi_M^* = \pi_M(T^*) = (\overline{p} - s) \int_0^{Q_M^*(T^*, \overline{p})} D\phi(D)\, dD$$

$$= (\overline{p} - s)g(\overline{p}) \int_0^{F^{-1}\left(\frac{\overline{p} - c}{\overline{p} - s}\right)} y f(y)\, dy \qquad (3.6)$$

and the resulting probability of meeting the profit target is

$$\theta^* = \theta(p_M^*(T^*), Q_M^*(T^*, p_M^*(T^*))) = 1 - F\left(\frac{T^*}{(\overline{p} - c)\, g(\overline{p})}\right) = \frac{c - s}{\overline{p} - s}.$$
$$(3.7)$$

So, the owner will set a profit target such that the manager will order exactly the quantity that will maximize expected retail profit *given a selling price of* $\overline{p}$. But since $\overline{p}$ is not the expected-profit-maximizing price, it is apparent that the delegation of operational decision-making creates inefficiency within the retail firm[5].

(3.7) has the following interesting interpretation. Recall that the owner sets T so that the optimal newsvendor quantity results. Since the manager's order is such that the profit target is attained when the retail demand is no less than the order quantity, the resulting likelihood of meeting the profit target is the same as that of stocking out.

3.3.   The manufacturer's problem

The manufacturer seeks to maximize his own profit, which we denote as $\Lambda_M(c)$, by appropriately setting the wholesale price c. Since the manufacturer's choice of c will influence the retailer, we henceforth parametrize the decisions and outcomes for the retailer as functions of c [i.e., $T^*(c)$, $\overline{p}(c)$, $Q_M^*(c)$, $\pi_M^*(c)$, and $\theta^*(c)$]. So, $\Lambda_M(c) = (c - m)Q_M^*(c)$, which is a deterministic function since the retail order quantity is determined prior to the realization of the stochastic market demand, and the manufacturer produces exactly to that order. In fact, $Q_M^*(c)$ is precisely the demand curve perceived by the manufacturer, and the choice of c simply boils down to the standard exercise of trading off profit margin against sales volume. Insight into this decision may be obtained by examining the effect of c on the retailer's decisions. These comparative statics results are presented in Theorem 3.3.

Theorem 3.3 *As c increases, the equilibrium for system M changes in the following ways:*

(i) the retail manager will increase the retail price, but by no more than half the increase in the wholesale price ($0 \leq \frac{d\overline{p}(c)}{dc} \leq \frac{1}{2}$), which implies that the unit contribution to net profit decreases ($\frac{d(\overline{p}(c) - c)}{dc} \leq -\frac{1}{2}$),

(ii) the retailer's order will decrease ($\frac{dQ_M^*(c)}{dc} \leq 0$),

(iii) the retail owner will lower the manager's profit target ($\frac{dT^*(c)}{dc} \leq 0$),

(iv) the retail manager's probability of meeting the profit target will increase ($\frac{d\theta^*(c)}{dc} \geq 0$), and

(v) the retail owner's expected profit will decrease ($\frac{d\pi_M^*(c)}{dc} \leq 0$).

Having studied the impact of c on the retailer, we turn our attention to the manufacturer's preferences towards c.

Theorem 3.4 *There is some $c \in (m, p_{max})$ for which the manufacturer's profits are maximized.*

While increasing c increases the manufacturer's profit per unit, eventually this is overwhelmed by the decrease in the quantity demanded by the retailer. So, there is some threshold below p_{max} at which the manufacturer would prefer not to increase the wholesale price. However, the optimal c, which we refer to as c_M^*, cannot be established analytically without assuming further structure on the form of the market demand distribution. The method for doing this is illustrated for a numerical example in Figure 3.2. The manufacturer's profit is $\Lambda_M^* = \Lambda_M(c_M^*)$, and then the expected total profit for the system will be denoted by $\Omega_M^* (= \pi_M^* + \Lambda_M^*)$.

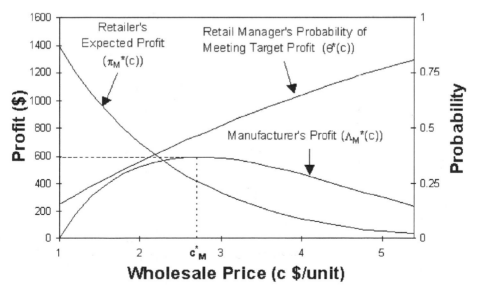

Figure 3.2. Manufacturer's Profit, Retailer's Expected Profit, and Retail Manager's Probability of Meeting Profit Target vs. Wholesale Price (Control System M)

3.4.   Discussion of control system M

Our study of this system reveals two key sources of inefficiency. The first is due to the delegation of authority within the retail entity. Because the retail manager and the owner have different criteria for success, the resulting price and quantity differ from those that the owner himself would select. Certainly, because the owner can influence the quantity decision through the specification of T, a quantity will result that maximizes expected profit for *whatever retail price is chosen*. However, the retail price chosen will, in general, not be optimal with respect to expected profit since this is not the retail manager's objective, and T has no leverage over the price decision. A second source of inefficiency comes from the interplay with the negotiation of the wholesale price. We show above that, all else being equal, there is a disparity between the owner's preferences for c and the manager's. Part (v) of Theorem 3.3 indicates that the retail owner prefers as low a wholesale price as possible (ideally $c = m$), and will oppose any increase the manufacturer might propose. However, surprisingly the retail manager actually prefers as high a c as possible, since increasing c increases the probability of attaining the profit target (part (iv) of Theorem 3.3). (This is because raising c causes the retail manager to lower the target.) So, different outcomes will result depending on who on the retail side negotiates the contract. If the owner negotiates c, all we know is that the wholesale price must be sufficiently low that his reservation profit is attained. If the retail manager negotiates, he will acquiesce to whatever the manufacturer proposes, no matter how much expected retail profit is sacrificed in the process.

4.   Analysis of Control System R

In this first benchmark, the manufacturer sets a wholesale price and the retail owner determines the retail price and order quantity, each party wishing to maximize individual expected profits. The retail manager plays no role.

4.1.   The retail owner's problem

Here the retailer's expected profit is

$$\pi_R(p, Q) = \int_0^Q (pD - cQ + s(Q - D))\phi(D)\,dD + \int_Q^\infty (p - c)Q\phi(D)\,dD$$

$$= (p - s)\widehat{\mu} - (c - s)Q - (p - s)g(p)\int_{Q/g(p)}^\infty [1 - F(y)]\,dy. \quad (3.8)$$

The retailer chooses p and Q to jointly maximize this, as indicated in the following theorem.

Theorem 3.5 *The retailer's behavior in system R has the following properties:*

(i) for any given selling price, the order quantity is

$$Q_R^*(p) = \Phi^{-1}\left(\frac{p-c}{p-s}\right) = g(p) \cdot F^{-1}\left(\frac{p-c}{p-s}\right). \qquad (3.9)$$

(ii) the selling price, denoted as p_R^*, may be obtained as the solution to the following condition,

$$\frac{(p-c)\,g'(p) + g(p)}{(c-s)\,g'(p)} + 1 = \frac{\nu(p)}{X(p)}, \quad where \qquad (3.10)$$

$$\nu(p) = F^{-1}\left(\frac{p-c}{p-s}\right) \quad and \quad X(p) = \int_0^{\nu(p)} [1 - F(y)]\,dy < \nu(p)$$

(iii) for any given c, p_R^* is greater than the price that results under system M, i.e. $c < p_M^* < p_R^* < p_{max}$.

(iv) the resulting retail profit is

$$\pi_R(p_R^*, Q_R^*(p_R^*)) = (p_R^* - s)\,g(p_R^*) \int_0^{F^{-1}\left(\frac{p_R^* - c}{p_R^* - s}\right)} y f(y)\,dy \qquad (3.11)$$

This theorem indicates that for a given wholesale price, system M provides end customers with a lower purchase price and more product to buy than does system R. A framework for evaluating the net benefit to consumers is illustrated in §6.

4.2.   The manufacturer's problem

The manufacturer prefers a c that will maximize his profits, denoted as $\Lambda_R(c) = (c - m)Q_R^*(p_R^*(c))$. As before, we parametrize all retailer decisions and outcomes to make explicit the influence of c [e.g., $p_R^*(c)$, $Q_R^*(p_R^*(c)), \pi_R^*(p_R^*(c))$].

Establishing analytically the properties of the manufacturer's preferred wholesale price, which we denote c_R^*, is difficult because $c_R^* = argmax_c[(c - m)g(p_R^*(c))F^{-1}((p_R^*(c) - c)/(p_R^*(c) - s))]$, and characterizing $p_R^*(c)$ in closed form is itself problematic. (See [16] for elaboration on the confounding factors.) However, c_R^* can be determined numerically, and its properties will be studied in the context of a numerical example later. The manufacturer's equilibrium profit is $\Lambda_R^* = \Lambda_R(c_R^*) = (c_R^* - m)Q_R^*(p_R^*(c_R^*))$, and the retailer's is

$$\pi_R^* = \pi_R(p_R^*(c_R^*), Q_R^*(p_R^*(c_R^*))).$$

We will denote the expected total supply chain profit as $\Omega_R^* = \Lambda_R^* + \pi_R^*$.

5.   Analysis of Control System C

In this, the first-best case, the selling price and quantity are set to maximize expected total supply chain profit. This is essentially a standard newsvendor problem with the pricing decision included (cf. [27] for a review, and [26] for additional details).

For a given selling price p and order quantity Q, the supply chain expected profit is

$$\Lambda_C(p, Q)$$
$$= \int_0^Q (pD - mQ + s(Q - D))\, \phi(D)\, dD + \int_Q^\infty (p - m)Q\phi(D)\, dD$$
$$= (p - s)\widehat{\mu} - (m - s)Q - (p - s)g(p) \int_{Q/g(p)}^\infty [1 - F(y)]\, dy. \qquad (3.12)$$

The following theorem establishes the behavior under this control system.

Theorem 3.6 *The optimal outcome in system C has the following properties:*

(i) for any given selling price, the order quantity is the following newsvendor solution:

$$Q_C^*(p) = \Phi^{-1}\left(\frac{p - m}{p - s}\right) = g(p) \cdot F^{-1}\left(\frac{p - m}{p - s}\right)$$

(ii) the selling price, denoted as p_C^*, may be obtained as the solution to the following condition,

$$\frac{(p - m)\, g'(p) + g(p)}{(m - s)\, g'(p)} + 1 = \frac{\nu(p)}{X(p)}, \quad where, \qquad (3.13)$$

$$\nu(p) = F^{-1}\left(\frac{p - m}{p - s}\right) \quad and \quad X(p) = \int_0^{\nu(p)} [1 - F(y)]\, dy < \nu(p).$$

(iii) the resulting total supply chain profit is

$$\Omega_C^* = \Lambda_C(p_C^*, Q_C^*(p_C^*)) = (p_C^* - s)g(p_C^*) \int_0^{F^{-1}\left(\frac{p_C^* - m}{p_C^* - s}\right)} y f(y)\, dy \qquad (3.14)$$

For reasons outlined in the previous section, closed forms for p_C^* and $Q_C^*(p_C^*)$ are not available.

6.   Comparing the Control Systems

In this section we compare the behavior of the decision makers across the three control systems. Since only limited analytical results are available for the general model, to enable explicit illustration of key properties we further specify the demand model as follows:

number of customers	N is exponentially distributed[6] with mean μ, so $F(y) = 1 - \exp(-y/\mu)$
demand per customer	$g(p) = a - bp^2 \quad (a, b > 0)$[7]

The resulting expressions for the equilibria under each control system are presented in Tables 3.3-3.5 in the Appendix. These indicate that a and b affect the outcome for each system only through the ratio a/b. Further, while the equilibrium wholesale price c is a function of the manufacturer's m, for any given c the retailer's behavior and outcomes are independent of m. Consequently, we lose no generality by using $b = 1$ and $m = \$1.00/$unit in our numerical analysis, which reduces the number of free parameters to be considered. For the base case example, we set $a = 60$ and $s = \$0.30/$unit. The choice of μ is immaterial since it is merely scales the size of the market, and we set it to 15. In the discussion that follows we will be very careful to note the extent to which the conclusions depend on the specific assumptions used for this numerical analysis. The main purpose of this exercise is to demonstrate a methodology for making such comparisons, and call attention to the possibility of certain non-obvious outcomes.

Figures 3.2 and 3.3 illustrate the results that were established analytically in Theorems 3.3 and 3.4. Specifically, Figure 3.2 shows the profits for each firm, along with the retail manager's probability of achieving his profit target, all as functions of c. Notice that as c increases, the retail profit continuously decreases, whereas the manager's likelihood of meeting his profit target increases. This reiterates one of the sources of inefficiency under such a system, as described earlier. If the retail manager were to negotiate the contract, he would allow a high wholesale price even though it opposes the interests of the retail organization. On the other hand, the owner would prefer as small a wholesale price as possible. Figure 3.3 shows the impact of c on the various retail decisions (p, T, and Q), based on the behaviors derived in §3. Consistent with the analysis presented in Theorem 3.3, as the wholesale price increases, the retail owner is forced to reduce the profit target for the retail manager, who in turn makes fewer units available to the market, and at a

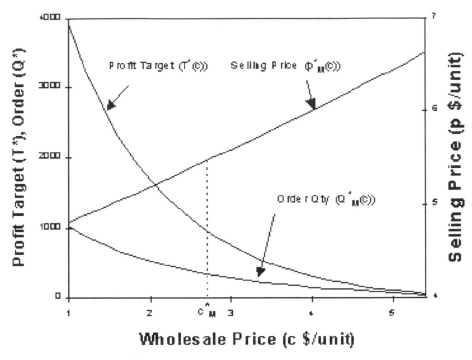

Figure 3.3. Profit Target, Order Quantity, and Selling Price vs. Wholesale Price (Control System M)

higher price. Thus, as the manufacturer increases the wholesale price, the end customer's buying price increases, which reduces the demand in the price-sensitive market.

The next two figures examine system R. Figure 3.4 shows how the retailer's expected profit and the manufacturer's profit vary with c, as established in §4.1. Figure 3.5 shows the corresponding order and retail price chosen by the retailer. As the wholesale price increases, the retail owner raises the retail price, orders less and makes less profit. The manufacturer's pricing policy therefore trades off the increased revenue per unit against the lower number of units sold. c_R^*, and hence the resulting system equilibrium, may be obtained from Figure 3.4.

Finally, we illustrate the dynamics driving the equilibrium for system C. Figure 3.6 shows how the order quantity and the system profit vary with p. The downward-sloping curve of mean customer demand $(g(p))$ is displayed for comparison. When the retail price is low, increasing it in conjunction with the order size increases profits. Although mean demand always decreases with price increases, this is more than offset by the amount collected per sale. However, beyond a certain threshold,

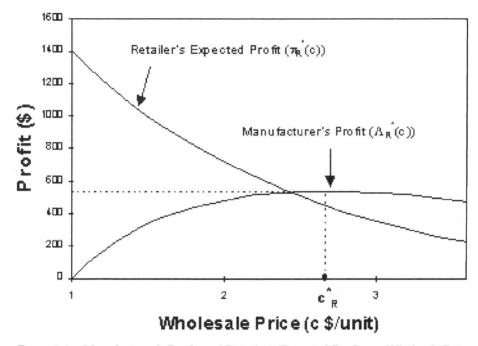

Figure 3.4. Manufacturer's Profit and Retailer's Expected Profit vs. Wholesale Price (Control System R)

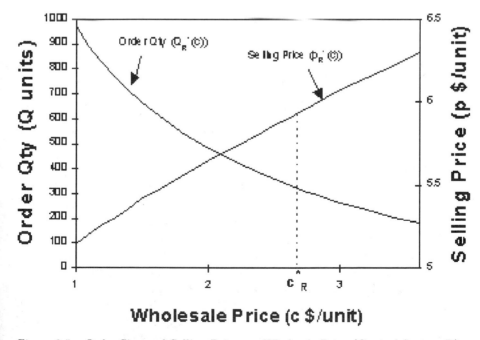

Figure 3.5. Order Size and Selling Price vs. Wholesale Price (Control System R)

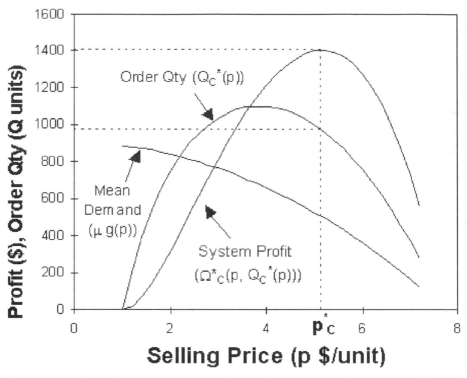

Figure 3.6. Manufacturer's Profit and Order Quantity vs. Selling Price (Control System C)

the reduction in customer demand dominates, leading to lower order quantities and profits. p_C^* is illustrated in the figure as the price at which the Ω_C reaches its peak.

We now turn our attention to comparing the three control systems. Table 3.1 summarizes the equilibrium behaviors and outcomes for all players in each case.

As expected, total expected profit is highest under the centralized control of system C. The inefficiency of system R is due to double marginalization (cf. [30]). That is, the retailer's benefit from making a sale is less than that which accrues to the supply chain as a whole, since part of the profit margin goes to the manufacturer via the wholesale price. Hence, the retailer stocks too little and prices too high relative to the supply chain optimum. System M has the additional issue of intrafirm conflict. Interestingly, though, while the retailer is less profitable in system M than in system R ($417.58 vs. $451.13), the manufacturer is better off by more than an offsetting amount ($589.20 vs. $537.95), making total profit lowest in system R[8]. This raises the possibility that reducing the extent of coordination within a supply chain need not damage the

Table 3.1. Comparison of Behavior and Performance Across Control Systems

Equilibrium Outcome	Control System		
	M	R	C
Wholesale Price (c)	$2.70	$2.67	–
Retail Manager's Profit Target (T)	$956.00	–	–
Retail Price (p)	$5.46	$5.93	$5.14
Retailer's Order Quantity (Q)	346.20	322.44	973.37
Expected Demand ($\hat{\mu} = \mu g(p)$)	452.46	372.22	503.21
Probability of Meeting All Demand	0.53	0.58	0.85
Retailer's Expected Profit (π)	$417.58	$451.13	–
Manufacturer's Profit (Λ)	$589.20	$537.95	–
Total Supply Chain Expected Profit (Ω)	$1006.78	$989.07	$1403.71
Manufacturer's Share of Total Exp. Profit	58.5%	54.4%	–

system-wide performance. Additional numerical analysis (details omitted due to space considerations) over a broad range of combinations of a (range: 10 to 100) and s (range: 0–0.8 m) has demonstrated this result to be robust to the specific values assumed in the base case.

The discussion thus far has focused on each supply chain entity's financial outcome. However, although difficult to describe using traditional constructs, the implications for the end customer should not be overlooked. A simplistic view might be that the lower the retail price, the better off are the customers. This would be reasonable if all demand were to be served. However, the stochastic premise of the newsvendor setting highlights the possibility that some demand will go unfilled, to an extent that is determined jointly by the demand realization and the stocking level. Indeed, a lower price does not necessarily make customers better off if the service level is drastically lessened as well (which is plausible, since newsvendor analysis recommends stocking less when the unit profit margin is reduced). This is demonstrated by Table 3.1, which reports that the three control systems may be ranked one way with respect to price (R > M > C) and differently with respect to the fraction of demand filled (C > R > M).

Our method for reconciling this derives from the classical economic notion of consumer surplus (cf. [33]). This measures the benefit derived by customers given a particular price outcome, and is typically used to inform public policy concerns such as antitrust policy. However, we can also view consumer surplus as a proxy for customer goodwill, which relates to long-term profits. Hence, this becomes relevant to profit-minded supply chain managers as well. Since the reality for such managers usually includes stochastic demand, we must modify the construct to account for the number of customers actually satisfied at a given price-quantity combination, as described below.

Since all customers are assumed to be identical, we can define the total consumer surplus for a given price, quantity, and demand realization, denoted by $TCS\,(p, Q, N)$, as follows:

$$TCS\,(p, Q, N) = CS(p) \cdot \min(N, Q/g\,(p))\,.$$

Here $CS(p)$ is the consumer surplus per customer $[= \int_p^{p_{max}} g(z)\,dz]$, and $\min(N, Q/g(p))$ gives the number of customers fully satisfied. Then, the expected total consumer surplus can be computed as:

$$E_N[TCS(p, Q, N)] = CS(p) \cdot E_N[\min(N, Q/g(p))].$$

Note that when demand is deterministic (in which case $Q = N \cdot g(p)$ is the appropriate quantity), the traditional consumer surplus construct is recovered. Alternatively, when the retail price is held fixed as in the basic newsvendor model, this metric is proportional to the fill rate. Table 3.2 summarizes the calculations comparing the three control systems. It suggests two observations. First, note that coordinating the supply chain in a way that maximizes total supply chain profit need not be detrimental to customers, as expected consumer surplus is highest for C. This extends a property observed in deterministic economic models, where eliminating double marginalization unambiguously improves consumer welfare since more customers are satisfied, and each is paying less ([31]). In our example the system with the lowest efficiency, R, turns out to provide the lowest consumer surplus. Since much of the economic discussion regarding the merits of vertical integration does not explicitly incorporate the intrafirm decision hierarchies and incentive systems, and the resulting consequences for consumer surplus, we believe that this observation suggests a possible topic for further investigation, both theoretical and empirical. Second, comparing the surplus

Table 3.2.   Comparison of Consumer Welfare Across Control Systems.

	Control System		
Equilibrium Outcome	M	R	C
Retail Price (p^*)	$5.46	$5.93	$5.14
Product Availability (Q^*)	346.20	322.44	973.37
Probability of Meeting All Demand $(\mathrm{Prob}\{N \cdot g\,(p^*) \le Q^*\})$	0.53	0.58	0.85
Consumer Surplus Per Customer Served $(CS\,(p^*))$	36.60	23.60	46.80
Expected Number of Customers Served $(E[\min(N, Q^*/g(p^*))])$	2.68	3.23	8.63
Expected Consumer Surplus $(E_N[TCS(p^*, Q^*)])$	**97.80**	**76.40**	**403.20**

per customer to the fraction of demand met from stock reinforces the idea that a higher service level need not be in the best interests of the customer base. Here, a greater proportion of interested customers' demand is met under R, but the surplus per consumer is lower than under M. The net effect is that M provides a higher total consumer surplus than R. So even though the retail organization suffers from internal inefficiencies under M (leading to lower expected profits for the retailer), under the assumptions of this study the supply chain is more efficient and the consumers are better off under this regime than under R. In this case the cost of supply chain inefficiency is borne by the retailer, not the manufacturer or the consumer.

7. Conclusion

In this chapter, we have developed a more detailed representation of decision making in supply chains than has appeared in the operations management literature. We have incorporated the very real premise that not all decision-making is guided by the maximization of expected profit (or minimization of expected cost). By examining the behavioral consequences of intrafirm incentives, we have been able to shed light on certain key inefficiencies. This, we believe, is a significant contribution to the emerging operations management literature on supply contracts.

Our analysis reveals that when goal incongruence exists within any organization, the outcome of any contract negotiation will be highly dependent upon who is involved in the negotiations. In our model, for example, while the retail owner would be expected to pressure the manufacturer for a lower wholesale price, the retail manager has no such agenda. A knowledge of such a possibility is obviously crucial to the owner of any business organization.

In contrast to extant supply chain literature, we have also examined the consequences for the end customer by extending the classical economic notion of consumer surplus to a stochastic environment. This composite measure is useful since standard metrics (selling price, stock level, or the fraction of customer demand filled) cannot individually characterize whether a particular control system makes customers better or worse off. Our analysis suggests that increasing the profitability of a supply chain need not be at the expense of the end customer. This notion has been developed in the economics literature, but primarily for deterministic settings.

Comparison across the three control systems revealed a counterintuitive observation. While the effect of double marginalization on supply chain efficiency is as expected, the effect of goal incongruence within an organization (the retail entity) is not. While, as would be expected, we obtained a higher expected retail profit in system R than in system M, this was not the case for the total system profit. We found that the internal conflict of goals within the retail organization can potentially counter the effect of double marginalization, since supply chain efficiency can be higher in system M than in system R. In other words, coordination of goals within an organization does not necessarily improve the efficiency of the supply chain. Whether this phenomenon results in more general settings will be, we believe, an important area of future research.

Clearly, the profit target criterion is only one alternative to expected profit. Our analysis could be further extended by considering behavioral responses to other intrafirm incentive scheme such as profit-sharing (reward as a percentage of profits), commission (reward as a percentage of sales), or multi-tiered compensation plans. The question of how to remedy the intrafirm goal incongruence remains open.

Finally, the manufacturer-retailer supply contract could take on more general structure. We have assumed only a linear wholesale price contract, and have not pursued the question of optimal contract design.

8.   Appendix: Proofs of Theorems

PROOF OF THEOREM 3.1. To obtain the p and Q that maximize $\theta(p, Q)$, we first compute the Q that is optimal for a given p. This result has been developed in several papers, including [19], [25] and [29]. We provide a slightly different proof here.

Clearly, if $Q \leq T/(p - c)$, $\theta(p, Q) = 0$. Therefore, to reach the target profit at all we must consider only values of Q such that $Q \geq T/(p - c)$. Then, $Z_M(p, Q) = T$ when $pD - cQ + s(Q - D) = T$, or $D = (T + (c - s)Q)/(p - s)$. Thus

$$\theta(p, Q) = Pr\left(D \geq \frac{T + (c - s)Q}{p - s}\right).$$

This is maximized by setting Q to the smallest allowable value, i.e., $Q_M^*(T, p) = T/(p - c)$.

The optimal objective function value for a fixed p, $\theta(p, Q_M^*(T, p))$, is then

$$\theta(p, Q_M^*(T, p)) = Pr\left(D \geq \frac{T + (c - s)Q_M^*(T, p)}{p - s}\right)$$

$$= Pr\left(D \geq \frac{T}{p - c}\right) \tag{3.15}$$

$$= 1 - \Phi\left(\frac{T}{p - c}\right) = 1 - F\left(\frac{T}{H(p)}\right) \tag{3.16}$$

where $H(p) = (p - c) g(p)$. This proves (i). Since $F()$ is monotone, the price that maximizes $H(p)$ does the same for $\theta(p, Q_M^*(T, p))$. Under assumptions (i)–(iv) on $g(p)$, $H(p)$ is a continuous function that is strictly positive everywhere in (c, p_{max}) and non-positive on the rest of its domain. Hence, an interior maximum exists within (c, p_{max}), and satisfies $H'(p) = 0$. Since $H''(p) = g(p) (p - c) + 2g'(p) < 0$ on this region, the solution to the first order conditions (FOC) must be unique. The FOC are precisely the conditions stated in (ii). (iii) and (iv) then follow from (i) and (ii).

PROOF OF THEOREM 3.2. By Theorem 3.1, the manager will set a retail price of $\bar{p}$ regardless of the T specified. From the newsvendor structure of the objective, for any p the profit maximizing quantity will be $\Phi^{-1}((p - c)/(p - s)) = g(p) \cdot F^{-1}((p - c)/(p - s))$. The value of T^* specified in (3.4) is then appropriate, since $Q_M^*(T, p) = T/(p - c)$ from (3.0), which then determines (3.5). (3.6) may be obtained by direct evaluation of $E[Z_M(\bar{p}, Q_M^*(T^*, \bar{p}))]$, and (3.7) follows from (3.3).

PROOF OF THEOREM 3.3. (i) Recall from Theorem 3.1 that $\bar{p}(c)$ is defined by the condition $(\bar{p}(c) - c) g'(\bar{p}(c)) + g(\bar{p}(c)) = 0$. By implicit differentiation,

$$\frac{d\bar{p}(c)}{dc} = \frac{1}{2 + (\bar{p}(c) - c) g''(\bar{p}(c)) / g'(\bar{p}(c))}. \tag{3.17}$$

Since $g'' \leq 0$ and $g' > 0$, $0 \leq d\bar{p}(c)/dc \leq 1/2$. This implies that

$$\frac{d(\bar{p}(c) - c)}{dc} \leq -\frac{1}{2}.$$

(ii) We know that

$$Q_M^*(c) = \Phi^{-1}\left(\frac{\bar{p}(c) - c}{\bar{p}(c) - s}\right). \tag{3.18}$$

Differentiating with respect to c, we get

$$\frac{dQ_M^*(c)}{dc} = -\frac{(\bar{p}(c) - s) - (c - s) \frac{d\bar{p}(c)}{dc}}{(\bar{p}(c) - s)^2 \phi(Q_M^*(T^*(c), \bar{p}(c)))} \leq 0.$$

The inequality follows since $0 \leq d\bar{p}(c)/dc \leq 1/2$ by part (i).

(iii) Since $T^*(c) = (\bar{p}(c) - c) Q_M^*(c)$, the previous two results indicate that $dT^*(c)/dc \leq 0$.

(iv) For a given c, the retail manager's optimal probability of achieving the profit target is

$$\theta^*(c) = 1 - \Phi\left(Q_M^*(c)\right) = \frac{c - s}{\overline{p}(c) - s}. \tag{3.19}$$

By the chain rule,

$$\frac{d\theta^*(c)}{dc} = -\phi\left(Q_M^*(c)\right)\frac{dQ_M^*(c)}{dc} \geq 0 \tag{3.20}$$

where the inequality is due to part (ii).

(v) Equation (3.6) provides the owner's optimal expected profit, which we rewrite in the following way to highlight the dependence on c:

$$\pi_M^*(c) = (\overline{p}(c) - s)\int_0^{Q_M^*(c)} D\phi(D)\,dD,$$

where $Q_M^*(c) = \Phi^{-1}\left((\overline{p}(c) - c)/(\overline{p}(c) - s)\right)$. By taking the total derivative of $\pi_M^*(c)$,

$$\frac{d\pi_M^*(c)}{dc} = \frac{d\pi_M^*(c)}{d\overline{p}(c)}\frac{d\overline{p}(c)}{dc} + \frac{\partial\pi_M^*(c)}{\partial c}.$$

Now

$$\frac{\partial\pi_M^*(c)}{\partial c} = \frac{\partial\pi_M^*(c)}{\partial Q_M^*(c)}\frac{dQ_M^*(c)}{dc}$$

$$= \left[(\overline{p}(c) - s)Q_M^*(c)\phi\left(Q_M^*(c)\right)\right]\left[\frac{-1}{(\overline{p}(c) - s)\phi\left(Q_M^*(c)\right)}\right]$$

$$= -Q_M^*(c),$$

and

$$\frac{d\pi_M^*(c)}{d\overline{p}(c)} = (\overline{p}(c) - s)\left[Q_M^*(c)\phi(Q_M^*(c))\frac{dQ_M^*(c)}{d\overline{p}(c)}\right] + \int_0^{Q_M^*(c)} D\phi(D)\,dD.$$

Since

$$\frac{dQ_M^*(c)}{d\overline{p}(c)} = \frac{c - s}{(\overline{p}(c) - s)^2\,\phi\left(Q_M^*(c)\right)},$$

this simplifies to

$$\frac{d\pi_M^*(c)}{d\overline{p}(c)} = \frac{c - s}{\overline{p}(c) - s}Q_M^*(c) + \int_0^{Q_M^*(c)} D\phi(D)\,dD.$$

So,

$$
\begin{aligned}
\frac{d\pi_M^*(c)}{dc} &= \left[\frac{c-s}{\overline{p}(c)-s} Q_M^*(c) + \int_0^{Q_M^*(c)} D\phi(D)dD \right] \frac{d\overline{p}(c)}{dc} - Q_M^*(c) \\
&\le \frac{1}{2} \left[\frac{c-s}{\overline{p}(c)-s} Q_M^*(c) + \int_0^{Q_M^*(c)} D\phi(D)dD \right] - Q_M^*(c) \\
&\le \frac{1}{2} \left[\frac{c-s}{\overline{p}(c)-s} Q_M^*(c) + Q_M^*(c)\Phi(Q_M^*(c)) \right] - Q_M^*(c) \\
&= -\frac{1}{2} Q_M^*(c) \le 0.
\end{aligned}
$$

This first inequality follows from part (i) of this theorem. The second is obtained by noting that the integral is defined on $D \le Q_M^*(c)$, so that $\int_0^{Q_M^*(c)} D\phi(D)\,dD \le Q_M^*(c)\int_0^{Q_M^*(c)} \phi(D)\,dD$. The subsequent equality follows from the fact that $\Phi(Q_M^*(c)) = (\overline{p}(c)-c)/(\overline{p}(c)-s)$.

PROOF OF THEOREM 3.4. We examine how $\Lambda_M(c) = (c-m)\,Q_M^*(c)$ changes as c increases. By differentiation we obtain

$$
\frac{d\Lambda_M(c)}{dc} = (c-m)\frac{dQ_M^*(c)}{dc} + Q_M^*(c)
$$

At $c = m$, $d\Lambda_M(c)/dc = Q_M^*(c) > 0$, which verifies that the manufacturer prefers a wholesale price higher than the manufacturing cost. And as $c \to p_{max}$, $Q_M^*(c) \to 0$, while $dQ_M^*(c)/dc < 0$, so $d\Lambda_M(c)/dc < 0$. Hence, there exists some $c \in (m, p_{max})$ that maximizes manufacturer profit.

PROOF OF THEOREM 5. Part (i) follows from the newsvendor structure. For (ii) and (iii), to obtain p_R^* we characterize the shape of $\pi_R(p, Q_R^*(p))$, which by (3.8) has the form

$$
\pi_R(p, Q_R^*(p)) = (p-s)\widehat{\mu} - (c-s)Q_R^*(p)
$$
$$
-(p-s)g(p) \int_{Q_R^*(p)/g(p)}^{\infty} [1 - F(y)]\,dy \qquad (3.21)
$$

First, note that $\pi_R(p, Q_R^*(p)) = 0$ at both $p = c$ and $p = p_{max}$. This is intuitively obvious, and may also be obtained directly from (3.21). Then, applying the Envelope Theorem (cf. [33]) to (3.21) and using the notation $\nu(p)$ and $X(p)$ as specified, we get

$$\frac{\partial \pi_R(p, Q_R^*(p))}{\partial p} = -(c-s)\nu(p)g'(p) + [(p-s)g'(p) + g(p)]X(p),$$

(3.22)

which may be rewritten (by adding and subtracting $cg'(p)X(p)$) as

$$\frac{\partial \pi_R(p, Q_R^*(p))}{\partial p}$$
$$= (c-s)g'(p)[X(p) - \nu(p)] + [(p-c)g'(p) + g(p)]X(p) \qquad (3.23)$$

We know $g'(p) < 0$ for all p, and $[(p-c)\,g'(p) + g(p)] \geq 0$ on $p \leq \bar{p} = p_M^*$ by the definition of $\bar{p}$. And since $X(p) < \nu(p)$ by construction[9], we conclude that $\partial \pi_R(p, Q_R^*(p))/\partial p > 0$ on $c \leq p \leq \bar{p}$. Hence, it must be the case that the retailer's optimal price, denoted as p_R^* and obtained by setting (3.23) equal to zero, must lie somewhere in $(\bar{p}, p_{max})$ since the endpoints of this interval each yield exactly zero profit.

We note that the first order condition is necessary but not sufficient to uniquely characterize the p_R^* that will result. That is, without additional restrictions on the problem, the existence of multiple candidate selling prices cannot be ruled out. But we do know that all of these candidates will be greater than the price that occurs under system M.

(iv) follows by analogy to equation (3.6).

PROOF OF THEOREM 6. The proof is identical to that of Theorem 5, except that the cost of the product is the manufacturing cost m, rather than the wholesale price c.

OUTCOMES FOR EXPONENTIALLY DISTRIBUTED N.
The decisions and outcomes under systems M and R for any fixed wholesale price (c) are presented below.

Table 3.3. Decisions and Outcomes for a Given c in System M

Retail Price	$p_M^*(c) = \bar{p} = \left(c + \sqrt{c^2 + \frac{3a}{b}}\right)/3$
Retail Order Quantity	$Q_M^*(c) = \mu \cdot g(\bar{p}) \cdot ln\left(\frac{\bar{p}-s}{c-s}\right)$
Profit Target Set by Retail Owner	$T^*(c) = \mu \cdot (\bar{p} - c) \cdot g(\bar{p}) \cdot ln\left(\frac{\bar{p}-s}{c-s}\right)$
Expected Retail Profit	$\pi_M^*(c) = \mu \cdot g(\bar{p}) \cdot \left[(\bar{p}-c) - (c-s)\,ln\left(\frac{\bar{p}-s}{c-s}\right)\right]$
Prob{Retail Profit $\geq T^*$}	$\theta^*(c) = \frac{c-s}{\bar{p}-s}$ (this is independent of the distribution of N)
Manufacturer Profit	$\Lambda_M(c) = \mu \cdot g(\bar{p}) \cdot (c-m) \cdot ln\left(\frac{\bar{p}-s}{c-s}\right)$

Table 3.4. Decisions and Outcomes for a Given c in System R

Retail Price	$p_R^*(c)$ is the p that solves $\frac{3p^2 - 2ps - \frac{a}{b}}{2p(c-s)}$ $= \frac{\mu \cdot ln\left(\frac{p-s}{c-s}\right)}{1 - \mu \cdot \left(\frac{c-s}{p-s}\right)}$ (from (3.10))
Retail Order Quantity	$Q_R^*(c) = \mu \cdot g(p_R^*) \cdot ln\left(\frac{p_R^* - s}{c-s}\right)$
Expected Retail Profit	$\pi_R^*(c) = \mu \cdot g(p_R^*) \cdot \left[(p_R^* - c) - (c-s)ln\left(\frac{p_R^* - s}{c-s}\right)\right]$
Manufacturer Profit	$\Lambda_R(c) = \mu \cdot g(p_R^*) \cdot (c-m) \cdot ln\left(\frac{p_R^* - s}{c-s}\right)$

Table 3.5. Decisions and Outcomes in System C

Retail Price	p_C^* is the p that solves $\frac{3p^2 - 2ps - \frac{a}{b}}{2p(m-s)}$ $= \frac{\mu \cdot ln\left(\frac{p-s}{m-s}\right)}{1 - \mu \cdot \left(\frac{m-s}{p-s}\right)}$ (from (3.13))
Retail Order Quantity	$Q_C^* = \mu \cdot g(p_C^*) \cdot ln\left(\frac{p_C^* - s}{m-s}\right)$
Total Supply Chain Expected Profit	$\Lambda_C^* = \mu \cdot g(p_C^*) \cdot$ $\left[(p_C^* - m) - (m-s)ln\left(\frac{p_C^* - s}{m-s}\right)\right]$

Acknowledgments

The authors would like to thank Gerard Cachon, Sandra Chamberlain, Marty Lariviere, Steve Nahmias, Bill Sundstrom, and Garrett van Ryzin for comments that have greatly assisted in the refining of our ideas. Any errors remain the responsibility of the authors.

Notes

1. Decision-making that focuses on achieving some threshold level of utility (rather than, say, maximizing the expected utility) has been termed "satisficing" ([24]).

2. The reward for achieving the target may take the concrete form of a monetary bonus, or simply the retention of the owner's goodwill. An empirical account of the latter is provided by [2], and our analysis will focus primarily on this case. Non-monetary incentive systems are common. In fact, some of the popular business literature on workforce motivation suggests that employees are more effectively motivated by rewards that cost very little in real terms, such as public recognition, greater job responsibilities, etc. (e.g., [15]). Similarly, employee stock options are an increasingly common incentive tool that do not have an immediate cost consequence to the firm, at least per the current accounting standards ([5]). The value of the options is tied that of the company's stock, which oftentimes is greatly influenced by whether profit expectations have been met.

3. Sales quotas, common in sales and merchandising settings ([21]), are one manifestation of this concept, as the quota essentially maps into a profit figure.

4. In the newsvendor setting, the ex post profit is largest when the demand outcome exactly matches the available quantity, thus avoiding both shortage and excess. In such a scenario, the full profit margin is earned on every unit stocked, for a total profit of $(p-c)\,Q$.

5. It is relatively straightforward to show that if the owner uses a cash bonus payment to motivate the manager, the order quantity will go up, yet another source of inefficiency. This will be true even though any bonus is merely an internal transfer of funds within the retail entity, due to the distortion of individual incentives. We do not present these details here since they are not the main thrust of our analysis. They are available from the authors by request.

6. The assumption of exponential distribution is made strictly for convenience. Similar outcomes follow if, for example, the uniform distribution is used.

7. This specific form of $g(p)$ satisfies the conditions specified in §2, with $p_{max} = \sqrt{a/b}$.

8. Analytically, it can be shown that (i) $\pi_R^* \geq \pi_M^*$, and, (ii) $\Omega_C^* \geq \Omega_R^*, \Omega_M^*$. Since control system R does not suffer from goal incongruence between the retail owner and manager, its expected retail profit is higher than that of control system M. However, the effect of the retail manager's actions on the manufacturer's profit is indeterminate. Hence the relative magnitudes of Ω_R^* and Ω_M^* are indeterminate as well. Thus, while it is clear that the total system profit is the highest in the first-best case, it is difficult to analytically compare the system profits under systems R and M.

9. The strictness of the inequality in the statement of $X(p)$ is required to make some of the claimed inequalities strict. Strictness is reasonable for all $p > c$ because this requires only that there is some $y < \nu(p)$ for which $F() > 0$ on the neighborhood around y. Only at $p = c$ is strictness lost, for at this point $X(p) = \nu(p) = 0$.

References

[1] ATKINSON, A.A. (1979) Incentives, Uncertainty, and Risk in the Newsboy Problem. *Decision Sciences*, Vol. 10, 341–357.

[2] BLACKWELL, D.W., J.A. BRICKLEY, AND M.S. WEISBACH. (1994) Accounting Information and Internal Performance Evaluation. *Journal of Accounting and Economics*, Vol. 17, 331–358.

[3] BORDLEY, R. AND M. LICALZI. (1998) Decision Analysis Using Targets Instead of Utility Functions. *Working Paper, Knowledge Network, GM Research Development Center.*

[4] EMMONS, H. AND S.M. GILBERT. (1998) Note: The Role of Returns Policies in Pricing and Inventory Decisions for Catalogue Goods. *Management Science*, Vol. 44, No. 2, 276–283.

[5] FOX, J. (1997) The Next Best Thing to Free Money. *Fortune*, July 7, 52–62.

[6] GAVER, J.J., K.M. GAVER, AND J.R. AUSTIN. (1995) Additional Evidence on Bonus Plans and Income Management. *Journal of Accounting and Economics*, Vol. 19, 3–28.

[7] GALLEGO, G., AND G. VAN RYZIN. (1994) Optimal Dynamic Pricing of Inventories with Stochastic Demand Over Finite Horizons. *Management Science*, Vol. 40, No. 8, 999–1020.

[8] GEOFFRION, A.M. (1967) Stochastic Programming With Aspiration or Fractile Criterion. *Management Science*, Vol. 13, 672–679.

[9] HARLOW, W.V. (1991) Asset Allocation in a Downside Risk Framework. *Financial Analysts Journal*, Sept–Oct, 28–40.

[10] HEALY, P.M. (1985) The Effect of Bonus Schemes on Accounting Decisions. *Journal of Accounting and Economics*, Vol. 7, 85–107.

[11] HOLTHAUSEN, R.W., D.F. LARCKER, AND R.G. SLOAN. (1995) Annual Bonus Schemes and the Manipulation of Earnings. *Journal of Accounting and Economics*, Vol. 19, 29–74.

[12] ISMAIL, B.E. AND J.G. LOUDERBACK. (1979) Optimizing and Satisficing in Stochastic Cost-Volume-Profit Analysis. *Decision Sciences*, Vol. 10, No. 2, 205–217.

[13] KARLIN, S. AND C.R. CARR. (1962) Prices and Optimal Inventory Policies. *in* K.J. Arrow, S. Karlin, and H. Scarf (Eds.), Studies in *Applied Probability and Management Science*, Stanford University Press, Stanford, CA.

[14] KHOUJA, M. (1995) The Newsboy Problem Under Progressive Multiple Discounts. *European Journal of Operational Research*, Vol. 84, No. 2, 458–466.

[15] KOUZES, J.M. AND B.Z. POSNER. (1995) *The Leadership Challenge, Jossey-Bass, San Francisco.*

[16] LARIVIERE, M.A. (1999) Supply Chain Contracting and Co-ordination with Stochastic Demand. *in Quantitative Models for Supply Chain Management*, S. Tayur, R. Ganeshan, and M. Magazine (Eds.), Kluwer, Norwell, MA.

[17] LAU, A.H. AND H. LAU. (1988) The Newsboy Problem with Price-Dependent Demand Distribution. *IIE Transactions*, Vol. 20, No. 2, 168–175.

[18] LAU, A.H. AND H. LAU. (1988) Maximizing the Probability of Achieving a Target Profit in a Two-Product Newsboy Problem. *Decision Sciences*, Vol. 19, No. 2, 392–408.

[19] LAU, H. (1980) The Newsboy Problem Under Alternative Optimization Objectives. *Journal of the Operational Research Society*, Vol. 31, 525–535.

[20] LAU, H. (1980) Some Extensions of Ismail-Louderbacks's Stochastic CVP Model Under Optimizing and Satisficing Criteria. *Decision Sciences*, Vol. 11, No. 3, 557–561.

[21] LEE, H.L., P. PADMANABHAN, AND S. WHANG. (1997) The Bullwhip Effect in Supply Chains. *Sloan Management Review*, Vol. 38, No. 3, 93–102.

[22] LI, J., H. LAU, AND A.H. LAU. (1990) Some Analytical Results For a Two-Product Newsboy Problem. *Decision Sciences*, Vol. 21, No. 4, 710–726.

[23] LI, J., H. LAU, AND A.H. LAU. (1991) A Two-Product Newsboy Problem with Satisficing Objective and Independent Exponential Demands. *IIE Transactions*, Vol. 23, No. 1, 29–39.

[24] MARCH J.G. AND H.A. SIMON. (1958)*Organizations,* Wiley, New York, NY.

[25] NORLAND, R.E. (1980) Refinements in the Ismail-Louderback Stochastic CVP Model. *Decision Sciences*, Vol. 11, No. 3, 562–572.

[26] PETRUZZI, N.C. AND M. DADA. (1999) Pricing and the Newsvendor Problem: A Review with Extensions. *Operations Research*, Vol. 47, No. 2, 183–194.

[27] PORTEUS, E.L. (1990) Stochastic Inventory Theory. in D.P. Heyman and M.J. Sobel (Eds.), *Handbooks in Operations Research and Management Science,* Vol. 2 (Stochastic Models), Elsevier Science Publishing Company, New York, NY.

[28] SANKARASUBRAMANIAN, E. AND S. KUMARASWAMY. (1983) Note on 'Optimal Ordering Quantity to Realize a Pre-Determined Level of Profit.' *Management Science*, Vol. 29, No. 4, 512–514.

[29] SHIH, W. (1979) A General Decision Model for Cost-Volume-Profit Analysis Under Uncertainty. *The Accounting Review*, Vol. 54, No. 4, 687–706.

[30] SPENGLER, J.J. (1950) Vertical Restraints and Antitrust Policy. *Journal of Political Economy*, Vol. 58, 347–352.

[31] TIROLE, J. (1988) The Theory of Industrial Organization, The MIT Press, Cambridge, MA.

[32] TSAY, A.A., S. NAHMIAS, AND N. AGRAWAL. (1999) Modeling Supply Chain Contracts: A Review. *in Quantitative Models for Supply Chain Management,* S. Tayur, R. Ganeshan, and M. Magazine (Eds.), Kluwer, Norwell, MA.

[33] VARIAN, H.R. (1984) *Microeconomic Analysis,* Second Edition, W.W. Norton & Co., New York, NY.

[34] ZIMMERMAN, J.L. (1995) *Accounting for Decision Making and Control,* Irwin, Chicago.

Chapter 4

IMPACT OF MANUFACTURING FLEXIBILITY ON SUPPLY CHAIN PERFORMANCE IN THE AUTOMOTIVE INDUSTRY

Stephan Biller
Enterprise Systems Lab
GM Research and Development Center
Warren, MI 48090, USA
stephan.biller@gm.com

Ebru K. Bish
Department of Industrial and Systems Engineering
Virginia Tech
Blacksburg, VA 24061, USA
ebru@vt.edu

Ana Muriel
Mechanical and Industrial Engineering Department
University of Massachusetts Amherst
Amherst, MA 01003, USA
muriel@ecs.umass.edu

1. Introduction and Motivation

The basis of competition in the automotive industry is changing. While product innovation and styling remain the most important areas of competition, an almost equally fierce battle is now developing in the areas of customization and order fulfillment (Stalk, Stephenson and King [35]). Currently, all models of vehicle distribution are fundamentally inventory-driven and do not promote customized ordering. However, several vehicle manufacturers — most notably Ford and General Motors — have recently launched initiatives to transform their companies from predominantly make-to-stock to predominantly make-to-order producers. This will enable vehicle manufacturers and their dealers not only to dramatically reduce their finished goods inventory but also respond to challenges and threats from third party Internet companies.

In 1999, 40% of all new vehicle buyers used the Internet during their shopping; this number grew to 542003 (J.D Power Automotive and Associates [21]). In the last few years, we have seen an evolution in the e-pricing arena from information sites, where invoice prices can be found, to referral sites, where price quotes can be requested, and finally to sites that post transaction prices and sometimes allow purchasing on the web. While currently most customers use the Internet to find out invoice prices and buy from franchise dealers, there are many third-party internet companies, such as AutobyTel and Microsoft's CarPoint, trying to convince customers to use their buying services.

These developments are a threat to the traditional dealer franchise system and Original Equipment Manufacturers (OEMs). First, disintermediation might result in third parties rather than dealers "owning" customer relations and consumer data. Second, if a specific third party becomes very powerful, it could dictate prices and wipe out manufacturers' already low margins, effectively making OEMs not very influential first-tier suppliers. Recognizing the opportunity to reduce distribution cost and, at the same time, fighting the threat of third parties, OEMs have launched their own direct-to-customer e-commerce initiatives; General Motors even founded a new division called e-GM. The Wall Street Journal (February 22, 2000) discusses GM's plans to reinvent the way it designs, builds and ships cars: "... GM's three-year makeover has to succeed if the company is to meet its goal of allowing people to custom-order cars online, and have the vehicles show up at their doors in as few as four days." The status of this transformation is described in Welch [39]: "Limited build-to-order systems may be in place in the next couple of years. GM, for instance, is starting to build custom versions of its Celta compact in Brazil (and sell them through the Internet). ... it plans to offer one-week delivery, which is possible because the car has only a few combinations of options." Moreover, key automotive suppliers are moving in this direction. Lear Corp., the world's largest supplier of automotive interior parts, has announced that its modular and interchangeable components will let buyers customize their car interiors by 2002 (Brady et al. [8]).

As manufacturers are moving towards a make-to-order environment, they have to increase their *flexibility* to absorb variability in demand of vehicles and options (additional features that customers can choose, such as leather seats) previously buffered by inventory at dealer lots. Three different levers of flexibility can be pulled to handle the demand variability: (1) manufacturing flexibility, resulting from being able to build different products in the same plant at the same time, also referred to as process flexibility (Sethi and Sethi [32]), (2) price flexibility, i.e., dynamically changing price over time, and (3) delivery time flexibility, i.e., offering financial incentives for customers who are willing to wait longer for their products. The first lever corresponds to supply flexibility, while the other two represent demand flexibility. In this chapter, we focus on manufacturing flexibility, and discuss the integration of supply- and demand-side flexibilities as a promising future research direction in Section 5.

While process flexibility investments need to be made long before production starts, when product demand is highly uncertain, these investments have a large impact on the overall performance of the supply chain. The main purpose of this chapter is to determine the magnitude of this impact and to provide a capacity-planning framework that considers this impact when designing a make-to-order manufacturing system.

To put the situation analyzed in perspective, we start this chapter with a general description of the automotive supply chain, its evolution over the last century, and its future prospects. The remainder of the chapter then focuses on characterizing the impact of manufacturing flexibility on supply chain performance, and is based on the recent works by Bish, Muriel and Biller [5], and Muriel and Somasundaram [29]. We conclude with an extensive discussion of challenges for the automotive industry in its transition towards mass customization and describe the respective research opportunities.

2.   The Automotive Supply Chain

The automotive industry and its supply chain have undergone several major changes since their inception at the end of the 19th century. What started as a job-shop driven industry went through mass production and lean production, and will now come around almost full circle if the predicted wave of mass customization actually takes place — a nice illustration of Fine's concept of the business double helix[1] with a clockspeed of more than 100 years (Fine [17]). In this section, we provide a historic perspective, describe recent developments, and try to predict the future of the automotive supply chain. We focus on two parts of the supply chain: supplier-manufacturer and manufacturer-dealer interactions.

2.1.   Historic perspective

In 1894, Eveyln Henry Ellis, a wealthy member of the English Parliament, ordered an automobile from P&L, a noted Parisian machine-tool company (Drucker [13]). While almost all parts were designed and manufactured in tiny job-shops, P&L served as a systems integrator and supply chain coordinator. Since P&L behaved in a general contractor (or Hollywood-like[2]) fashion and the making of every part was unique requiring a highly skilled labor force, Ellis could totally customize his car – he had the transmission, brake, and engine controls transferred from the right to the left. This, of course, had consequences for the delivery time: Ellis could not take delivery of the car before the spring of 1895.

During the next two decades, many automobile manufacturers were founded since economic profits, as well as excitement about the new technology, were high. The manufacturing processes, however, were all still based on craftsmanship. Then came Henry Ford: he introduced and perfected the concept of interchangeable parts to automotive manufacturing in 1908 and the moving assembly line in 1913 (Womack et al. [40]). Both concepts were crucial to allow for mass production and, in turn, mass consumption. However, Henry Ford took mass production to an extreme level: he produced literally one identical car down to the color ("The customer can have any color as long as it is black"). The capital required for a "car factory" and the falling profits for automobiles led to an industry consolidation, as well as a vertical integration of the automotive industry. In fact, General Motors was founded by consolidation of 4 automotive companies and Ford even owned his own iron ore mines. During the 1920s, Alfred Sloan, GM's long-time CEO, developed a more customer focused product portfolio strategy ("a car for every purse and purpose", Sloan [33]), which enabled GM to overtake Ford Motor Company as the largest automobile manufacturer.

The manufacturer-dealer relationship (franchise system) has its roots also at the end of the 19th century. In fact, the first formal franchise agreement was signed in 1898 between a Pennsylvania bicycle dealer and the Winston Motor company (Dickinson et al. [12]). Franchise agreements assigned car dealers an exclusive territory in exchange for providing "adequate service and suitable facilities" (Sloan [33]). The manufacturer promised to sell vehicles at a discount to the dealer, who would then sell the vehicle to the customer at a price predetermined by the manufacturer. In addition, dealers would advance large sums of money to the manufacturer before the car was even produced. These franchise agreements allowed manufacturers to grow quickly since they received "venture capital" from their dealers, did not have to carry any finished goods inventory, and could invest all capital in their manufacturing systems. In addition, manufacturers could operate their factories without considering demand fluctuation since dealers absorbed the demand variability in their inventory. Dealers, on the other hand, also benefited from the exclusivity agreement for new products and aftersales services.

2.2.   Recent developments and short-term outlook

In the following 70 years, the automotive supply chain did not change substantially. Figure 4.1 depicts the current tier-type supply chain. What has changed is the relative number of operators and ownership within this supply chain. For example, both Ford and General Motors have recently spun off their internal suppliers, GM's Delphi and Ford's Visteon, and control only two parts of the supply: sheet metal and powertrain (i.e., engine and transmission) manufacturing. In the not so distant future, this could lead to a power shift from Original Equipment Manufacturers (OEMs) to suppliers: similar to the computer

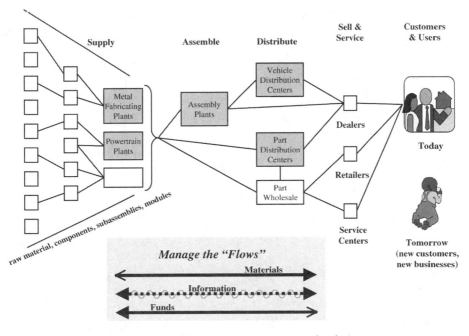

Figure 4.1. Current automotive supply chain

industry's "intel inside", customers might demand Lear seats or Bose stereo systems. Furthermore, we observe that declining margins have led to a consolidation of suppliers and OEMs alike (e.g., the Daimler-Chrysler merger, equity alliances of General Motors with at least four other OEMs, etc.) — again a development consistent with Fine's business double helix. In addition, all OEMs have successfully reduced the number of suppliers to lessen managerial complexity and most are moving towards modular assembly. GM's Blue Macaw and Volkswagen's Resende plants in Brazil exemplify the modularity aspect: doors are taken off the car after the body shop and suppliers install entire modules within the OEM's assembly plant. Lastly, on the dealer-side, we have seen a steady decline in the number of dealers over the last 30 years again caused by declining margins.

All the above changes, however, are small compared to the emerging changes caused by the advent of the Internet. Stalk [34], who first formalized the idea of time-based competition, believes that "after achieving remarkable gains in quality, manufacturing cost, and product development speed, order fulfillment will become the next battleground in the automotive industry" [35]; this development will be accelerated by the arrival of third parties trying to sell cars through the Internet.

Currently, there are 5 basic automotive Internet retail models (Biller et al. [4]): (1) A customer specs out a vehicle and a third party matches the request with a dealer who has this configuration available and offers the lowest price (e.g., AutobyTel). (2) A customer electronically searches a manufacturer's inventory (collection of dealer inventories within a specified distance from the customer's home), and contacts the dealer through a website to inquire the price and/or arrange for a test drive (e.g., GM Buypower). (3) A customer names a price and dealers participating in the third party's service can accept or reject it (e.g., Priceline). (4) A customer bids in an electronic auction (e.g., eBay). (5) A customer buys vehicles through third parties on the web at posted web-prices (e.g., CarsDirect, CarOrder). The main difference between models (1) and (3) is that in the latter the customer loses a deposit if he decides not to purchase the vehicle. As a result, model (3) has not enjoyed much success. Model (4), however, has been successful, especially for unique type of vehicles. Which of these models (if any) will prevail is uncertain: At the end of the year 2000, CarOrder has ceased to take orders, stock prices of third-parties are near all-time lows, and OEMs (in concert with their dealers) are vigorously trying to implement a click-and-mortar strategy. In fact, to improve internet retail model (2), GM has just partnered with Autobytel and its Washington D.C. Chevrolet dealers to market vehicles in dealer inventory on-line at a fixed e-price.

There are two main reasons for these fierce battles between OEMs and third parties in the Business-to-Consumer (B2C) space. First, OEMs and their dealers are afraid that they will lose the contact to their customer and hence, will not be able to predict future customer needs. Secondly, the cost savings due to on-line sales are huge: reductions in factory cost (field sales support, freight) and dealer cost (inventory, sales commissions, overhead) amount to 4% of total vehicle cost (Lapidus [27]).

The impact of the Internet on the manufacturer-supplier relationship (B2B) is as large or even larger. Much of an OEM's and its suppliers' activities are currently focused on lean manufacturing and/or cost reduction, leading to the optimization of each node rather than of the holistic supply chain and creating friction between the different parties. Breaking down these barriers through web-enabled processes will allow OEMs and their suppliers to significantly reduce their inventories. Other benefits, such as productivity gains, reduction in procurement cost, improved quality, alignment of specifications, etc. will also significantly contribute to the savings. Overall, the estimated savings amount to another 4% of total vehicle cost (Lapidus [27]). Considering these savings, it is no surprise that the B2B space was as contested as the B2C space. In this case, however, the Big Three (GM, Ford, and DaimlerChrysler) agreed to cooperate and set up an auto-exchange called Covisint. This exchange will allow the OEMs and their suppliers to be connected through a spider rather than a linear network, i.e. all tiers of suppliers and OEMs will be connected through one central data repository so that third tier suppliers could potentially access point-of-sale information of an OEM. Finally, any OEM that could implement an order-to-delivery process with significantly faster response times than the current 60 days, could realize another 6% savings in so-called "phantom cost" through fewer price discounts (inventory clearing), stock outs, and mix losses (Lapidus [27]). Overall, adding up savings from B2C, B2B, and build-to-order, we expect the savings to amount to 14%, or an average of $3650 per vehicle (Lapidus [27]).

2.3.   The future of the automotive supply chain

In this section, we will try to predict the developments of the next 20 years. Clearly, we expect to be wrong. However, we would like to take the opportunity to provide researchers with ideas as to where we believe the automotive industry is heading. We focus on three areas: (1) information technology, (2) organizational structure, and (3) product.

In the not-so-distant future, web-based real-time tracking and tracing capabilities at the component/order level throughout the supply chain will provide customers, service providers, shippers and producers "live" access to all material movements. This will lead to the automation of procurement activities and, in turn, to "brokered" transportation and warehousing services that will facilitate a "just-in-time" shipping mentality without significant impact on costs. Furthermore, activity based costing will become real-time and extendable throughout the supply chain. These developments will enable the use of smart agent decision support systems to make supply chain related decisions instantaneously.

Operators of or participants in the automotive supply chain will also undergo significant changes. We anticipate the transportation industry to become highly integrated (including multi-modes, freight payment, packaging, etc.), which will facilitate global one-stop shopping for an OEM's transportation needs. In addition, logistics service providers will become highly integrated within corporate culture and will form temporal (i.e., Hollywood-type) sub-organizations. How far the OEMs' outsourcing efforts will go is hard to predict; however, the authors strongly believe that outsourcing the entire supply chain would prove to be a fatal mistake for OEMs, but cost and earnings pressures may force some to do just that. Finally, we expect the consolidation in the logistics industry to result in dramatic increases in cross-company and cross-industry designed supply chains.

Clearly, the product itself will be significantly different. But rather than predicting flying cars, we will focus here on product related changes that will have an impact on the supply chain. We anticipate two key drivers of change: (1) mass customization, and (2) environmental regulations. The former will require the supporting supply chain to be flexible to adjust to customer specific requests in a timely fashion. On the other hand, due to cost pressures, we predict component production to become further extended throughout the globe and the assembly of modular components to be spread out over the supply chain, both suppliers and OEMs becoming dependent on time-based customer demand information. Finally, we will see OEMs becoming responsible for complete recycling of their products, a trend which has already started in Europe. Hence, balancing product design with respect to manufacturing and disassembly, as well as logistics and reverse logistics, will become crucial to accommodate environmental regulations.

Of the many research topics critical to the current evolution of the automotive industry towards mass customization, we address that of evaluating the costs and benefits of manufacturing flexibility in a make-to-order environment in the next section.

3.   Model Overview and Analytical Results

To assess the impact of flexibility on the performance of the supply chain, we quantify total sales, production variability, variability observed upstream, component inventory levels, and outbound transportation costs, under flexible and non-flexible settings.

In a make-to-order environment, capacity is allocated to final products after demand is realized and, thus, no finished goods inventory is held. However, some product components have long lead-times and need to be ordered (or made-to-stock) and kept in inventory before demand uncertainty is resolved. Specifically, we consider a manufacturing system with multiple plants and products, where flexible capacity allows some plants to produce more than one product, and the same product to be manufactured at several sites. We assume that a flexible plant can dedicate anywhere from 0 to 100% of its capacity to manufacture any of the products that it can produce in response to current demands. In addition, we consider that each product requires a unique (product-specific) component that is single-sourced. Hence, there is no component inventory pooling opportunity, and the variability observed by suppliers is determined by the total variability of production of the product whose component they provide. We study the case of multiple product-specific components via simulation in Section 4. Product-specific components are of particular interest to the automotive industry since they comprise over 70% of all components and have typically higher costs and longer lead-times.

The benefits of flexibility to hedge against demand uncertainty are well known: increased sales and higher capacity utilization (Jordan and Graves [23]). However, the impact of manufacturing flexibility on supply chain performance, including variability in upstream production and manufacturers' operational costs, such as component inventory holding and transportation costs, has not been explored. Adding flexibility results in increased variability in the production level of each product at each plant, because a surge in the demand of any product may require shifting the production of other products among different plants. Consequently, component inventory levels and associated inventory costs in a flexible production environment are generally higher. Moreover, in a flexible system, component suppliers will observe higher variability, and spread it upstream the supply chain, increasing production costs throughout. In addition, manufacturing the same product in different plants leads to lower volumes of components shipped from suppliers to each plant for single-sourced components. This, in turn, results in higher shipping costs and/or longer lead-times of the components. A simple EOQ analysis shows that, even in the case of constant demand, splitting volume between two plants, say a fraction of β to one plant and $1 - \beta$ to the other, total transportation set-ups and inventory costs would increase by a factor of $\sqrt{\beta} + \sqrt{1 - \beta}$ (as much as $\sqrt{2}$). On the other hand, outbound transportation cost will be lower, and order-to-delivery times will be faster due to the increased ability to produce closer to the customer. Consequently, a model that balances the benefits of flexibility not only against the initial equipment investment cost, but also against changes in operational costs, could offer new insights for a cost-benefit analysis of manufacturing flexibility.

The remainder of this chapter is organized as follows. In Section 3.1, we give an overview of the relevant literature. In Section 3.2, we introduce a multi-stage approach to study the impact of manufacturing flexibility on supply chain costs. Decisions pertaining to each stage are analyzed sequentially. First, we find, at each period, the production plan that will minimize shortage and lead to minimum production variability using a mathematical program. Then, we study simple analytical models to (1) quantify the increase in production variability associated with different capacity allocation policies in the flexible system, (2) determine the impact of the increased variability on component suppliers and on the manufacturer's component inventory levels, and (3) quantify the expected reduction in the manufacturer's outbound transportation cost in the flexible system. In Section 3.5, we illustrate how the prior analysis can be incorporated into the capacity planning decision process based on the multi-stage approach. A simulation study is presented in Section 4 to show the validity of the results obtained from the simple models, and to better understand the impact of flexibility on manufacturing systems with a larger number of plants and products, where products have multiple components. Finally, Section 5 draws conclusions based on our analysis, and discusses future research directions.

3.1.   Literature review

Manufacturing flexibility and the value of flexibility have been extensively studied in the past; see Beach et al. [3], De Toni and Tonchia [11], Kouvelis [26], and Sethi and Sethi [32] for extensive reviews on flexibility in manufacturing. In this section, we discuss the literature that considers investment decisions in flexible capacity and addresses the value of flexibility.

Much of the earlier work in flexibility comes from the field of economics (see, for instance, Jones and Ostray [22]). In particular, financial option theory has been successfully used to determine the net present value of the investment in flexibility (see Andreou [1] and Triantis and Hodder [36]).

Eppen, Martin, and Schrage [15] consider a scenario-based approach to determine capacity and flexibility investments over a planning horizon. They develop a decision support system that provides optimal net present value investments subject to an acceptable level of risk, where risk is defined as the expected deviation from a revenue target, considering all possible scenarios.

A scenario approach is also used in Fine and Freund [16]. They consider a single-period two-product model and assume that capacity acquisition and variable production costs are linear, while the revenue functions are strictly concave. Their model, a non-linear stochastic program, leads to the characterization of necessary and sufficient conditions for optimal investments in flexible and dedicated capacity. Considering a similar model with two products and a single-period, Van Mieghem [37] studies the sensitivity of flexibility investment decisions with respect to product margins, investment costs, demand uncertainty, and demand correlations for a two-product firm. He characterizes the optimal investment decision and shows that it can be advantageous to invest in flexible resources even when the two products are perfectly positively correlated, but have different profit margins. The common restrictive assumption of these papers is that they consider a single-period in which demand is satisfied or, otherwise, lost. To extend these findings to more general settings, Caulkins and Fine [9] consider a multi-period model in which the firm can hold inter-period seasonal inventories. They consider a two-plant two-product model, where all uncertainty is resolved before production and inventory decisions are made, and study how holding seasonal inter-period inventories affects the utility of flexible capacity.

Most relevant to our work is that of Jordan and Graves [23]. They analyze the benefits of manufacturing process flexibility and conclude that the level of expected sales and capacity utilization resulting from full flexibility (each plant is able to produce all products) can be achieved almost fully with limited or partial flexibility (each plant producing only a few of the products) by assigning products to plants such that the

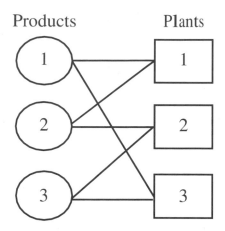

Figure 4.2.   A chain flexibility configuration

product-plant assignment graph is connected, see Figure 4.2. This strategy is referred to as *chaining*. Thus, most of the benefits of total flexibility can be obtained with a significantly lower capital investment in a chain configuration. To show this, they consider a single-period model in which production capacity is optimally allocated to the different products once demands are realized. They illustrate their model through a capacity planning problem at General Motors (see Section 4.4), and show that adding just 6 links to a 16-product 8-plant environment increases capacity utilization from 86.6% to 94.0% (compared to 94.7% for full flexibility). At this point, we briefly digress from the literature review to describe the impact of this research on General Motors capacity planning. This work "had a substantial impact on the way capacity planning is done at GM. The principles of chaining have become entrenched in the GM lexicon of manufacturing capacity planning." (Jordan and Graves [23]). This remains true today; manufacturing flexibility planning and the principles of chaining are even taught as part of GM's continuing education program. However, changing from a make-to-stock to make-to-order manufacturer obviously presents new challenges to capacity planning.

More recently, Graves and Tomlin [5] study the extent to which the findings of Jordan and Graves [23] apply to multi-stage supply chains, and develop insights to guide the choice of process flexibility in such settings. They identify potential inefficiencies of partial flexibility that lead to reduced system throughput due to the interaction of the various stages, and study how these inefficiencies can be mitigated by deploying chain flexibility configurations at each stage. Their simulation and analytical results show that partial flexibility still yields most benefits of full flexibility when configured appropriately across the supply network.

The value of flexibility in non-manufacturing settings, such as flexible workers, flexible service capacity and substitutable inventory systems, has also been widely studied. The results obtained in these areas can help better understand the value of flexible manufacturing capacity. For example, we can view cross-trained workers, who can perform different tasks on the shop-floor (see, for instance, Van Oyen, Senturk-Gel, and Hopp [38]), as flexible resources, who can be allocated to any demand in the system, just as the capacity of a flexible manufacturing plant can be allocated to the various products built.

There are systems where a fully flexible allocation may not be preferable or possible. As an example, consider a rental car company that needs to decide how many vehicles of various models to keep at each location. Although each customer will demand a specific car model, customers can be satisfied with an available higher value model (i.e., mid-size cars may be substituted for compact cars, etc.). Thus, the higher value models are partially flexible resources that can be used to satisfy demand for the lower value ones. Netessine, Dobson, and Shumsky [30] describe the case of a telecommunications company that must determine the number of technicians to hire and train. The technicians are required to perform functions such as equipment installation, testing, and repair, listed in order of increasing complexity. Installation technicians are limited to installation only, testing technicians can handle installation and testing, and repair technicians can handle all tasks. Thus, repair technicians are the fully flexible and most expensive resources; testing technicians are partially flexible resources that can satisfy only two functions; and installation technicians are dedicated resources, the cheapest to maintain. In order to obtain an analytically tractable model, Netessine, Dobson and Shumsky [30] approximate this service structure by a *more restrictive structure*, in which demand for a function (e.g., installation) can be fulfilled by either the corresponding resource (i.e., installation technicians), or the adjacent one with the next higher value (i.e., testing technicians, but not repair technicians). Studying this model, they show that an increase in correlation leads to an alternating pattern in adjustments to optimal capacity decisions. That is, if the optimal capacity of one resource increases, then the optimal capacity of the adjacent resource decreases.

A more general substitution structure has been studied in Bassok, Anupindi, and Akella [2] in the context of *substitutable inventory systems*, in which demand for one product can be satisfied by other related products. Specifically, they study a single-period multi-product inventory problem, where demand for a particular product can be substituted by any product of higher value, and derive properties of the optimal solution. In fact, substitutable inventory systems have received much attention in the operations management literature, see Khouja [25] for a review. Single-period, multi-product models with substitution have been also studied in Bitran and Dasu [6], Hsu and Bassok [20], and Pasternak and Drezner [31], among others.

Similar situations where higher value products or services are used to satisfy demand for lower value ones arise in the airline and hotel industries, in the allocation of airfare classes and different types of hotel rooms, respectively. These problems have been studied in the literature, mainly from a demand management (revenue management) perspective, (see Bitran and Gilbert [7], Carrol and Grimes [10], Edelstein and Melynk [14], Karaesmen and van Ryzin [24], and Geraghty and Johnson [18]). Research findings for such systems may provide insights on the value and management strategies of manufacturing flexibility. In fact, Biller et al. [4] have made attempts to apply demand management techniques to a make-to-order manufacturing environment.

In the aforementioned literature, a two-stage sequential model is always considered to determine the benefits of capacity flexibility: in the first stage, dedicated and flexible capacity investment decisions are made under uncertain demand, and in the second stage, demand is realized and capacity is allocated to products. Even the approach of Caulkins and Fine [9], who consider inter-period inventories, can be reduced to a two-stage model, since all uncertainty is resolved in the first period, before production and inventory decisions are made. As a result, the evaluation of different capacity and flexibility decisions cannot consider the impact of flexibility on safety stocks and transportation costs at the operational level. To fill this gap we introduce a new capacity investment decision framework in the next section.

3.2.   Model overview

To quantify the benefits and costs associated with a given flexibility configuration, we consider a three-stage process (see Figure 4.3): First, capacity investment decisions are made under high uncertainty. Second, component inventory and capacity allocation policies are determined when aggregate demand is fairly well known but there is a

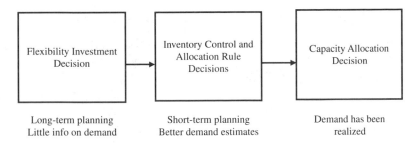

Figure 4.3. A three-stage approach

need to hedge against period-to-period demand variability. Recall that, although the final product is made to order, some components may need to be kept in stock to achieve acceptable customer lead-times. Third, product demand is allocated to different plants each period after orders have been received.

The introduction of an intermediate stage allows us to include tactical decisions, namely component inventory control and capacity allocation policies, in the decision framework. Observe that in a flexible manufacturing system in which each product can be built in several plants, there are multiple ways to allot plant capacities to product demands. For instance, each product can be *prioritized* at a particular plant that will satisfy as much as possible of its demand, while any excess demand is allocated to other plants. Alternatively, each product demand could be evenly *distributed* over several plants, or priorities could be assigned to products based on their *profit margin*. These various allocation schemes may all result in the same total shortage in the system; however, the variability in production of each product at each plant and the resulting inbound and outbound distribution costs will be different under each of them.

Given an allocation scheme and an inventory policy, we can determine the impact of flexibility on expected sales, production and upstream variability, inventory levels, and outbound distribution cost. The variability observed upstream and component inventory levels are driven by the variability of production of the corresponding products at the different plants, and are therefore highly dependent on the capacity allocation policy used in stage 3. Thus, in what follows, we take a bottom-up approach to examine decisions at each stage. We start by studying the capacity allocation decision at stage 3 and subsequently analyze stage 2. Finally, we illustrate how this analysis can be incorporated into the capacity planning decisions at stage 1.

3.3.   Determining optimal production levels

In the third stage of our model, we consider the allocation of available capacity after demand is observed each period. We determine the production levels of each product at each plant given orders received, available component inventories, chosen capacity allocation policy, and capacity constraints. Total shortage (or unsatisfied demand) can be minimized solving a linear program similar to that of Jordan and Graves [23]. However, we add a quadratic term to the mathematical program to ensure minimum variability in the period-to-period production levels. As mentioned above, this is important, since the need for component safety stock is mainly driven by the variability in production. In the following, we formulate the resulting quadratic program.

Let n be the number of plants, each with capacity C_j, $j = 1, 2, \ldots, n$, and m the number of products. Let A denote a given flexibility configuration, where $(i, j) \in A$ if product i can be built in plant j, $i = 1, 2, \ldots, m$, and $j = 1, 2, \ldots, n$. We define variables s_i to represent the shortage of product i, $i = 1, 2, \ldots, m$, and variables x_{ij} to be the quantity of product i to be produced at plant j, $\forall (i, j) \in A$. Finally, let ϵ be a constant and let μ_{ij} denote the *target production level* of product i at plant j, $\forall (i, j) \in A$. These target production levels depend on the long-term capacity allocation policy, which is chosen at stage 2 along with the inventory control policy, and are, therefore, inputs to our mathematical program. We will further discuss this in Section 4. Given each product's demand realization, d_i, $i = 1, 2, \ldots, m$, and the available inventory of components for product i at plant j, I_{ij}, $\forall (i, j) \in A$, the minimum possible shortage for configuration A can be determined by solving the following mathematical program. Later, we will refer to its solution as the *quadratic allocation*.

$$\text{QP}: \qquad \text{Min} \quad \sum_{i=1}^{m} s_i + \epsilon \sum_{(i,j) \in A} (x_{ij} - \mu_{ij})^2$$

$$s.t.$$

$$\sum_{j:\ (i,j) \in A} x_{ij} + s_i = d_i, \quad i = 1, \ldots, m, \qquad (4.1)$$

$$\sum_{i:\ (i,j) \in A} x_{ij} \leq C_j, \quad j = 1, \ldots, n, \qquad (4.2)$$

$$0 \leq x_{ij} \leq I_{ij}, \quad \forall (i, j) \in A, \qquad (4.3)$$

$$s_i \geq 0, \quad i = 1, \ldots, m.$$

We choose ϵ to be small so that the quadratic program minimizes the total shortage, while selecting the production quantities that will result in the minimum deviation from the target production levels, and hence, the minimum possible production variability. Observe that constraints (4.1) define the shortage of each product as the difference between its demand

and overall production. Constraints (4.2) and (4.3) ensure that production does not exceed plant capacity or available component inventory.

Note that transportation cost could easily be added into this model. For this purpose, we would consider the service area divided in multiple regions, define variables X_{ij}^r to be the production of product i in plant j that is destined for region r, and include associated inbound and outbound transportation costs in the objective function.

Different capacity allocation schemes correspond to different target production levels in the quadratic program. We will test the performance of flexible systems under various such allocations using simulation in Section 4. In the next section, however, we study a two-plant two-product model that is analytically tractable, to get insight into the effect of flexibility and allocation policies on supply chain performance. In this stylized model, simple allocation policies, rather than the quadratic program, will be employed to minimize shortages.

3.4.   Evaluating supply chain performance

In this section, we determine the impact of capacity flexibility on supply chain performance for various capacity allocation policies. In particular, we consider the following metrics: total sales, production variability, variability observed upstream, component inventory levels, and outbound transportation costs. Results presented in this section are based on Bish, Muriel, and Biller [5].

Production levels leading to minimum shortage and variability can be determined by solving the quadratic program presented above. However, it is difficult to analytically determine the variability of the production levels resulting from this particular allocation, because the optimal production level of each product at each plant in a certain period depends on the realized demands for all the products, component inventory levels, the capacity allocation policy, and the selected flexibility configuration. Thus, to gain insight into the impact of manufacturing flexibility and different capacity allocation policies on supply chain performance, we study a two-plant two-product scenario and analytically determine the resulting production variability for simple allocation rules.

This section is organized as follows. We start by introducing the details of the analytical model and various allocation policies. In Sections 3.4.2–3.4.5, we quantify the impact of flexibility on (i) sales and production variability, (ii) variability observed by component suppliers, (iii) manufacturer's inventory levels, and (iv) manufacturer's outbound cost. We consider the non-flexible system as the base case throughout. We analyze two cases: (1) demands for both products follow independent and identically distributed normal distributions with mean equal to plant capacity; (2) demands for both products follow independent but general normal distributions. For case (1), we report exact analytical results, whereas for case (2), we make use of numerical integration. Although the assumption of mean demand of each product equal to plant capacity in case (1) may seem restrictive, this case is important because flexibility is most valuable when capacity is neither too tight nor too loose (Jordan and Graves [23]).

3.4.1   Stylized model and allocation rules.   We consider a two-plant two-product model, where product i is the *main product* produced in plant i, $i = 1, 2$, and, therefore, the only one in the absence of flexibility. Recall that, since we consider a make-to-order environment, there is no final product inventory (i.e., sales at any period are equal to the amount produced in that period). In addition, we assume that backlogging is not allowed and hence, excess demand at any period is lost. Finally, as mentioned previously, we consider that each product requires a unique major component (e.g. transmission) that is single-sourced. We extend this analysis to multiple components in Section 4.

In what follows, we let C denote the capacity of each plant, and D_i denote the demand for product i, $i = 1, 2$. We assume that product demands in stage 2 are independently normally distributed with mean μ_{D_i}, standard deviation σ_{D_i}, probability distribution function $f_i(.)$, and cumulative distribution function $F_i(.)$ for $i = 1, 2$. Observe that the normal demand distribution considered is a continuous approximation of the discrete demand. In addition, the probability of negative values in this normal distribution is assumed to be negligible. In our analysis, this is justified for demand coefficients of variation $(\frac{\sigma_{D_i}}{\mu_{D_i}})$ of at most 0.25, which are typical of the period-to-period variability observed at the second stage by the automotive industry.

Let X_{ij}^A denote the production quantity of product i at plant j under allocation policy A, for $i, j = 1, 2$. For the case of full flexibility, where both plants can produce both products, we investigate the performance of three different capacity allocation policies:

1 *Prioritized Allocation Policy:* production of each product is concentrated in its main plant. As much demand for product i as capacity permits is allocated to plant i, $i = 1, 2$; that is,

$$X_{ii}^{F-Prior} = \min\{D_i, C\}$$
$$X_{ij}^{F-Prior} = \min\{D_i - X_{ii}^{F-Prior}, C - X_{jj}^{F-Prior}\}, \quad \text{for } i \neq j.$$

2 *Distributed Allocation Policy:* production of each product is distributed among the two plants. Half of the demand for product i is allocated to plant i, $i = 1, 2$, and the remaining demand to the spare capacity of the other plant; that is,

$$X_{ii}^{F-Dist} = \min\left\{\frac{1}{2}D_i, C\right\}$$

$$X_{ij}^{F-Dist} = \min\left\{\frac{1}{2}D_i, C - X_{jj}^{F-Dist}\right\}, \quad \text{for } i \neq j.$$

3 *Profit Based Allocation Policy:* the product with higher profit margin gets allocated first. This policy is identical to the prioritized allocation policy, except that the product of higher profit margin, say product 1, also gets priority on the second plant. That is,

$$X_{11}^{F-Profit} = \min\{D_1, C\},$$
$$X_{12}^{F-Profit} = \min\{D_1 - X_{11}^{F-Profit}, C\}$$
$$X_{22}^{F-Profit} = \min\{D_2, C - X_{12}^{F-Profit}\},$$
$$X_{21}^{F-Profit} = \min\{D_1 - X_{11}^{F-Profit}, C - X_{22}^{F-Profit}\}.$$

Similarly, we could consider a distributed-like profit-based allocation policy, in which the production of the higher profit margin product, product 1, is evenly split between the two plants. However, we show in [5] that all results derived below for the distributed policy given in (2) also hold for such a distributed profit-based policy; thus, the latter will not be mentioned any further.

As discussed in Section 3.2, each of these policies is attractive in practice for different reasons. Prioritized policies concentrate production of each product mostly in a single plant, leading to lower manufacturing and inbound distribution costs; distributed policies may be attractive because they allow producing closer to the customer's location, thus resulting in shorter delivery times to customers; and profit based policies are economically attractive when the products have significantly different profit margins.

3.4.2   Sales and production variability.   In this section, we determine the mean and variability of production of each product at each plant given a flexibility configuration and a capacity allocation policy, regardless of component inventory constraints (since the amount of component inventory to be held will be derived from these parameters).

We first consider independent, identically distributed normal product demands with mean equal to plant capacity. In our analysis, we assume that the demand for a single product never exceeds the overall capacity in the system and is always positive; that is, $P(0 < D_i < 2C) \approx 1$, $i = 1, 2$. This is justified for $\mu_{D_i} = \mu_D = C$, $i = 1, 2$, and demand coefficients of variation no larger than 0.25, which are reasonable parameters in the automotive industry.

With no manufacturing flexibility in the system, each plant will only produce its main product. Let X_i^{NF} be the production quantity of product i, $i = 1, 2$, in the non-flexible case. Observe that $X_i^{NF} = \min\{D_i, C\}$, and we can determine its expected value and variability as,

$$E\left(X_i^{NF}\right) = \int_0^C D_i f_i(D_i)\, dD_i + C[1 - F_i(C)]$$

$$E\left[\left(X_i^{NF}\right)^2\right] = \int_0^C D_i^2 f_i(D_i)\, dD_i + C^2[1 - F_i(C)],$$

and $Var(X_i^{NF}) = E[(X_i^{NF})^2] - [E(X_i^{NF})]^2$. Solving these integrals, we derive the mean and variability of the production level of product i at plant i, $i = 1, 2$, for the non-flexible case, and obtain the following result (Bish, Muriel, and Biller [5]).

Theorem 4.1 *Considering independent and identically distributed normal product demands with mean $\mu_D = C$ and standard deviation σ_D, the mean and variance of the production level of product i, $i = 1, 2$, in the non-flexible environment is given by:*

$$E\left(X_i^{NF}\right) = C - \frac{1}{\sqrt{2\pi}}\, \sigma_D, \quad Var\left(X_i^{NF}\right) = \left[\frac{1}{2} - \frac{1}{2\pi}\right]\sigma_D^2$$

Similarly, we analytically determine the expected value and variance of production levels in the flexible system, under the prioritized, distributed, and profit-based allocation policies. For example, under the prioritized policy, we observe that $E(X_{ii}^{F-Prior}) = E(X_i^{NF})$, and determine the expected value of production of product i in plant j, $i \neq j$, as

$$X_{ij}^{F-Prior} = \left\{ \begin{array}{ll} C - D_j, & \text{if } D_j < C \text{ and } D_i - C > C - D_j \\ D_i - C, & \text{if } D_i > C \text{ and } D_i - C < C - D_j \\ 0, & \text{otherwise.} \end{array} \right\}$$

Thus, assuming $\mu_D = C$ and $P(0 < D_i < 2C) \approx 1$, we can write:

$$E\left(X_{ij}^{F-Prior}\right) = \int_C^{2C} \int_{2C-D_i}^C (C - D_j) f_j(D_j) f_i(D_i)\, dD_j\, dD_i$$

$$+ \int_0^C \int_C^{2C-D_j} (D_i - C) f_i(D_i) f_j(D_j)\, dD_i\, dD_j$$

Solving the required integrals leads to the following result.

Theorem 4.2 *Considering independent and identically distributed normal product demands with mean $\mu_D = C$ and standard deviation σ_D, adding flexibility results in an increase in expected sales (or, equivalently, expected production) per period of at most*

$$2\left[\frac{1}{\sqrt{2\pi}} - \frac{1}{2\sqrt{\pi}}\right] \sigma_D \approx 0.23\ \sigma_D$$

for the prioritized, distributed, and profit-based allocation policies.

Observe that the above expression is an upper bound on the increase in expected sales, since we are not considering inventory constraints.

Through numerical integration, we also evaluate the case where mean demand is not necessarily equal to plant capacity (see [5] for details). Our results indicate that:

1 As demand for product 1 and product 2 becomes more unbalanced, flexibility leads to a higher increase in sales.

2 In general, flexibility has a higher impact on sales in cases of high demand variability.

3 However, when product demands are unbalanced and capacity is tight, lower variability leads to a higher increase in sales. Observe that this is quite intuitive. For instance, if the sum of the mean product demands equals total capacity, and one mean is much larger than plant capacity, whereas the other one is much smaller, then flexibility is most advantageous when demands are exactly equal to their means.

Furthermore, in [5] we characterize the resulting production variability under the prioritized, distributed, and profit-based policies in a flexible system, and obtain the following results.

Theorem 4.3 *Considering independent and identically distributed normal product demands with mean $\mu_D = C$ and standard deviation σ_D, the* *production variability of product i, $i = 1, 2$, in the flexible environment is given by:*

- *Under the prioritized allocation policy,*

$$Var\big(X_{ii}^{F-Prior}\big) = Var\big(X_i^{NF}\big) = \left[\frac{1}{2} - \frac{1}{2\pi}\right]\sigma_D^2$$

$$Var\big(X_{ij}^{F-Prior}\big) = \left[\frac{1}{4} - \frac{(5\sqrt{2} - 4)}{4\sqrt{2\pi}}\right]\sigma_D^2, \quad \text{for } i \neq j$$

- *Under the distributed allocation policy,*

$$Var\big(X_{ii}^{F-Dist}\big) = \frac{1}{4}\,\sigma_D^2,$$

$$Var\big(X_{ij}^{F-Dist}\big) = \frac{0.6817}{4}\,\sigma_D^2, \quad \text{for } i \neq j$$

- *Under the profit-based allocation policy,*

$$Var\big(X_{11}^{F-Profit}\big) = Var\big(X_{12}^{F-Profit}\big) = \left[\frac{1}{2} - \frac{1}{2\pi}\right]\sigma_D^2$$

$$Var\big(X_{21}^{F-Profit}\big) = \left[\frac{1}{4} - \frac{(5\sqrt{2} - 4)}{4\sqrt{2\pi}}\right]\sigma_D^2$$

$$Var\big(X_{22}^{F-Profit}\big) = \left[\frac{3}{4} - \frac{(2\sqrt{2} + 1)}{4\pi}\right]\sigma_D^2$$

The next section studies the impact of this production variability on component suppliers.

3.4.3   Variability observed by component suppliers.   For each product, we consider a single vendor that supplies a major component of that product system-wide and study the variability observed by this supplier in the flexible system versus that in the non-flexible setting. In what follows, we refer to the variability observed by the supplier as *system variability:* this is the variability associated with the total production quantity of the corresponding product.

The *system variability* associated with the production of product i in the non-flexible setting is simply given by the variability in production at the only plant that can manufacture it; that is, $Var(System_i^{NF}) \equiv Var(X_i^{NF})$, $i = 1, 2$. In the fully flexible setting, however, the *system variability* associated with product i, $i = 1, 2$, and a particular allocation policy A, $A = Prior, Dist$ or $Profit$, is given by $Var(System_i^{F-A}) \equiv Var(\sum_{j=1}^2 X_{ij}^{F-A})$. Thus, system variability can be calculated as

$$Var\big(System_i^{F-A}\big) = Var\big(X_{ii}^{F-A}\big) + Var\big(X_{ij}^{F-A}\big) + 2Cov\big(X_{ii}^{F-A}, X_{ij}^{F-A}\big),$$

where $Cov(X_{ii}^{F-A}, X_{ij}^{F-A})$ is the covariance term between X_{ii}^{F-A} and X_{ij}^{F-A}. Deriving analytical expressions for system variability in [5], we obtain the following results.

Theorem 4.4 *Considering independent and identically distributed normal product demands with mean $\mu_D = C$ and standard deviation σ_D, system variability associated with product i, $i = 1, 2$, in the flexible system is given by:*

1. $Var(System_i^{F-Prior}) = [\frac{3}{4} - \frac{3}{4\pi}] \sigma_D^2 \approx 0.51 \ \sigma_D^2$, *under the prioritized allocation policy;*

2. $Var(System_i^{F-Distr}) = \frac{2.6817}{4} \sigma_D^2 \approx 0.67 \ \sigma_D^2$, *under the distributed allocation policy;*

3. $Var(System_1^{F-Profit}) = \sigma_D^2$ *and* $Var(System_2^{F-Profit}) = [1 - \frac{1}{\pi}] \sigma_D^2 \approx 0.68 \ \sigma_D^2$, *under the profit-based policy.*

Observe that the profit-based policy is not symmetric and, consequently, the resulting system variability is product-dependent. Under this policy, the total production variability of the first product is simply equal to demand variability since all the demand for product 1 will be satisfied. The high system variability of product 2 is due mainly to the variability of its production in plant 2.

Using Theorem 4.4, we can quantify the *change in system variability* associated with each product in the flexible versus non-flexible settings.

Corollary 4.5 *Considering independent and identically distributed normal product demands with mean $\mu_D = C$ and standard deviation σ_D, the standard deviation of demand for components associated with product i, $i = 1, 2$, in the flexible setting increases by*

1. 22%, *under the prioritized allocation policy;*

2. 40%, *under the distributed allocation policy;*

3. 71% *and* 41% *for products 1 and 2, respectively, under the profit-based allocation policy.*

The highest system variability is observed by the supplier of product 1 components when the manufacturer uses the profit-based policy. In this case, the system variability is identical to that of product demand since no shortages for the product can occur under our assumptions. However, under the prioritized and distributed policies, as well as in the non-flexible system, plant capacity limits the production of each product, leading to lower production variability than demand variability. The distributed policy leads to the second highest system variability for both suppliers: this is because under this allocation, total production of a product i is $\min\{D_i, \frac{1}{2}D_i + C - \frac{1}{2}D_j\}$, that is, it varies between $1/2D_i$ and D_i depending on the magnitude of the demand for the other product.

The results show that suppliers observe significantly higher variability in the flexible system, as compared to the non-flexible system. The increase in the standard deviation of demand for any component is at least 22% under any of the three allocation policies studied. In addition, notice that the observed variability is highly dependent on the allocation policy implemented. The next section discusses the impact of flexibility on the manufacturer's inventory levels.

3.4.4   Manufacturer's component inventory levels.   In this section, we evaluate the effect of flexibility on the manufacturer's component inventory levels, using the results from Section 3.4.2, and focusing on parameters typical for OEMs. Similar results could be derived for parameters of other industries. We consider a major component with a long procurement lead-time, L, that needs to be kept in inventory, while the final product is made-to-order.

Different inventory control policies could be devised for these components. However, due to its simplicity and practical use, we limit our analysis to order-up-to inventory policies. Observe that demand for each component is triggered by the production of the final product, which is constrained by plant capacity. For this reason, we consider a *modified order-up-to inventory policy* in which we limit the safety stock of components of product i at plant j to the average unused plant capacity during the lead-time. That is, the order-up-to level of the component of product i at plant j is given by $L\mu_{ij} + \min\{z\sqrt{L}\sigma_{ij}, L(C - \mu_{ij})\}$, where μ_{ij} and σ_{ij} are the mean and standard deviation of the production level of product i at plant j, respectively, and z is a safety factor corresponding to a specified stock-out probability (for normally distributed demands). Thus, the average level of component inventory is given by

$$\frac{\mu_{ij}}{2} + \min\{z\sqrt{L}\sigma_{ij}, L(C - \mu_{ij})\}. \tag{4.4}$$

First, we consider the case of product demands independent and identically distributed according to a normal distribution with mean $\mu_D = C$ and standard deviation σ_D. Figure 4.4 shows the percentage increase in sales and inventory as the coefficient of variation of period-to-period demand varies, under each of the allocation policies considered for a lead-time of $L = 2$ weeks.

Our analysis leads to the following insights:

1 The higher the variability of product demand, the higher the benefits of flexibility, since inventory increases at a lower rate than

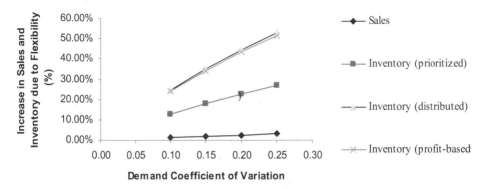

Figure 4.4. Percentage increase in sales and inventory under the various allocation policies considered

sales as variability grows. Thus, the benefits of adding flexibility increase as product demand becomes more variable, but due to the higher inventory levels, not as much as one would predict.

2 Component inventory levels are highly dependent on the allocation policy implemented: the prioritized allocation policy results in significantly lower component inventory levels than the distributed and profit-based allocation policies.

These results demonstrate the significant increase in component inventory levels — approximately 10% for each percentage point increase in sales in the case of mean demand equal to capacity — when flexibility is added to the system, and suggest that the prioritized allocation policy should be used in a flexible environment, since it leads to lower inventory levels.

A natural question at this point is whether these properties continue to hold for more general cases. In [5], we evaluate the case where mean demand is not necessarily equal to plant capacity and obtain the *increase in expected component inventory level per percentage point increase in sales* for different values of the mean product demands, for each allocation policy. This increase in inventory relative to sales decreases substantially as product demands become more unbalanced. The results are somewhat surprising. In the perfectly balanced case, in which both mean product demands are equal to plant capacity, the prioritized policy always dominated the distributed policy. However, in the general case no allocation policy is dominant over the others. In fact, which allocation policy to use depends on the relation of the mean demands of the various products to each other, and to plant capacity. In particular, the distributed policy leads to significantly higher component inventory than the prioritized policy when capacity is tight; however, it outperforms the prioritized policy otherwise. The practical implication of this insight is quite significant: *As demand patterns change over time, tactical planning decisions, such as capacity allocation and inventory policies, should be continuously revisited and changed, if appropriate.*

Finally, other issues need to be considered when determining a capacity allocation policy. For example, in practice it may not always be desirable to use a prioritized allocation policy under which each plant allocates most of its capacity to one product, and only residual capacity to the other product. The resulting low volumes of production of the second product will lead to higher manufacturing costs due to the higher number of setups, slow learning, and other lost benefits of scale. In addition, delivery times in the distributed case might be considerably shorter, due to the ability to produce closer to the customer, making this policy more attractive. Quantifying the associated reduction in outbound transportation cost for the manufacturer is the focus of the next section.

3.4.5   Manufacturer's outbound distribution.

Adding flexibility is also likely to result in an increase in inbound transportation cost, and a reduction in outbound transportation cost and shipping times to customers. These distribution costs and times can be significant, and should be considered when allocating production capacity to customer orders. For example, it may not be desirable to satisfy a customer order from a far away plant in the current period, if the order can be satisfied from the local plant in the next period. In addition, some customers may be willing to wait for capacity to become available at the local plant when given adequate financial incentives.

To evaluate the effect of added capacity flexibility on outbound distribution, we consider a two-plant two-product model, similar to that of the previous section, except that demand is now "location-based". We assume that customer and plant locations are uniformly distributed over a bounded region. As before, we assume independent normal product demands.

In a non-flexible environment, each plant satisfies as many customers as possible who demand its product, independent of the location of the customer. In a fully flexible environment, we consider an allocation policy, referred to as the *location-based allocation policy*, in which each plant gives priority to customers in its "production area"; that is, to customers who are closer to it than to the other plant. When capacity for one plant is exceeded by the demand in its production area, surplus orders are assigned to the other plant, if it has spare capacity after serving customers in its area. Our objective is to compare the expected unit shipping cost in the non-flexible and fully flexible environments. For that purpose, we assume that the total shipping cost is proportional to the total distance traveled to satisfy each customer directly from her corresponding plant, and determine the expected distance traveled in the two manufacturing environments using rectilinear (right-angle) distances. Such a distance metric is often used due to the approximate rectilinear patterns of city streets and the U.S. interstate highway system (Larson and Odoni [28]). However, if rectilinear distances are not appropriate, a linear adjustment factor may be used to approximate the correct distance function from the rectilinear (for example, the ratio of Euclidean distances to rectilinear distances is derived in Larson and Odoni [28]).

Following the location-based allocation policy in the flexible manufacturing system, we find that demand coming from customers in each plant's production area can be approximated by a normal distribution (Bish, Muriel and Biller [5]), and derive analytical expressions and bounds on the expected distance covered by each plant to serve customers assigned to it under no flexibility and full flexibility. Figure 4.5 gives lower bounds on the percent reduction in expected unit shipping cost in the flexible versus dedicated environments when both product demand distributions are identical. This is done for different

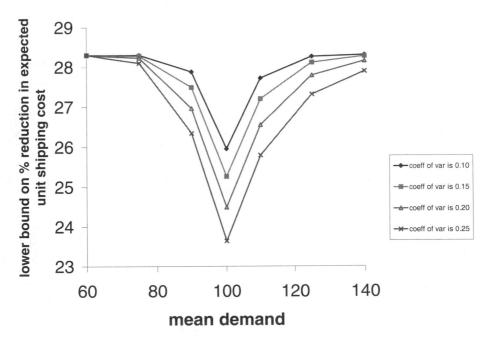

Figure 4.5. Lower bounds on the percent reduction in the expected unit shipping cost in the flexible system when product demands are identically distributed

demand coefficients of variation and means, for the case where each plant has a capacity of 100 units.

This analysis yields the following insights:

1 For all scenarios studied, full flexibility leads to a significant reduction (at least 23%) in expected unit shipping cost.

2 This reduction is more pronounced when the average overall demand is much smaller or much larger than total capacity (i.e., cases with $\mu_D = 60$ and $\mu_D = 140$). This is intuitive; the flexible system yields lower unit shipment cost in these cases, because capacity of each plant is almost entirely allocated to orders within its production area in the flexible environment.

3 The lower the variability of product demand, the higher the reduction in expected unit shipping cost. This is true because as demand variability increases, the demand that each plant faces from customers closest to it becomes more variable. As a result, the average number of customers satisfied from the further plant in the flexible system increases, leading to higher shipping cost.

We obtained similar results for the general case with different mean product demands (see [5]). A natural question at this point is how this allocation policy, which focuses on the outbound transportation cost only, affects inventory levels and other performance measures in the supply chain. Yet another interesting research direction is to evaluate the performance of different allocation policies which select customers and shipping routes so as to minimize the total shipping distance. We will discuss these questions in Section 5.

Having exploited our simple analytical models to understand the implications of flexibility and different capacity allocation policies at stage 2, we will now complete the analysis of our planning framework with stage 1.

3.5.   Capacity and flexibility planning

Our analysis demonstrates the significant impact of flexible capacity on the operational performance of the supply chain. While manufacturing flexibility allows the company to satisfy more demand and produce closer to the customer's location, it requires significantly higher levels of component inventory and results in higher upstream variability. Consequently, the impact on operational performance needs to be considered when making capacity investment decisions to more accurately determine the benefits of flexibility. We suggest the following type of cost-benefit analysis to incorporate the impact of flexibility on future supply chain performance at the capacity investment stage.

We first determine the expected sales and operational cost in the following way. Let m be the number of products that the company manufactures. In stage 1, given little information on demand, we use historic sales and market research data to determine a joint probability distribution for the future mean product demands $M_{D_1}, M_{D_2}, \ldots, M_{D_m}$, given by $P(M_{D_1} = \mu_{D_1}, M_{D_2} = \mu_{D_2}, \ldots, M_{D_m} = \mu_{D_m})$, for any possible realization $\mu_{D_1}, \mu_{D_2}, \ldots, \mu_{D_m}$. This reflects the point forecast and its error. Observe that this approach allows including correlation among the mean product demands in the analysis. Given plant capacities, the proposed flexibility configuration, and the level of period-to-period demand variability in the system, the analysis in Section 3.4 can be used to determine, for any possible realization of the mean demands, $\mu_{D_1}, \mu_{D_2}, \ldots, \mu_{D_m}$, the associated supply chain operational cost, which we denote by $Cost(\mu_{D_1}, \mu_{D_2}, \ldots, \mu_{D_m})$. Thus, the expected operational cost corresponding to the given investment decision can be written as:

$$\sum_{\mu_{D_1}, \mu_{D_2}, \ldots, \mu_{D_m}} P\big(M_{D_1} = \mu_{D_1}, \ldots, M_{D_m} = \mu_{D_m}\big)\, Cost\big(\mu_{D_1}, \ldots, \mu_{D_m}\big)$$

$$(4.5)$$

Expected sales can be calculated in a similar fashion. This analysis is repeated with different capacity and flexibility configurations. In this way, the investment in flexibility can be traded off not only against the increased sales revenue, but also against the associated operational costs. To illustrate this approach, we present the following example.

3.5.1   An example.   Consider the two-plant two-product scenario analyzed in Section 3.4. Suppose that in stage 1, the future mean demand of each product is estimated to be 60, 75, 90, 100, 110, 125 and 140 units with certain probabilities. We consider various discrete probability distributions over those values, each with the same expectation of 100 (the point forecast), but with different coefficients of variation, to capture different forecast inaccuracies in stage 1. To estimate the impact of adding flexibility on operational performance, we determine the expected sales and supply chain cost corresponding to each possible realization of the mean product demands (there are 7×7 of them). These values are then used to determine the overall expected value, as given in Equation (4.5).

For this specific example, Table 4.1 presents the expected percent increase in sales and component inventory levels resulting from the flexible capacity investment, when assuming different levels of forecast error in stage 1 (expressed as coefficients of variation of demand). In stage 1, we

Table 4.1. Impact of flexible capacity investment on supply chain performance as a function of forecast error in stage 1

Coefficient of Variation of Forecast Error	Sales	Percent Increase in		
		Inventory		
(Stage 1)		Prioritized	Distributed	Minimum
0.10	2.11	15.93	21.10	15.48
0.20	2.41	16.32	22.87	15.65
0.30	2.74	16.80	24.74	16.01
0.40	2.97	17.10	25.88	16.28

consider normal distributions with mean 100 and standard deviations of 10, 20, 30, and 40, truncated to the interval [40, 160]. As before, plant capacity is 100, the coefficient of variation of the period-to-period demand distribution (at stage 2) is 0.15, and the component procurement lead-time is 2 weeks. Three different allocation policies are considered. The first two utilize the same policy, either the prioritized or the distributed policy, for all possible realizations of mean demands at stage 2. The third one (Minimum), however, builds upon our previous insight, and chooses, for each realization of average demands at stage 2, the one that leads to minimum inventory for each particular realization of the mean demands.

This example shows how the expected benefits of flexibility increase as the uncertainty of the demand forecast in stage 1 increases, but they are diminished by the increase in component inventory levels. With the appropriate data for flexibility investment, inventory holding costs, transportation costs as well as profit margins, the proposed method can be used to determine the optimal level of flexibility. In addition, sensitivity analysis of investment cost versus demand uncertainty proves to be very helpful for capacity planning decision makers.

4.   A Simulation Study

To extend our findings and explore more general settings with multiple plants and products and different flexibility configurations, we analyze three-plant three-product, five-plant five-product and full-scale automotive scenarios using simulation. The results presented in this section are based on Muriel and Somasundaram [29].

Building on the results of Jordan and Graves [23], we consider manufacturing systems with limited capacity flexibility following chain configurations such as the one depicted in Figure 4.2. As mentioned earlier, most benefits of full flexibility can be achieved in this way with a significantly reduced flexibility investment cost. In the five-plant five-product case, four different single-chain configurations are possible; we consider chains with two, three, four, and five (i.e., full flexibility) products built at each plant (see Figure 4.6). Our objectives are: (1) to quantify the costs and benefits of flexibility in these more complex settings, (2) to evaluate the benefits of the optimization-based allocation policy (quadratic allocation) presented in Section 3.3 relative to the

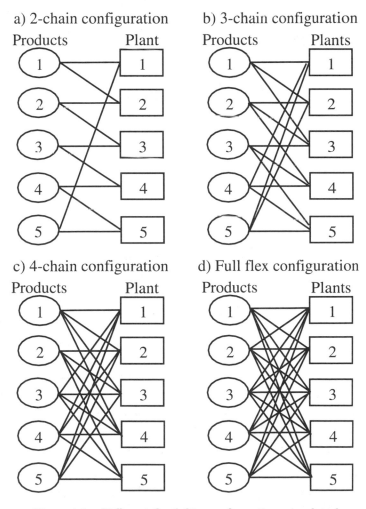

Figure 4.6. Different flexibility configurations simulated

naïve allocation policies (prioritized and distributed policies) considered in Section 3.4, (3) to compare the performance of the partially flexible systems to identical systems under full and no flexibility, and (4) to study the impact of multiple product-specific components on the value of manufacturing flexibility. In what follows, we describe the simulation experiments (see Muriel and Somasundaram [29] for a more extensive simulation analysis).

We first generate 50 realizations of the mean demands for the different products using independent normal distributions, each with mean 100 and standard deviation 40, truncated to the interval [20, 180]. This simulates the first stage of our approach (see Section 3.2).

For each of these 50 realizations of the mean demands, in the second stage we consider independent normal period-to-period demands for the three products, each with a coefficient of variation of 0.15. In this stage, capacity allocation policies and inventory control policy parameters need to be determined. Recall that, in the case of quadratic allocation policies, this also includes setting a target production level for each product at each plant (see Section 3.3). Given an allocation policy (quadratic or naïve allocation), we simulate 1000 dummy periods, not considering inventory restrictions, to obtain estimates of the mean and variance of the production levels associated with that allocation policy. These mean and variance of the production levels are then used to determine the target production levels and the order-up-to inventory control policy for the corresponding allocation. In the case of the quadratic policies, we estimate the initial target production levels using a naïve allocation policy on the mean demands to run the dummy periods.

In the third stage, we simulate 1000 new periods, identical to the previous ones, except that we include inventory constraints in the allocation decision, and, if a quadratic policy is used, we update the target production levels to the ones determined after the dummy periods have been run. Clearly, available component inventories in this stage will be highly dependent on the corresponding inventory control policy parameters determined in the second stage. Based on these last 1000 runs, we obtain estimates on the total sales and average component inventory levels in the system.

Specifically, we analyze the performance of the following four capacity allocation policies:

1 Q-Prior: The quadratic program is used to allocate plant capacities to orders each period. The initial target production levels are determined based on the prioritized allocation policy (see

Section 3.4), which produces as much as possible of each product at its associated plant.

2 Q-Dist: Identical to the previous one, except that the initial target production levels are determined based on the distributed allocation policy which evenly assigns products to the plants that can produce them.

3 Naïve-Prior: Period-to-period orders are allocated using the prioritized allocation policy.

4 Naïve-Dist: Period-to-period orders are allocated using the distributed allocation policy.

We rank the 50 mean demand realizations tested in order of "unbalancedness" to observe how the benefits and costs of flexibility increase as demands for the different products become more unbalanced. We measure unbalancedness through $\sqrt{\sum_{i=1}^{m}(\mu_i - \bar{\mu})^2}$, where μ_i, $i = 1, \ldots, m$, are the mean demands of the m products, and $\bar{\mu} = \frac{\sum_{i=1}^{m}\mu_{D_i}}{m}$.

The remainder of this section is organized as follows. In Section 4.1, we compare the performance of naïve allocation policies with the optimization based quadratic allocation policies for partially flexible systems. Section 4.2 focuses on comparing the performance of partially flexible systems to fully flexible and non-flexible systems. Section 4.3 studies the impact of multiple product-specific components on the value of flexibility. Finally, we conclude by simulating a real example from the automotive industry in Section 4.4.

4.1.   Performance of naïve vs. quadratic allocation policies

We first analyze a three-plant three-product manufacturing system with the following chain configuration (see Figure 4.2): plant 1 can produce products 1 and 2, plant 2 products 2 and 3, and plant 3 products 1 and 3. We assume a component procurement lead-time of 2 weeks.

Figures 4.7 and 4.8 show how sales and component inventory increase under each of the four allocation policies in this partially flexible system, compared to the dedicated system. We observe that, on average, the optimization-based allocation policies (quadratic allocation) lead to 3% higher sales than the naïve policies. The naïve distributed policy has the poorest performance, leading to the lowest sales and highest inventories. This is due to the amount of variability induced in the system as a result of allocating half of the demand of one product to one plant and

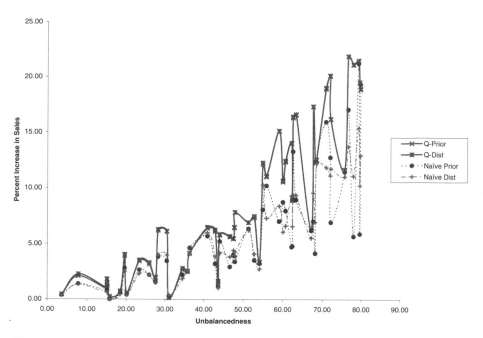

Figure 4.7. Percent increase in sales in the partial flexibility configuration under the different allocation policies considered

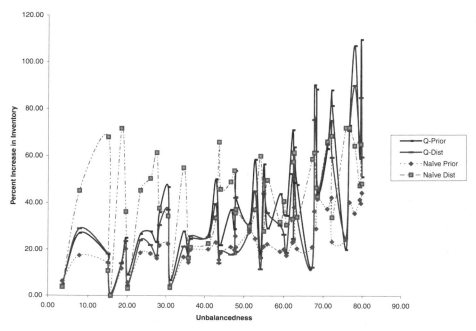

Figure 4.8. Percent increase in inventory in the partial flexibility configuration under the different allocation policies considered

the rest to the unfilled capacity of the other plant. Surprisingly, the Q-Dist allocation policy, which uses the quadratic program and starts out with distributed target production levels, achieves slightly higher sales than the Q-Prior policy, while keeping lower inventories (7% lower on average); see Figures 4.7 and 4.8. This is mainly due to the fact that in the Q-Prior policy, the production levels of the secondary products assigned to plants are highly variable, resulting in higher inventory and more frequent stock-outs (despite having the same safety factor z; this is because the distribution of production levels is skewed and not normal). Thus, although production of the main products under this policy is less variable due to plant capacity constraints, the resulting variability in the system is higher.

In addition, Figures 4.7 and 4.8 show that as demand for the three products becomes more unbalanced, both sales and inventory increase significantly for all allocation policies. As unbalancedness grows, the benefits of flexibility tend to be more variable. This suggests that there are other important factors that limit the benefits of the added flexibility. We find the major decisive factors to be: (1) total system demand relative to total capacity, and most importantly, (2) demand for the different products being below and above capacity.

Finally, Figure 4.9 depicts the percent increase in both sales and inventory for the best performing allocation policy (Q-Dist). We observe that while the average increase in sales is 9% and never higher than 22%, the average increase in inventory is 34% and it can be as high as 85%.

4.2.   Performance of partial vs. full flexibility

In this section we compare the benefits of full flexibility (all plants can manufacture all products) with those obtained for the partially flexible system following a chain configuration. As in the previous section, the non-flexible configuration is the base case, and we consider the percent increase in sales and inventory over the base case. We only consider the more realistic quadratic allocation policies: Q-Prior and Q-Dist. We assume a component procurement lead-time of 2 weeks.

We start with the three-plant three-product scenario (see Figure 4.2). The results, summarized in Figure 4.10, are striking. While sales remain almost the same as more flexibility is added into the system (as shown by Jordan and Graves [23]), *component inventory levels are reduced under full flexibility!* The reason is that the added flexibility allows for allocations that are closer to the target production levels, resulting in lower variability. Thus, although a limited flexibility

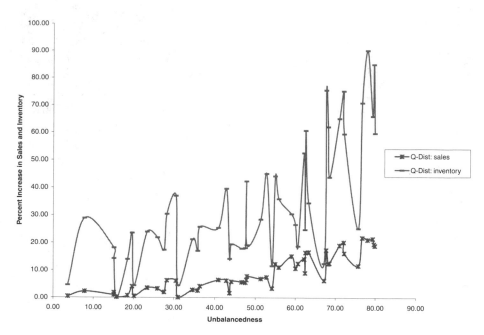

Figure 4.9. Percent increase in sales and inventory in the partial flexibility configuration under the Q-Dist allocation policy

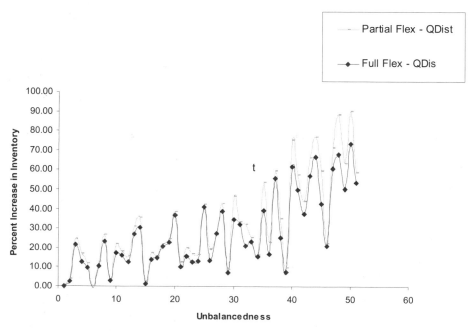

Figure 4.10. Percent increase in inventory levels in the full flexibility and partial flexibility configurations under the Q-Dist allocation policy, with respect to the non-flexible system

Table 4.2. Average percent increase in sales and inventory under the various chaining configurations in Figure 4.6, relative to the non-flexible case

Configuration	Policy	Percent Increase in	
		Sales	Inventory
2-Chain	Q-Prior	10.14	59.17
2-Chain	Q-Dist	10.19	46.33
3-Chain	Q-Prior	10.24	40.87
3-Chain	Q-Dist	10.30	38.61
4-Chain	Q-Prior	10.27	37.25
4-Chain	Q-Dist	10.33	35.96
5-Chain	Q-Prior	10.28	35.59
5-Chain	Q-Dist	10.35	35.41

investment following a chain configuration extracts most of the benefits in sales, it leads to higher operational costs due to the larger safety stocks required. Observe that this effect is more significant as demands become more unbalanced, leading to a reduction in component inventory levels of over 15% in some cases.

Next, we compare the various chaining configurations possible in the five-plant five-product system, including the fully flexible configuration, depicted in Figure 4.6. Table 4.2 shows that adding more flexibility to the 2-chain configuration barely impacts sales. However, inventories are significantly reduced as we add more flexibility in the system (Figure 4.11). This generalizes the result for the three-plant three-product case.

4.3. Impact of component variety

In this section, we analyze the impact of manufacturing flexibility on sales and component inventory levels as the number of components with different lead-times required to build the product grows. Observe that, without loss of generality, product-specific components with the same lead-time can be bundled as a single component in our model. We consider the three-plant three-product system with the following two variations: In the first one, each product requires two components with lead-times of 2 and 4 periods. In the second one, products are composed of five components each, with respective lead-times of 1, 2, 3, 4, and 5 periods. These components are assumed to be unique to each vehicle (no inventory pooling). Again, partial flexibility and full flexibility configurations are considered and compared to the non-flexible

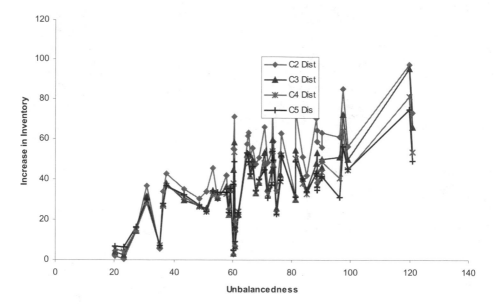

Figure 4.11. Percent increase in inventory of each of the four chain configurations considered in the five-plant five-product setting, under the Q-Dist allocation policy

system, our base case. The results over the 50 simulated mean demand scenarios are presented in Table 4.3. We only report the simulation results for the Q-Dist allocation policy, since again we found it to perform consistently better than the Q-Prior policy. The average inventory for components associated with each lead-time is listed separately. To allow

Table 4.3. Impact of partial (PF) and full (FF) manufacturing flexibility on sales and inventory levels as the number of components required to assemble each product grows

Number of Components	Lead-Time	Percent Increase in			
		Sales		Inventory	
		PF	FF	PF	FF
1	2	8.52	8.56	32.25	31.05
1	4	8.51	8.57	37.41	36.64
2	2	8.52	8.59	32.49	31.66
	4			37.12	36.22
5	1	8.90	9.00	29.55	28.62
	2			33.92	32.84
	3			36.78	35.53
	4			39.01	37.74
	5			40.72	39.46

for comparison, we include cases with a single component and different lead-times.

Table 4.3 shows that one additional component barely changes the effect of flexibility on sales and inventory levels. As more components are required, the increases in sales and inventory grow an extra 0.4% and 1.5% (considering the components with lead-time 2 and 4 for comparison), respectively, and the benefits of full versus partial flexibility slightly increase. We conclude that the effect of multiple components on the value of flexibility is quite weak.

4.4. A full-scale automotive industry scenario

To conclude our simulation study, we consider an example described in Jordan and Graves [23], which is based on a real set of vehicles and assembly plants. The manufacturing system consists of 8 plants and 16 products, which are originally arranged in six chains (groups of products and plants that are connected directly or indirectly by product-plant assignments); see Figure 4.12a).

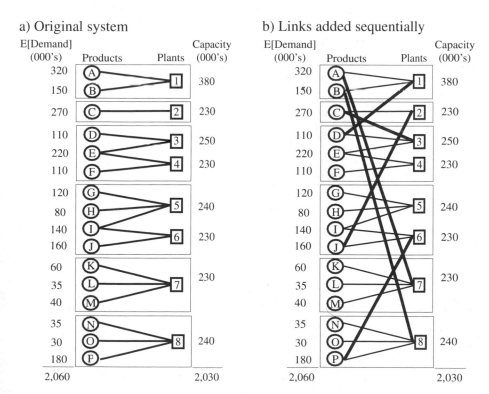

Figure 4.12. The 8 plant 16 product example considered in Jordan and Graves [23]

In our model, each planning period corresponds to one week. Therefore, we adjust the annual demands and capacities given in Figure 4.12 by dividing by the number of production weeks in a year. While demands of products within the same chain are positively correlated (correlation of 0.3), products in different chains are independent of each other. Overall, total average demand and capacity are fairly balanced (41,200 versus 40,600 per week). However, this is not the case for each of the chains, some of which have serious capacity deficits or surplusses. Thus, significant benefits can be obtained by linking the chains that have deficits with those that have a surplus and whose product demands are not correlated. We study the effect of adding six such links, (A-2), (A-3), (C-7), (D-6), (J-8), and (P-1), to the configuration, one at a time and in this order; see Figure 4.12 b). Figure 4.13 illustrates the changes in average weekly sales and component inventories resulting from adding these links sequentially to the original system. As before, in our simulation, we have assumed the forecast error at stage 1 to have a coefficient of variation of 0.4, truncated the average demand at stage 2 to $\pm 2\sigma$ of the point forecast, taken the coefficient of variation of period-to-period demand in stage 2 to be 0.15 and a component lead-time of 2 weeks.

We observe that while adding one link results in a 2.8% and 3.4% increase in sales and inventory, respectively, adding six links provides a 7.4% increase in sales accompanied by a 9.6% increase in inventory. Full flexibility, which requires adding 110 links, leads to increases of 9.6%

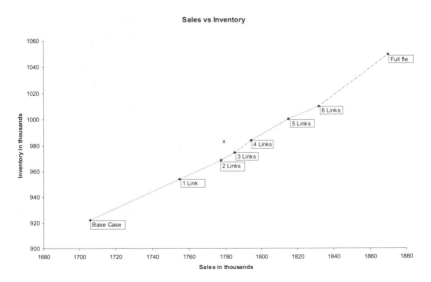

Figure 4.13.   Sales vs. inventory levels for a full-scale automotive industry scenario

and 13.8% in sales and inventory, respectively. In this case the percent increases in sales and component inventory levels are close in magnitude and grow at similar rates as we add more links. This is not surprising since demands are quite unbalanced in the system and we are adding strategic links to balance demands and capacities within each chain. As in Jordan and Graves [23], we conclude that "a small amount of flexibility configured in the right way can have virtually all the benefits of full flexibility."

5.   Conclusions and Directions for Future Research

Delivery speed and on-time performance reliability are becoming competitive battle-grounds in the automotive industry. Since capacity and flexibility decisions determine the agility of manufacturing companies, strategic capacity planning, which has always been a determining factor of financial success in the automotive industry, has become even more important in a make-to-order environment. In this chapter, we showed that manufacturing flexibility has a large impact not only on sales but also on the operational cost of the supply chain, and introduced a capacity planning framework that considers this effect. We used a multi-stage approach that considers strategic capacity investment decisions at the first stage, tactical capacity allocation and inventory control decisions at the intermediate stage, and the assignment of customer orders to available capacity at the final (operational) stage. Through a simple two-plant two-product model, we quantified the effects of adding flexibility on sales, component inventory levels, outbound distribution cost and demand variability observed upstream. Simulation studies complement this analysis considering more general models. Our results show that while flexibility is beneficial in increasing sales, it leads to a significant increase in variability observed by suppliers and requires higher levels of component inventory to be kept at the manufacturer. However, a flexible system would allow producing closer to the customer, thus considerably reducing the manufacturer's outbound transportation cost. This suggests that when evaluating investments in flexible capacity at the strategic level, not only the future increase in sales, but also the impact on operational costs of the supply chain must be weighed against the required capital outlay. Finally, we illustrated the usefulness of the proposed multi-stage approach as a capacity planning tool through a simple example and quantified the impact of adding different levels of flexibility for a real automotive capacity planning problem.

Our analysis provided additional managerial insights, such as how the benefits and costs of flexibility increase as the variability of period-to-period demand grows and as the mean demands for the different products become more unbalanced. In addition, we showed that the benefits and costs of flexibility are highly dependent on the allocation policy used to allocate different product demands to plants in a flexible system. In fact, as demand patterns change over time, the allocation policy initially implemented should be continuously revised, since its performance may deteriorate relative to that of other allocation policies. Finally, simulation of partial and full flexibility multi-plant multi-product models show that although partial flexibility results in almost the same sales as full flexibility, it requires significantly higher component inventory levels.

Numerous extensions of the models studied deserve further attention. In our previous analysis, product demands were assumed to be independent and each product required a single component, which is unique to that particular product. However, in reality, product demands are correlated over time and among products, and multiple components — some unique, some shared among different products — are required to assemble the final product. Moreover, some of the demand not satisfied in a particular period could be backlogged and satisfied in the following production period. In addition, other capacity allocation policies, such as policies that consider both inbound and outbound distribution costs when allocating capacity to customers in different locations, might be attractive in practice. Finally, the effect on supply chain performance of different flexibility configurations in production systems with multiple plants and products needs to be further studied.

Furthermore, we considered demand for "basic" vehicles with one component. This model is easily extendable to multiple components if the components are part of the basic vehicle, e.g. steering wheel. However, if the components are options a customer can choose (e.g. sunroof), other issues have to be addressed. First, the option capacity decision is dependent on the capacity decision for the vehicle model. Secondly, demand for the different options might be correlated (e.g., sports package). Thirdly, customers have a different attitude towards different options. For example, they might not buy a car without an automatic transmission, but accept that other options, such as a vanity mirror, are missing. Furthermore, an option might be shared among vehicles but the option price might be different depending on the vehicle. This leads to very different risks of over — and under-capacitizing for each particular option.

While we focused on supply-side flexibility, this is not the only potential lever to hedge against uncertainty. Price flexibility and delivery time flexibility can be used to smooth demand at the operational level and mitigate the effects of a poor capacity planning decision. We show elsewhere that flexible or dynamic pricing can be a useful lever to balance supply and demand if demand is high at the beginning of the planning horizon but declines over time. However, price is not a good lever if demand is initially low and increases over time (Biller et al. [4]). Therefore, strategic capacity investment decisions could improve if price and demand patterns as well as delivery time flexibility are considered as levers that can be pulled at the tactical or operational levels. Thus, future work that focuses on capacity planning, by considering the impact of demand management techniques on the optimal capacity decisions, would provide new insights into capacity planning. This analysis would be further complicated when considering demand management for vehicles and options.

Finally, we could reduce demand uncertainty by better predicting what customers want based on the two main sources of consumer information available: traditional sales data and web-user clicks. How could these different data be combined into a reliable forecasting system? How could web-user clicks be used to predict short-term demand and to suggest product enhancements that will make it more appealing to the consumer? Information of actual web-user product configurations could be evaluated using conjoint analysis. The ultimate in customer relationship management would be to link the data gathered from the web with actual purchasing behavior on an individual basis. However, privacy issues should be a premier concern for companies gathering data.

The transformation of vehicle manufacturers from a build-to-stock to a make-to-order producer is progressing rapidly. While other manufacturers, such as Dell Computers, have perfected the make-to-order model for modular assembly, much work needs to be done in non-modular industries to improve capacity investment decision making, choosing allocation and inventory policies, and pricing. We expect to see more OR models in these areas trying to integrate the four levers — capacity, component inventory, pricing and delivery flexibility — in a make-to-order environment.

Acknowledgments

The authors would like to thank Dr. Jeff Tew, Ted Costy, and Lynn Truss of the GM R&D Center for their continuous support of this project, feedback, and suggestions. Special thanks are due to Professor Muriel's students, Anand Somasundaram and Farhad Munshi, for their contribution to the simulation and numerical integration results.

Notes

1. Analogous to the DNA double helix, Fine illustrates how industry and/or product structure evolves from vertical/integral to horizontal/modular and back.

2. Today, Hollywood producers hire sub-contractors on an as needed basis (actors, directors, stage builders, special effects people, etc.) after they have read the script. This is fundamentally different from the big-studio model of the 1930s through 1960s where all people involved in producing a movie were employed by the one studio.

References

[1] ANDREOU, S.A. (1990) A Capital Budgeting Model for Product-Mix Flexibility. *Journal of Manufacturing Operations Management*, Vol. 3, 5–23.

[2] BASSOK, Y., R. ANUPINDI, AND R. AKELLA. (1999) Single-period Multiproduct Inventory Models with Substitution. *Operations Research*, Vol. 47, 632–642.

[3] BEACH, R., A.P. MUHLEMANN, D.H.R. PRICE, A. PATERSON, AND J.A. SHARP. (2000) A Review of Manufacturing Flexibility. *European Journal of Operational Research*, Vol. 122, 41–57.

[4] BILLER, S., L.M.A. CHAN, D. SIMCHI-LEVI, AND J. SWANN. (2000) Dynamic Pricing and the Direct-to-Consumer Model in the Automotive Industry. In preparation.

[5] BISH, E.K., A. MURIEL, AND S. BILLER. (2000) Capacity and Flexibility Planning in a Make to-Order Environment. Under review for *Management Science*.

[6] BITRAN, G.R. AND S. DASU. (1992) Ordering Policies in an Environment of Stochastic Yields and Substitutable Demands. *Operations Research*, Vol. 40, 999–1017.

[7] BITRAN, G.R. AND S. GILBERT. (1996) Managing Hotel Reservations with Uncertain Arrivals. *Operations Research*, Vol. 44, 35–49.

[8] BRADY, D., K. KERWIN, D. WELCH, L. LEE, AND R. HOF. Customizing for the Masses. *Business Week*, March 20, 2000.

[9] CAULKINS, J.P. AND C.H. FINE. (1990) Seasonal Inventories and the use of Product-Flexible Manufacturing Technology. *Annals of Operations Research*, Vol. 26, 351–375.

[10] CARROLL, W.J. AND R.C. GRIMES. (1995) Evolutionary Change in Product Management: Experiences in the Car Rental Industry. *Interfaces*, Vol. 25, 84–104.

[11] DE TONI, A. AND S. TONCHIA. (1998) Manufacturing Flexibility: A Literature Review. *International Journal of Production Research*, Vol. 36, 1587–1617.

[12] DICKINSON, J., S. DEMBKOWSKI, C. SHAH, AND R. MORRISON. (1998) The Future of Automotive Distribution. *Financial Times Automotive*, London, UK.

[13] DRUCKER, P. (1946) *The Concept of a Corporation*. John Day, New York, NY.

[14] EDELSTEIN, M. AND M. MELYNK. (1977) The Pool Control System. *Interfaces*, Vol. 8, 21–36.

[15] EPPEN, G.D., R.K. MARTIN, AND L. SCHRAGE. (1989) A Scenario Approach to Capacity Planning. *OR Practice*, Vol. 37, 517–527.

[16] FINE, C.H. AND R.M. FREUND. (1990) Optimal Investment in Product-Flexible Manufacturing Capacity. *Management Science*, Vol. 36, 449–466.

[17] FINE, C.H. (1998) *Clockspeed*. Harper Collins, New York, NY.

[18] GERAGHTY, M.K. AND E. JOHNSON. (1996) Revenue Management Saves National Car Rental. *Interfaces*, Vol. 27, 1-7-127.

[19] GRAVES, S.C. AND B.T. TOMLIN. (2000) Process Flexibility in Supply Chains, under review for *Management Science*.

[20] HSU, A. AND Y. BASSOK. (1997) Random Yield and Random Demand in a Production System with Downward Substitution. To appear in *Operations Research*.

[21] J.D POWER AUTOMOTIVE AND ASSOCIATES 2000 J.D. Power Automotive and Associates (2000) 2000 New Autoshopper.com Study, in *The Power Report*, October, Augora Hills, CA.

[22] JONES, R.A. AND J.M. OSTROY. (1984) Flexibility and Uncertainty. *Rev. Economic Studies*, Vol. 51, 13–32.

[23] JORDAN, W.C. AND S.C. GRAVES. (1995) Principles and Benefits of Manufacturing Process Flexibility. *Management Science*, Vol. 41, 577–594.

[24] KARAESMEN, I. AND G. VAN RYZIN. (1998) Overbooking with Substitutable Inventory Classes. Working paper, Columbia University, NY.

[25] KHOUJA, M. (1999) The Single-period (News-vendor) Problem: Literature Review and Suggestions for Further Research. *Omega*, Vol. 27, 537–553.

[26] KOUVELIS, P. (1992) Design and Planning Problems in Flexible Manufacturing Systems: A Critical Review. *Journal of Intelligent Manufacturing*, Vol. 3, 75–99.

[27] LAPIDUS, G. (2000) Gentlemen, Start Your Engines. *Goldman Sachs Research Report*, New York, NY.

[28] LARSON, R.C. AND A.R. ODONI. (1981) *Urban Operations Research*. Prentice-Hall, Inc., New Jersey.

[29] MURIEL, A. AND A. SOMASUNDARAM. (2000) Simulation Study of the Impact of Manufacturing Flexibility on Sales, Inventory, and Transportation Costs. Working paper, University of Massachusetts, Amherst, MA.

[30] NETESSINE, S., G. DOBSON, AND R.A. SHUMSKY. (2000) Flexible Service Capacity: Optimal Investment and the Impact of Demand Correlation. To appear in *Operations Research*.

[31] PASTERNAK, B. AND Z. DREZNER. (1991) Optimal Inventory Policies for Substitutable Commodities with Stochastic Demand. *Naval Research Logistics*, Vol. 38, 221–240.

[32] SETHI, A.K. AND S.P. SETHI. (1990) Flexibility in Manufacturing: A Survey. *The International Journal of Flexible Manufacturing Systems*, Vol. 2, 289–328.

[33] SLOAN, A.P. (1963) My Years with General Motors. *Currency and Doubleday*, New York, NY.

[34] STALK, G. (1988) Time — the next source of competitive advantage. Harvard Business Review July–August 1988, Vol. 66, No. 4.

[35] STALK, G., S. STEPHENSON, AND T. KING. (1996) Searching for Fulfillment: Breakthroughs in Order-to-Delivery Process in the Auto Industry. The Boston Consulting Group.

[36] TRIANTIS, A.J. AND J.E. HODDER. (1990) Valuing Flexibility as a Complex Option. *The Journal of Finance*, Vol. 45, 549–565.

[37] VAN MIEGHEM, J.A. (1998) Investment Strategies for Flexible Resources, *Management Science*, Vol. 44, 1071–1078.

[38] VAN OYEN, M., E. SENTURK-GEL, AND W. HOPP. (2000) Performance of Workforce Agility in Collaborative and Non-Collaborative Work Systems. To appear in *IIE Transactions*.

[39] WELCH, D. Where's My Dream Car? *Business Week*, November 27, 2000.

[40] WOMACK, J.P., D.T. JONES, AND D. ROOS. (1990) *The Machine that Changed the World*. Harper Collins, New York, NY.

Chapter 5

OPTIMAL USE OF DEMAND INFORMATION IN SUPPLY CHAIN MANAGEMENT

Guillermo Gallego

Industrial Engineering and Operations Research
Columbia University, 324 Mudd Bldg, New York, NY 10027
ggallego@ieor.columbia.edu

Özalp Özer

Management Science and Engineering
Stanford University, 336 Terman, Stanford, CA 94305
ozalp.ozer@stanford.edu

1.   Introduction

Alan Greenspan was puzzled by the data on his computer screen. Capital expenditures in high technology were rising sharply, unemployment was declining, prices were holding steady and profits were rising. At the same time, the Labor Department's statistics showed that productivity had decreased by one percent during the second quarter. Greenspan did not agree with the productivity data. He found the missing link in the Survey of Current Business from the Bureau of Labor Statistics, Woodward [44], which indicated that business inventories were shrinking significantly while the economy was growing. This suggested that new computer technology was allowing just-in-time orders. Instead of stocking products weeks or months in advance, businesses could keep detailed track of what was needed and order within days, while competitive pressures were forcing quality control. Greenspan eventually became convinced by this argument and voiced it publicly in the *State of the Economy* address before the Committee on Ways and Means, U.S. House of Representatives, January 20, 1999.

Are companies really using demand information optimally to manage inventories? If so, what models are driving these decisions? Anecdotal evidence suggest that the implementation of Enterprise Resource Planning (ERP) software, such as SAP, Baan has allowed companies to integrate data bases and greatly improve the quality of the data that resides therein. In addition, the purchase and implementation of numerous ERP bolt-ons, such as i2 and Manugistics, were justified, at least in part, in integrating manufacturing and replenishment decisions with key suppliers and basing these decisions on customers' needs.

Capital investments in information technology have certainly provided members of supply chain networks with more timely and better quality data than ever before. As members of supply networks, inventory managers need to decide what data to collect, request, or buy from their customers, and what data to pass on, or sell to their suppliers. In addition, inventory managers need to develop effective replenishment policies that make full use of demand information.

This chapter is concerned with models where current demand information is used to drive inventory replenishment policies in distributed decision making settings and with advance demand information models where customers place orders in anticipation of future requirements. The problem of using current demand information on single echelon systems has been studied extensively and reviewed by Scarf [38], Veinott [43], and Porteus [35]. Recent papers in this area are concerned with the use of current demand information in a supply chain context. A typical example is a retailer having point of sale (POS) information that is valuable to his supplier. Questions that arise include: What is the optimal policy for the supplier when he only sees orders from the retailer? What is the optimal policy for the supplier when in addition to retail orders he has access to POS information? What is the value of POS information to the supplier? Is the retailer better off by sharing POS information? If not, can POS information be traded so that both the retailer and the supplier are better off? Authors that have studied different versions of this problem include Bourland et al. [4], Chen [6], Gallego et al. [16], Gavirneni et al. [21], and Lee et al. [30].

Models of advance demand information assume that the total demand for a period can be expressed as the sum of orders placed over a certain horizon prior to the period. For example, the total demand for period five may be the sum of orders placed in periods two, three, four, and five. There are two motivations for studying demand models of this form. First, under long term contracts, downstream supply chain partners often agree to forecast and update future requirements and to freeze orders within a certain time window. Ford Motor Company, for example, issues orders and updates these orders for the next few weeks to its catalytic converter supplier as discussed in the Harvard Business School teaching case "Corning Glass Works" (1991). Long term contracts, however, are not the only way companies get advance demand information. Advance demand information also occurs when customers place orders in advance, perhaps through the Internet, for customized products such as a upscale furniture and computer equipment.

The second reason to study models with advance demand information is that there is a growing consensus, see Appell et al. [1], that manufacturers can benefit from having a portfolio of customers with different demand leadtimes[1]. A portfolio of customers with different demand leadtimes can enable better capacity utilization, higher and more consistent revenues and better customer service.

Customers often need to be induced to place advance orders. Chen [7] provides an example where market segmentation is used to gain advance demand information. The problem of finding price structures to induce advance bookings has a revenue management flavor, Gallego and van Ryzin [20], and is outside the scope of this chapter. Advance demand information inventory models, however, are very complementary to the problem of finding effective incentives for customers to book in advance. Indeed, the advance demand information models studied in this chapter are about finding effective inventory policies and about evaluating the cost benefits of advance demand information. Without the ability to assess the cost benefits of advance demand information it is very difficult to design cost effective incentives to induce advance bookings. Authors who have studied models related to the concept of advance demand information include Gallego and Özer [17], [18], Gallego and Toktay [19], Graves et al. [22], Güllü [23], [24], Hariharan and Zipkin [26], Heat and Jackson [28], Toktay and Wein [42], Özer [33] Schwarz, et al. [40], and Sethi et al. [41].

The aim of this chapter is to summarize the result in Gallego, et al. [16], Gallego and Özer [17], [18], and Özer [33]. In the course of summarizing these papers we also review the work of other authors that has contributed to our understanding of these problems. Section 2 deals with models using current demand information, while Section 3 deals with models using advance demand information.

2.   Using Current Demand Information

This section will deal mostly with multi-location models using current demand information. However, we start our discussion with a single location model due to Herbert Scarf. His model is an example where demand information is used to decide whether or not to exercise an option. After presenting Scarf's model we will discuss papers by Bourland et al. [4], Chen [6], Gavirneni et al. [21], Gallego et al. [16], and Lee et al. [30]. These authors find that sharing *current* demand information in supply chains is valuable when order cycles are asynchronous, when information is centralized in a vertically integrated system, when capacity is limited, when the retailer orders in batches to take advantage of economies of scales, and when the demand process is autocorrelated.

2.1.   Optimal inventory policies when sales are discretionary

Herbert Scarf is well know for his early contributions to inventory theory, see Clark and Scarf [11], Scarf [37], and [38]. After a 40 years hiatus, Scarf [39] returns to the field to investigate the nature of optimal policies when sales are discretionary, i.e., when the inventory manager can choose to meet a fraction of the demand that arises during any given period. Inventory managers may exercise the discretionary sales option in anticipation of higher prices or higher costs.

In Scarf's model the horizon is assumed to be finite, ordering costs are fixed plus linear, lead times are zero, and unsatisfied demands are lost. The demand during the period is a random variable D. The special feature of this model is that the manager may choose to sell any quantity z with $0 \leq z \leq \min[y, D]$ at the market price p where y is the period's starting inventory after ordering. The ordering cost, selling price and demand distributions can be time varying aside from the requirement that the set-up costs decrease monotonically over time. The objective is to maximize expected discounted profit.

Scarf's analysis makes use of a novel property of K-concave functions[2]. Scarf shows that if $d \geq 0$ is a constant, f is a K-concave function defined on $[0, \infty)$ with a finite number of local maxima,

$$g(y) = \max_{0 \leq z \leq \min[y,d]} f(y - z),$$

and $z(y)$ is the largest minimizer of $f(y - z)$ subject to $0 \leq z \leq \min[y, d]$, then g is K-concave and $y - z(y)$ is monotonically increasing in y.

Scarf uses this property to show that policy that maximizes the expected discounted profit is of the (s, S) form. Under this policy the inventory manager observes the inventory x and places an order for $S - x$ units if $x \leq s$ and does not order otherwise. After observing the demand, say d, during the period, the inventory manager sells $z(y) \in [0, \min(y, d)]$

and keeps $y - z(y)$ units for the next period. While speculating with inventories is rarely seen in stationary environments, this behavior is typical in inflationary environments and in environments where there is significant price and or cost volatility.

2.2.   The value of demand information when order cycles are asynchronous

Bourland et al. [4] consider a two stage serial system where each stage follows a periodic review base-stock policy. Each stage is assumed to have ordering periods of equal length. The authors investigate the benefits of sharing customer demand information through EDI when the ordering cycles are *not* synchronized. As an example, consider a supplier and a retailer making ordering decisions on a weekly basis. Assume the retailer places orders every Thursday and the supplier places orders every Monday. Assume that the supplier incurs linear holding and backorder costs. In the absence of demand information the retailer faces a periodic review inventory problem with iid demands. Under this condition the supplier's expected cost is proportional to standard deviation of the weekly demand. Suppose, instead, that on Mondays the supplier obtains information from the sales experienced by the retailer since Thursday. Thus, at the time of placing the order, the supplier only faces uncertainty about the demand for Monday, Tuesday, and Wednesday. The cost to the supplier is now proportional to the standard deviation of the demand over these three days. If the demand is uniform over the week then the expected cost under demand information sharing is 65.4% ($\sqrt{3/7}$) of the expected cost in the absence of demand information sharing.

2.3.   The value of demand information when capacity is limited

Gavirneni et al. [21] study the holding and penalty cost of a finite capacity supplier facing demands from a single retailer following an (s, S) policy. The authors assume that the retailer's lead time is zero and that the retailer can immediately procure units from an alternative source if the supplier is not capable of filling an order. The supplier absorbs the assumed linear cost of procuring such units. Under these assumptions, the authors compare the costs with and without information sharing, and report significant savings for the supplier from demand information sharing. The savings come from the supplier's increased ability to anticipate the timing and the magnitude of the next order from the retailer.

2.4. The value of demand information when the retailer batches orders

Gallego et al. [16] assess the benefits of sharing demand information in a supply chain consisting of a single supplier and a single retailer. The retailer faces Poisson demands at rate λ, economies of scale in ordering, and places orders from the supplier according to a (Q, r) policy where Q is fixed.[3] The supplier orders from an outside source with ample stock and incurs linear holding and backlogging costs at rates h and p respectively. Units ordered from the outside source arrive at the supplier after L units of time, where L is a known constant. The supplier's problem is to minimize the expected holding and backorder penalty costs with or without retail demand information. The authors show that the supplier's optimal policy under demand information sharing calls for monitoring the retailer's inventory position and results in a cost that is independent of the order size Q. In the absence of demand information they show that a modified base-stock policy where the supplier introduces a random delay after receiving the order from the retailer is optimal. The authors then investigate whether or not the retailer is better off by voluntarily sharing demand information. When this is not the case, the authors identify conditions under which information can be traded, i.e., purchased by the supplier.

In the absence of demand information the supplier only observes orders of size Q from the retailer. The time between orders observed by the supplier is an Erlang random variable, say T_Q, with parameters Q and λ. The authors assume that the supplier knows the distributional form of T_Q or equivalently that the supplier knows, or can accurately estimate, the demand rate λ. The expected cost of the supplier's optimal base-stock policy is given by $\min H(mQ)$ where the minimization is done over integer values of m, $H(y) = hE(y - N_L)^+ + pE(N_L - y)^+$ and N_t denotes a Poisson random variable with parameter λt. They show that the largest optimal base stock level is $S = m_o Q$ where m_o is given by

$$m_o = \min\left\{ m : m \in \mathcal{Z}^+, \frac{1}{Q} \sum_{k=mQ+1}^{(m+1)Q} P(N_L < k) > \frac{p}{h+p} \right\}. \quad (5.1)$$

The case $Q = 1$ is of special interest, since in this case the retail orders coincide with the demand seen by the retailer. Thus, in this case, the supplier has full demand information. The optimal base stock level, say

S^*, is given by the smallest integer S satisfying

$$P(N_L \leq S) > \frac{p}{h+p}. \tag{5.2}$$

Obviously, $H(S^*) \leq H(m_o Q)$ for all $Q > 1$. In fact $H(S^*)$ is a lower bound on the optimal with respect to all policies for all Q, even if the supplier has full demand information.

2.4.1   Optimal policies under demand information sharing. When the retailer shares demand information with the supplier, the supplier's optimal policy consists of monitoring the retailer's inventory position and placing an order of size Q when the retailer's inventory position drops to $r + n$ where $n = S^* - (m-1)Q$ and m is an integer such that $(m-1)Q < S^* \leq mQ$. The expected cost to the supplier under this policy is shown to be equal to $H(S^*)$, and is independent of Q. A numerical comparison shows that the cost of this policy can be significantly lower than the cost of the best base-stock policy. For example, for $\lambda = 20, L = 1, h = 1$, and $p = 9$, $S^* = 26$, and $H(26) = 8.19$. On the other hand, for $Q = 15$, $m_o = 2$ and $H(30) = 10.32$ while for $Q = 20$, $m_o = 1$ and $H(20) = 17.77$.

2.4.2   Optimal   policies   without   demand   information sharing.   The optimal policy under demand information sharing *delays* the placement of orders until $Q - n$ units are demanded. This suggest that expanding the class of base-stock policies by allowing the supplier to delay orders. We first consider base-stock policies with fixed and then random delays. Gallego et al. show that random delay base-stock policies are optimal.

If the supplier delays orders arriving from the retailer by τ units of time then the system behaves as if the lead time was $L + \tau$ instead of L. We can write the cost of delayed base-stock policies as $H(mQ, \tau) = hE[y - N_{L+\tau}]^+ + pE[N_{L+\tau} - y]^+$. Fixed delay base-stock policies were recently proposed by Moinzadeh [31]. He computes the optimal delay τ for a fixed order size mQ for the case where the inter-order distribution is normal and for the case where only the mean and variance of the interorder distribution is known. Let

$$m_1 = \min \left\{ m \in \mathcal{Z}^+ : P(N_L < mQ) = P(t_{mQ} > L) > \frac{p}{h+p} \right\}. \tag{5.3}$$

Corresponding to every integer $m \geq m_1$ there is a unique $\tau_m > 0$ such that (mQ, τ_m) is a stationary point. The existence of a countable number of stationary points lead Moinzadeh to an extensive computational search for an optimal solution. Gallego et al. show that it is enough to consider only the points $(m_o, 0)$ and (m_1, τ_1) where τ_1 is the unique positive root of $P(t_{mQ} > L + \tau) = P(N_{L+\tau} < mQ) = \frac{p}{h+p}$.

2.4.3 Random delay base-stock policies.

The authors consider a policy that terminates the delay if the next order from the retailer arrives before the fixed delay. Let ν be the fixed delay and T_Q the time until the next order from the retailer. Under the random delay base-stock policy, the supplier delays her order by $\min(\nu, T_Q)$. By the reward renewal theorem the average cost of this policy can be written as

$$G(mQ, \nu)$$
$$= h\lambda E\left(T_{(m-1)Q} - L + (T_Q - \tau)^+\right)^+ + p\lambda E\left(L - T_{(m-1)Q} - (T_Q - \tau)^+\right)^+$$
$$= hE[(m-1)Q - N_L + (Q - N_\nu)^+]^+ + pE[N_L - (m-1)Q - (Q - N_\nu)^+]^+$$

Notice that if $\nu = 0$ then

$$G(mQ, 0) = H(mQ, 0) = hE[mQ - N_L]^+ + pE[N_L - mQ]^+.$$

Thus, the largest optimal m for $\nu = 0$ is m_o as defined by (5.1). We need to compare the cost $G(m_oQ, 0) = H(m_oQ, 0)$ to the cost of the best policy with a positive delay. Recall the definition of m_1 in equation (5.3). Let ν_1 be the unique solution to

$$P(N_L + N_\nu < m_1 Q \mid N_\nu < Q) = \frac{p}{h + p}. \tag{5.4}$$

The authors show that the pair (m_1Q, ν_1) is the only strictly interior stationary point of $G(mQ, \nu)$. Thus, to find the optimal random delay base-stock policy we only need to compare $G(m_oQ, 0)$ and $G(m_1Q, \nu_1)$. Let (m^*Q, ν^*) denote the pair with lower cost among these two candidate solutions. The authors also show that policy (m^*Q, ν^*) minimizes the supplier's long run average holding and penalty cost among all possible policies by using arguments due to Katircioglu [25].

Although the random delay base-stock policy has the virtue of being provably optimal, our computations indicate that the savings relative to the fixed delay base-stock policy are almost always negligible. The only case where there was a small, but a significant, difference in cost was for $Q = 20$ where $\nu_1 = 0.6279$. This case resulted in an average cost of \$10.19, which is a slight improvement over the cost \$10.33 of using the fixed delay $\tau_1 = 0.6069$. As a final observation, note that random delay policies are unlikely to be significantly better than fixed delay policies for small and large values of Q. Indeed, for small values of Q we almost have retail demand information, whereas for large Q values T_Q becomes large and almost deterministic.

Q	ν^*	m^*Q	$G(m_oQ,0)$	$G(m^*Q,\nu^*)$	$\frac{G(m_oQ,0)-G((m^*Q,\nu^*)}{G(m_oQ,0)}$
1	0	26	8.19	8.19	–
5	0	25	8.31	8.31	–
10	0.1615	30	10.32	8.83	14.44%
15	0.1615	30	10.32	8.83	14.44%
20	0.6279	40	17.77	10.19	42.66%
25	0	25	8.31	8.31	–
30	0.1615	30	10.32	8.83	14.44%
40	0.6069	40	20.00	10.33	48.35%
50	1.059	50	30.00	11.64	61.21%

2.4.4   To share or not to share demand information.

Suppose it takes l units of time for a shipment to get from the supplier to the retailer. If we take the view that stockouts at the supplier cause shipment delays to the retailer, then the retailer's lead time demand under demand information sharing is

$$N_l + (N_L - S^*)^+,$$

where S^* is given by (5.2). On the other hand, the retailer's lead time demand without information sharing is

$$N_l + (N_L - (m-1)Q - (Q - N_{\nu^*})^+)^+,$$

where (m^*Q, ν^*) minimizes G.

The retailer needs to compute his cost under each of these lead time demands to decide whether or not it is in his best interest to share demand information. For example, when $Q = 20$, the retailer would be willing to share demand information. This is a win–win situation since both the retailer and the wholesaler benefit from sharing demand information. On the other hand, for $Q = 10$ the retailer will be unwilling to share demand information. A similar situation arrises when $Q = 15$ and when $Q = 30$. If the retailer is worse off sharing demand information, the supplier may be willing to buy this information from the retailer if the expected gain to the supplier exceeds the expected cost to the retailer. Although we are not aware of any practical instance where the retailer sells demand information for a fee, there may be an implicit cost paid by the supplier in the form of lower unit prices. Moreover, as retailers become more powerful, they may soon be in a position of actually demanding payment for demand information. Finally, if neither selling nor sharing demand is jointly profitable, the supplier and the retailer may attempt to jointly optimize the sum of their costs and then find a way to share the savings. We note, however, that moving from distributed to centralized decision making is nontrivial since it entails sharing cost structures which may be misrepresented.

2.5.   The value of centralized demand information in vertically integrated firms

Chen [6] considers a continuous–time model of a vertically integrated serial supply chain where the demand process at the downstream stage is compound Poisson. Each stage is assumed to follow a (nQ, r) replenishment policy with given batch sizes. Reorder points are computed based on echelon (centralized) and on local (decentralized) information with the objective of minimizing system wide expected costs. It is known that the cost difference is zero when demand is Poisson and the batch sizes are all one, but Chen's is the first comprehensive numerical study to investigate the cost difference under compound Poisson and arbitrary, but fixed, batch sizes. Surprisingly, the cost difference is fairly small, 1.75% on the average, and a maximum of 9%. The cost seems to be larger the longer the lead time, the larger the batch sizes, and the greater the number of echelons in the chain. We remark that the policies found by Chen are optimal among the class of (nQ, r) policies he considers. It is quite possible that the idea of delaying orders, described in the previous section, may provide a strict improvement over the class of (nQ, r) policies, but more research is needed to assess the extent to which delay policies are beneficial in this setting.

2.6.   The value of demand information when demand is autocorrelated

Lee et al. [30] study the benefit of demand information sharing to the manufacturer in a two stage supply chain consisting of a single manufacturer and a single retailer facing an autocorrelated demand process. The manufacturer and the retailer incur linear holding and backlogging costs, experience constant lead times, and follow base-stock policies. The authors assume that the manufacturer immediately ships retail orders, instantly procuring from an alternative source at a linear penalty cost in the case of a shortfall. The authors report large savings for the manufacturer when the autocorrelation coefficient $\rho \in (-1, 1)$ is high. These savings are computed under the assumption that the manufacturer knows the parameters of the demand process, observes only retail orders, and makes forecasts based only on the *last* retail order.

The demand process is assumed to be given by

$$D_t = d + \rho D_{t-1} + \epsilon_t,$$

where $d > 0$, $-1 < \rho < 1$, and ϵ_t is iid normally distributed with mean zero and variance σ^2.

Let S_t denote the order-up-to level in period t. At the end of time t the retailer orders

$$Y_t = D_t + (S_t - S_{t-1})$$

which represents the demand during period t plus the change made in the order-up-to level. The authors acknowledge that it is possible to have $Y_t < 0$ but they assume $\sigma \ll d$ so that $P(Y_t < 0)$ is negligible. The authors show that the conditional expectation and the conditional variance of the total lead time demand are given by

$$m_t = a + bD_t,$$

and

$$\nu_t = c^2 \sigma^2,$$

where a is a constant that depends on d, ρ and l, while b and c are constants that depend only on ρ and the length of the lead time l. Consequently,

$$S_t = m_t + kc\sigma,$$

and

$$Y_t = D_t + b(D_{t+1} - D_t),$$

where k is the $p/(h+p)$ percentile of the standard normal distribution.

The authors then solve the manufacturers problem under both information sharing and no information sharing. The authors assume that, in the case of no information sharing, the manufacturer estimates the conditional mean and variance of his lead time demand based on Y_t only, as opposed to D_t in the case of demand information sharing. It is assumed that the manufacturer knows d, ρ, and σ. The conditional mean and variance of the manufacturer's lead time demand is then used to set the manufacturer's order-up-to levels. The authors attribute very large savings to the manufacturer's long run average cost, especially for $\rho > 0.5$, to demand information sharing.

We find, however, that the manufacturer can estimate D_t as follows: Let $\hat{D}_o = \frac{d}{1-\rho}$, and for $t > 0$ let

$$\hat{D}_t = \frac{Y_t + b\hat{D}_{t-1}}{1 + b}.$$

Let $e_t = D_t - \hat{D}_t$ be the estimation error in period t. A little algebra reveals that

$$D_t - \hat{D}_t = \frac{b}{1+b}(D_{t-1} - \hat{D}_{t-1})$$

$$= \frac{b}{1+b}e_{t-1}$$

$$= A^{t-1}e_1,$$

where $A = \frac{b}{1+b} = \frac{\rho(1-\rho^{l+1})}{1-\rho^{l+2}} \in (-1,1)$, which implies that $e_t \to 0$ as $t \to \infty$. Since $\hat{D}_o$ is the unconditional mean of D_t, it follows that $e_1 = \epsilon_1$ is normal with mean zero and variance σ^2. Thus the absolute value of the initial error is, with very high probability, bounded by $3\sigma \ll d$. As an example, assume that $\rho = 0.6$, $d = 100$, $l = 5$, and $\sigma = 10$. Then a large initial error of $3\sigma = 30$ is 15% of the mean demand. After 5 weeks the error is 1.8% of the mean demand, and after 12 weeks the error is negligible. Armed with the estimate $\hat{D}_t$ the manufacturer can do as well as under demand information sharing after 12 periods. This analysis shows that the savings in average costs reported by the authors are only transient and are zero in the long run. Demand information sharing is valuable when the manufacturer cannot infer current demand from the order history. This would be the case, for example, if demand were driven, in part, by pricing and promotion activities.

3.   Using Advance Demand Information

Hariharan and Zipkin [26] incorporate advance demand information in a single echelon continuous review system by introducing the concept of *demand leadtimes*. In their model customers place orders l periods in advance of their requirements. As a consequence, the inventory manager has perfect information about future demand. For fixed demand lead time, L_D, they prove the optimality of base stock policies and show that demand lead times directly offset supply leadtimes. As an example, consider the case where the supply lead time is L. If $0 \leq L_D \leq L$ the system with demand leadtimes behaves as a system with supply lead time $L - L_D$. As L_D approaches L the system moves from make-to-stock to make-to-order. If $L_D > L$, then the system is make-to-order and it is optimal for the inventory manager to delay orders by $L_D - L$ units of time.

There have been a number of attempts to generalize the results of Hariharan and Zipkin to the case where advance demand information is not perfect. Schwarz et al. [40] consider a periodic review model where a retailer has *imperfect* demand information over the demand leadtime L_D. To better understand their model, consider the special case $(L_D, L) = (1, 2)$. At the beginning of a period, the inventory manager observes the potential demand $Z = z$ for the next period. The demand, say X, that materializes in the following period depends on z. For example, Z may be Poisson with parameter $\lambda/(1-p)$ and X binomial with parameters $Z = z$ and $1 - p \in [0, 1]$ where p is the cancelation rate. Schwarz et al. consider the problem of maximizing the total expected discounted profit over an infinite horizon for a problem with stationary cost parameters and a stationary demand process under the assumption that inventory manager can order or dispose units at a linear cost. They show that a state dependent base stock policy is optimal where the state is the vector of demand signals over the demand leadtime. The authors compare systems with imperfect information, $0 < p < 1$ to the extreme cases of perfect and no information, and provide insights about the nature of the optimal policies and the corresponding expected discounted profits. Notice that the sequence of demand forecasts namely, λ, $Z(1-p)$ and X forms a martingale since $E[Z(1-p)] = \lambda$ and $E[X|Z(1-p)] = Z(1-p)$ with $E[X] = E[E[X|Z]] = E[Z(1-p)] = \lambda$. Viewed this way, the work of Schwarz et al. is best examined in the light of earlier results dealing with demand forecast revisions.

Work on demand forecast revisions include Hausman [27], Heat and Jackson [28], Graves et al. [22], and Güllü [23], [24], Gallego and Toktay [19], and Toktay and Wein [42]. Hausman [27] models the evolution of forecast as a quasi-Markovian or Markovian process. He suggests that ratios of successive forecasts can be modeled as independent lognormal variates and incorporated into sequential decision problems. Heat and Jackson [28] model the evolution of forecast using martingales. They name their model the Martingale Method of Forecast Evolution (MMFE) and use it to generate forecasts for a simulation model and to analyze economic safety stock levels for a multi-product multi-facility production system. Graves et al. [22] independently model a single item version of MMFE and use it to analyze a two stage production planning system. Güllü [23] determines the form of the optimal policy that arises under MMFE for a capacitated single item/single facility inventory system with zero set up cost and instantaneous delivery. He shows that the system, which employs demand forecast one period in the future, attains lower expected minimum cost than a system that does not incorporate future demand information. This result is a precursor to Schwarz et al. in the context of cost minimization that does not make the disposal assumption. Güllü [24] studies the behavior of the optimal order up to policy with respect to capacity levels and the forecast state. Gallego and Toktay [19] study a production problem with demand updates where capacity is limited and the fixed cost is large enough to justify an all or nothing production policy. Under such a policy, production in a given period is either zero or the maximum capacity. They show that a state dependent threshold policy is optimal. Toktay and Wein [42] model a setting where the production stage is modeled as a single-server discrete-time continuous-state queue. They use heavy traffic and random walk theory to obtain a closed form approximation for the forecast adapted base stock policies where forecasts are updated as in MMFE. Sethi et al. [41] study a model of forecast evolution with zero setup costs and multiple delivery modes and show that a state dependent policy is optimal.

The work that we review in this section, Gallego and Özer [17], Gallego and Özer [18], and Özer [33], deal respectively, with single location, serial, and distribution systems under a model of advance demand information. The advance demand information model is a discrete time generalization of Hariharan and Zipkin, and fits into the MMFE framework in a very explicit way. The idea is that total demand for any given period is the sum of customer commitments made over a certain horizon. For example, the total demand for period five is the sum of commitments made by customers in periods three, four, and five. This generalizes the idea of Hariharan and Zipkin in that part of the demand is known in advance. Our model, unlike that of Schwarz, Petruzzi and Wee, does not allow for cancelations. While this is a limiting factor, as stated in the introduction, there are two reasons to study these models. First, some manufacturers have long term price agreements with customers that place firm orders in advance. Second, there is a growing consensus that manufacturers can benefit from having a portfolio of customers with different demand lead times. Such portfolios lead naturally to inventory systems with advance demand information.

3.1.   Advance demand information model

In this section we present the advance demand information model in detail. To model advance demand information, we assume that customers place orders during a period. Such orders may be either for immediate delivery or to be delivered at a specified period in the near future. To be more precise, we assume that during period t we observe the demand vector

$$D_t = (D_{t,t}, \ldots, D_{t,t+N}),$$

where $D_{t,s}$ represents orders placed by customers during period t for periods $s \in \{t, \ldots, t+N\}$, where N represents the length of the *information horizon*. $N = 0$ represents the classical case of no advance demand information.

At the beginning of period t, the demand to prevail in period $s \geq t$ can be divided into two parts: The *observed* part that is known to us

$$O_{t,s} \equiv \sum_{r=s-N}^{t-1} D_{r,s}, \qquad (5.5)$$

and the part that is *unobserved* and not yet known to us

$$U_{t,s} \equiv \sum_{r=t}^{s} D_{r,s}. \qquad (5.6)$$

We define $O_{t,s} \equiv 0$ for $s \geq t+N$. At the beginning of period t, we know

$$(O_{t,t}, \ldots, O_{t,t+N-1}).$$

Example: Assume that $N = 2$. At the beginning of period t, $O_{t,t} = D_{t-2,t} + D_{t-1,t}$, $O_{t,t+1} = D_{t-1,t+1}$ and $O_{t,t+2} = 0$, while $U_{t,t} = D_{t,t}$, $U_{t,t+1} = D_{t,t+1} + D_{t+1,t+1}$, and $U_{t,t+2} = D_{t,t+2} + D_{t+1,t+2} + D_{t+2,t+2}$. Notice that $O_{t,s} + U_{t,s} = \sum_{j=0}^{2} D_{s-j,s}$ for $s \in \{t, t+1, t+2\}$. $D_{s-i,s} \equiv 0$ for $i = 1, 2$ models the case of no advance demand information, while the case $D_{s,s} \equiv 0$ models the case where all the demand information is obtained at least one period in advance.

We assume throughout that there is a centralized decision maker who has the opportunity to decide at the beginning of each period how much to order/produce and when and where to ship goods taking into account the advance demand information at hand. The objective is to establish an optimal control mechanism to minimize the expected discounted cost of managing the production/distribution system over a finite or infinite horizon. Since decisions are sequential and made under uncertainty, the problems will be formulated using dynamic programming.

3.2.   Single location models

In this section, we consider a single location inventory model with fixed lead time L under advance demand information. At the beginning of each period the inventory manager has to decide whether or not to order and how much to order from its supplier in light of the advance demand information.

At the beginning of period t, the inventory manager knows

I_t : inventory on hand

B_t : number of backorders

z_s : pipeline inventory $s \in \{t - L, \ldots, t - 1\}$

$O_{t,s}$: observed part of demand for periods $s \in \{t, \ldots, t + N - 1\}$.

We assume that the unsatisfied demands are backordered and satisfied as inventory is available. After observing the inventory on hand, the number of backorders, the pipeline inventory, and the advance demand information the manager places an order of size $z_t \geq 0$. This order will arrive at the beginning of period $t + L$. We assume that the cost of ordering $z_t \geq 0$ units in period t is given by $K_t \delta(z_t) + c_t z_t$ where $K_t \geq 0$ is the set-up cost and $\delta(z_t) = 1$ if $z_t > 0$ and zero otherwise. This cost is realized whenever the order is placed. This assumption can be easily modified to incorporate other cases, including that where the cost is realized at the time of delivery. After the ordering decision is made, the demand vector D_t is realized. Demand is satisfied from on-hand inventory given priority to existing backorders, if any. At the end of each period, holding cost is charged based on the inventory on hand, if there is any. Otherwise, a penalty cost is charged based on the number of backorders.

The net inventory at the end of period $t + L$ is given by

$$I_t + \sum_{s=t-L}^{t-1} z_s - B_t + z_t - D_t^L,$$

where D_t^L is the total demand realized during the periods $t, t+1, \ldots, t + L$. We use the term *protection period demand* to refer to the demand over these periods. Some authors follow a different convention and refer to D_t^L as the lead time demand. Notice that viewed from the beginning of period t, we can divide the protection period demand into the *observed* part

$$O_t^L \equiv \sum_{s=t}^{t+L} O_{t,s},$$

and the *unobserved* part

$$U_t^L \equiv \sum_{s=t}^{t+L} U_{t,s}.$$

This suggests a way to summarize the state space. Let

x_t : modified inventory position <u>before</u> the ordering decision is made $\equiv$

$$I_t + \sum_{s=t-L}^{t-1} z_s - B_t - O_t^L,$$

y_t : modified inventory position <u>after</u> the ordering decision is made $\equiv$
$x_t + z_t$

We use the term *modified* to distinguish the definition from the classical definition of inventory position, which does not subsume the observed part of the protection period demand. In addition to x_t, we have to keep track of the observations beyond the protection period,

$$O_t = (O_{t,t+L+1}, \ldots, O_{t,t+N-1}).$$

To summarize, the state of the system is given by (x_t, O_t). Notice that the vector O_t is meaningful only if $N > L+1$. Consequently, if $N \leq L+1$ the state space is given solely by the modified inventory position.

The single period cost charged to period t is the discounted expected holding and penalty cost at the end of period $t + L$. This is given by

$$\tilde{G}_t(y_t) = \alpha^L E g_t(y_t - U_t^L),$$

where $\alpha < 1$ is the discount factor, $g_t(\cdot)$ is the holding and penalty cost function, and the expectation is taken with respect to the unobserved part of the protection period demand U_t^L. For each t, we assume that (*i*) g_t is convex, (*ii*) $\lim_{|x| \to \infty} g(x) = \infty$ and (*iii*) $E(D_{t,s})^\rho < \infty$ for some $\rho > 1$. These are classical assumptions in inventory literature. All these assumptions are satisfied when the holding and backorder penalty costs are linear, e.g., at rates h_t and p_t respectively.

After observing $D_t = (D_{t,t}, \ldots, D_{t,t+N})$ the *modified* inventory position is updated by

$$x_{t+1} = x_t + z_t - D_{t,t} - \sum_{s=t+1}^{t+L+1} D_{t,s} - O_{t,t+L+1} \qquad (5.7)$$

and the vector of observed demand beyond the protection period by

$$O_{t+1} = (O_{t+1,t+L+2}, \ldots, O_{t+1,t+N}) \qquad (5.8)$$

where $O_{t+1,s} = O_{t,s} + D_{t,s}$.

A rigorous proof of the state space reduction is given in Özer [34]. $N = 0$ is the classical case studied extensively in inventory theory, see Scarf [37], Veinott [43], and Zheng [45].

For $1 \leq N \leq L + 1$, although there is non-trivial information about future demands, it is subsumed in the modified inventory position so *all* the classical results described for the case $N = 0$ apply! In addition, if ordering takes place in a period then the order quantity is increasing in the observed protection period demand. Brown et al.[5] consider this case and restrict the information horizon to be, at most, the length of the protection period.

We let T denote the length of the horizon when it is finite. We assume that inventory leftovers at the end of the planning horizon are salvaged at unit rate c_{T+1}. Final backorders (if there are any) are satisfied by a final procurement at unit rate c_{T+1}. Our final assumption is that discounted set-up costs are non-increasing. More formally, we assume that $\alpha K_{t+1} \leq K_t$ holds for all t where α is the discount rate. None of these assumptions are stronger than the assumptions of classical inventory problems, see Scarf [37], Veinott [43].

3.2.1   Inventory problems with positive set-up costs.

This section focuses on problems with positive set up costs, $K_t > 0$, where the information horizon satisfies $N > L + 1$. We first characterize the form of optimal policies for the finite horizon case and then for the infinite horizon case. Recall that O_t is known at the beginning of period t. From now on we denote random variables or random vectors by lower case letters when their realization is known, e.g., $O_t = o_t$ when O_t is known.

The optimal cost-to-go function satisfies the dynamic program

$$J_t(x_t, o_t) = \min_{y_t \geq x_t} \{ K_t \delta(y_t - x_t) + V_t(y_t, o_t) \} \qquad (5.9)$$

where $J_{T+1}(\cdot, \cdot) \equiv 0$,

$$V_t(y_t, o_t) = G_t(y_t) + \alpha E J_{t+1}(x_{t+1}, O_{t+1}), \qquad (5.10)$$

and $G_t(y) = (c_t - \alpha c_{t+1})y + \alpha^L E g_t(y - U_t^L)$. The expectation in (5.10) is with respect to the vector $D_t = (D_{t,t}, D_{t,t+1}, \ldots, D_{t,t+N})$. An intuitive explanation of this dynamic program is as follows: If the decision is to order $y_t > x_t$ than we incur fixed and variable ordering costs plus cost of managing the system for a single period *plus* cost of managing this system starting from the next period to the end of the planning horizon T. Notice that we have subsumed the terminal condition into the dynamic program itself which results in a dynamic program with zero terminal cost. A formal construction of this DP can be found in Gallego and Özer [17].

Equation (5.9) can be expressed as

$$J_t(x_t, o_t) = V_t(x_t, o_t) + \min\{H_t(x_t, o_t), 0\}$$

where $H_t(x_t, o_t) := K_t + \min_{y_t \geq x_t} V_t(y_t, o_t) - V_t(x_t, o_t)$. If $H_t(x_t, o_t) \leq 0$, then it is optimal to order. On the other hand, if $H_t(x_t, o_t) > 0$, it is not optimal to order. If $H_t(\cdot, o_t)$ has a *unique* sign change from $-$ to $+$ for every o_t then the policy has a simple form: an interval in which ordering is optimal followed by an interval in which ordering is not optimal. It can be shown, by contradiction and induction arguments and the definition of K-Convexity, that $H_t(\cdot, o_t)$ has a unique sign change form $-$ to $+$. The following result establishes the optimality of state dependent (s, S) policies.

Theorem 5.1 *The following statements are true for any fixed vector o_t:*

1 $V_t(\cdot, o_t)$ is K_t-convex and $\lim_{|x| \to \infty} V_t(x, o_t) = \infty$.

2 An optimal policy is defined by a state dependent $(s_t(o_t), S_t(o_t))$-policy where

$$S_t(o_t) = \min\{y : V_t(y, o_t) \leq V_t(x, o_t) \text{ for all } x\},$$
$$s_t(o_t) = \max\{x : H_t(x, o_t) \leq 0\}.$$

3 $J_t(\cdot, o_t)$ is K_t-convex and $\lim_{x \to \infty} J_t(x, o_t) = \infty$, $\lim_{x \to -\infty} J_t(x, o_t) = V_t(s_t(o_t), o_t)$.

Thus, it is optimal to order up to $S_t(o_t)$ only if the modified inventory position is below $s_t(o_t)$. Notice that these critical levels depend on the observation beyond the lead times.

3.2.2   Stationary problems.   We refer to an inventory problem as stationary if the demand and the cost parameters are stationary, i.e. $c_t = c$, $g_t = g$, $\alpha_t = \alpha$ and $K_t = K$ and we also drop the subscript from single period cost function G. Let us define for stationary problems

$$S^* = \min\{y : G(y) \leq G(x) \text{ for all } x\}$$
$$s^* = \max\{y \leq S^* : G(y) \geq K + G(S^*)\}$$
$$\bar{S} = \min\{y > S^* : G(y) > G(S^*) + \alpha K\}$$

The pair (s^*, S^*) is a myopic policy for the positive set-up cost case. Such a policy ignores the impact of current decisions to future and also does not depend on advance demand information. Myopic policies for non-stationary problems are defined similarly but they are time dependent. The points above exist since G is convex with respect to y and $\lim_{|y| \to \infty} G(y) = \infty$.

The following Lemma shows that optimal policies are, in a sense, bounded by myopic policies.

Lemma 5.2 *For all t and any fixed vector o_t, $S^* \leq S_t(o_t) \leq \bar{S}$ and $s^* \leq s_t(o_t)$.*

We remark that our proofs do not require the underlying functions to be continuous and differentiable. Our arguments are based on the study of first differences and allow us to cover demand processes with integer domains.

The next Theorem provides a horizon result that sheds more light on the structure of optimal policies under advance demand information.

Theorem 5.3 *For finite horizon stationary problems, if $(\bar{S} - s^*) \leq o_{t,t+L+1}$, then $S_t(o_t) = S^*$.*

This result shows that once the observed demand for period $t + L + 1$ exceeds $\bar{S} - s^*$, the myopic order-up-to level is optimal for the stationary positive set-up cost problems. The threshold level is a function of the lower bound for the reorder point and the upper bound for the order-up-to level. Tighter bounds result in a lower threshold level. As the set-up cost increases the observed demands for the immediate period beyond the protection period need to be higher for the horizon result to hold. This result has both managerial and computational implications. Management can ignore advance demand information beyond period $t + L + 1$ if the observed demand for period $t + L + 1$ is sufficiently high. In particular, management should concentrate on ordering, if needed, to satisfy the demand for period $t + L$, knowing that a new order will be placed in period $t + 1$. The horizon result limits the need to search for state dependent policies, when the observed demand for period $t + L + 1$ is sufficiently large, making it easier to compute optimal policies.

Theorem 5.4 *The results obtained for the stationary finite horizon problem with positive set-up cost also hold for stationary infinite horizon problems.*

3.2.3 Inventory problems with zero set-up costs.
The functional equation for the zero set-up cost case is given by equation (5.9) with $K_t = 0$ for all t. Using similar arguments as in the case of positive set up cost, we obtain the following results for the finite horizon zero set-up cost case.

Theorem 5.5 *The following statements are true for any vector* o_t:

1 $V_t(\cdot, o_t)$ *is convex and* $\lim_{|x| \to \infty} V_t(x, o_t) = \infty$.

2 *An optimal ordering policy is a state dependent base-stock policy where the order-up-to level is given by the smallest minimizer of* $V_t(\cdot, o_t)$, *i.e.*

$$y_t(o_t) = \min\Big\{y : V_t(y, o_t) = \min_x V_t(x, o_t)\Big\}. \tag{5.11}$$

3 $J_t(\cdot, o_t)$ *is increasing convex.*

4 $V_t(x, o_t)$ *has decreasing differences[4] in* (x, o_t).

5 $y_t(o_t)$ *is increasing in* o_t.

6 $J_t(x, o_t)$ *has decreasing differences in* (x, o_t).

This indicates that an optimal policy is to order whenever the modified inventory position falls below a state dependent base stock level. The fifth statement shows that systems maintain higher order-up-to levels, hence higher average inventory levels, as the observed demand beyond the protection period increases.

3.2.4 Stationary policies. The following result is necessary to establish optimal policies for the stationarity zero set-up cost infinite horizon case.

Lemma 5.6 *For all* t *and any vector* (x, o), $V_{t-1}(x, o) \geq V_t(x, o)$, $y_{t-1}(o) \leq y_t(o)$, $J_{t-1}(x, o) \geq J_t(x, o)$.

A myopic policy ignores the effect of upcoming periods and focuses on minimizing the expected cost for the current period. Thus a myopic policy for the zero set-up cost case is any minimizer of G_t, i.e., any $y \in (y_{min}^*, y_{max}^*)$ where

$$y_{min}^* = \min\Big\{y : G_t(y) = \min_x G_t(x)\Big\},$$
$$y_{max}^* = \max\Big\{y : G_t(y) = \min_x G_t(x)\Big\}.$$

Notice that the range collapses into a unique point when $G_t(\cdot)$ is strictly convex.

Theorem 5.7 *For a stationary problem any base stock policy, where the base stock level* y^* *is* $\in (y_{min}^*, y_{max}^*)$, *is optimal for the finite horizon problem.*

This result shows that information beyond the protection period does not affect the order-up-to level when we assume stationary costs and demand distributions. Intuitively, it makes sense to order only to cover for the protection period demand in the absence of fixed costs. This result significantly reduces the computational effort since the state space collapses to a single dimension. It also implies that management does not need to obtain advance demand information beyond the protection period for inventory control purposes. As with the case of positive set up cost the results carry over to the infinite horizon.

3.2.5   Numerical study.   We use a backward induction algorithm to solve the functional equation (5.9). For the purpose of our numerical study we assume $L = 0$ and $N = 2$. This is the simplest case for which the problem is non-trivial and is general enough to capture the main ideas. We model $D_{t,t+i}$ as Poisson with parameter $\lambda_i, i = 0, 1, 2$. Notice that $o_t = D_{t-1,t+1}$.

Figure 5.1 depict the relationship between x_t and y_t with respect to $o_t = D_{t-1,t+1}$, the observed demand information beyond the protection

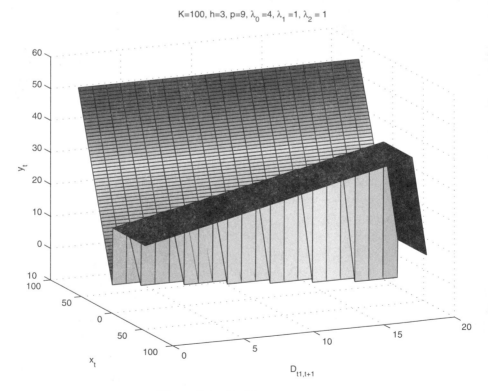

Figure 5.1.   Positive set-up cost

period. Notice that order-up-to level increases as the level of observed demand increases for large set-up costs. We find counter examples, however, which show that this monotonistic behavior is not a general property. On the other hand, our extensive experiments indicate that reorder point $s_t(D_{t-1,t+1})$ decreases as $D_{t-1,t+1}$ increases. We found this observation surprising, because intuition suggests that the reorder point, $s_t(D_{t-1,t+1})$, should be increasing in $D_{t,-1,t+1}$ making it more likely to place an order to cope with large observed demands. Careful thought, however, reveals a more complete story. First, notice that if x_t is not too low, the holding and penalty cost of not ordering may be lower than the cost of ordering and carrying $D_{t-1,t+1}$ for one period. This suggests that at high values of $D_{t-1,t+1}$, it may be better to incur a shortage cost now rather than to place an order and carry inventory for the next period. On the other hand, for sufficiently low values of x_t and very high values of $D_{t-1,t+1}$ it is best to place two consecutive orders, which is shown in Theorem 5.3. In this case, it is optimal to raise the modified inventory position of the first order to minimize current costs, i.e., $S_t(D_{t-1,t+1}) = S^*$ when $D_{t-1,t+1} \geq \overline{S} - s^*$. A sharp decline of order-up-to-level in Figure 5.1 depicts this result.

We now illustrate how our model quantifies the trade off between the benefits of advance demand information and the cost of implementing a pricing strategy that induces advance bookings. Recall that the total s demand for period s is given by $\mathcal{D}_s = D_{s-2,s} + D_{s-1,s} + D_{s,s}$. In Table 5.1, we fix the expected total demand for a period to be $6(= \lambda_0 + \lambda_1 + \lambda_2)$ and increase λ_2 while decreasing λ_0 (This allows us to model when the inventory manager has more advance demand information). Assume that a brand manager is trying to acquire advance demand information through pricing strategies. She is willing to reduce the price of the product if customers are willing to book early. Strategies one

Table 5.1. Optimal (s, S) levels with respect to advance demand information

$K=100, h=1, p=9, s_t(D_{t-1,t+1}), S_t(D_{t-1,t+1})$ for $D_{t-1,t+1} \in \{0,\dots,15\}, T=12$									
#	$(\lambda_0, \lambda_1, \lambda_2)$		0	1	$\dots$	14	15	$(s^*, S^*, \overline{S})$	$J_1(0,0)$	$J_1(0,10)$
1	(5,1,0)	$S(\cdot)$	35	36	$\dots$	49	50	$(-7,8,110)$	418.81	427.80
		$s(\cdot)$	2	2	$\dots$	0	0			
2	(4,1,1)	$S(\cdot)$	33	34	$\dots$	47	48	$(-8,7,108)$	415.81	420.12
		$s(\cdot)$	1	1	$\dots$	-1	-1			
3	(1,1,4)	$S(\cdot)$	27	28	$\dots$	41	42	$(-11,2,104)$	379.77	386.85
		$s(\cdot)$	-2	-2	$\dots$	-3	-3			
4	(0,1,5)	$S(\cdot)$	25	26	$\dots$	39	40	$(-12,0,100)$	353.91	375.67
		$s(\cdot)$	-3	-3	$\dots$	-4	-4			

through four in Table 5.1 model different levels of aggressiveness in the pricing strategy to induce advance bookings. The last two columns in Table 5.1 show the reduction in costs for two initial states. Notice that expected cost decreases as more and more customers are induced to book early. More examples can be found in Gallego and Özer [17]. It is also evident from this table that the order-up-to level and the reorder point decrease as more customers place advance orders, suggesting a reduction in average inventory level.

3.3.   Multi-stage serial systems

In this section, we incorporate advance demand information for a multi-stage serial system and establish that state dependent base stock policies are optimal. Consider a production/distribution system with J stages. External demand occurs only at stage J. Stage $j \geq 2$ satisfies its requirements from Stage $j - 1$. The first stage orders from an outside supplier with ample stock. Shipments arrive after exogenous, stage specific, lead times. Replenishment decisions are centralized and based on system wide inventory information. The objective is to minimize the expected discounted cost of managing the system over a finite horizon. A two stage serial system can be interpreted as a manufacturer and a retailer. The retailer procures from the manufacturer to satisfy an uncertain demand process. The manufacturer satisfies its own requirement from an outside supplier. A centralized decision maker has to decide when and how much to procure from the outside supplier and when and how much to ship to the retailer to minimize the holding and penalty cost of managing the system over a finite horizon in light of advance demand information. The infinite horizon problem is essentially a limiting case of the finite horizon problem. More details can be found in Gallego and Özer [18].

Clark and Scarf [11] show that the problem without advance demand information decomposes into single stage problems and prove the optimality of base stock policies. Federgruen and Zipkin [13] extend these results for stationary infinite horizon problems. Chen and Zheng [9] establish lower bounds on the cost of managing the infinite horizon inventory problems for the average cost criteria and construct feasible policies that achieve these lower bounds. Chen and Song [8] is the only paper to establish the form of optimal policies for the infinite horizon case with non-stationary demands. In their model, demand is governed by a Markov modulated Poisson process. They prove the optimality of state dependent echelon base stock policies and provide an algorithm to calculate these policies. Rosling [36] shows the equivalence of series and assembly systems under a mild assumption–long run balance–on the initial stock levels. We refer the reader to Federgruen [12] for a further discussion on the relation among series, distribution and assembly systems.

3.3.1   Impact of advance demand information.

The demand process is as described in section 3.1. At the beginning of each period, after receiving previously placed orders and/or scheduled shipments, the decision maker decides how much to order, say $z_1 \geq 0$, from an outside supplier and how much to ship, say $z_j \geq 0$, to Stage $j \in \{2, \ldots, J\}$ from Stage $j - 1$. A linear ordering/shipping cost $\sum_{j=1}^{J} c_{jt} z_j$ is charged to period t. An order from the outside supplier placed at the beginning of period t arrives at Stage 1 at the beginning of period $t + L_1$. Similarly, a shipment to Stage $j \in \{2, \ldots, J\}$ placed at the beginning of period t arrives at the beginning of period $t + L_j$.

At the beginning of period t, in addition to the on hand inventory I_{jt}, at stage $j = 1, \ldots, J$, and the backorders B_t at Stage J, the decision maker also knows

$$\vec{O}_t = (O_{t,t}, \ldots, O_{t,t+N-1})$$
$$\vec{z}_{jt} = \left(z_{j1}, \ldots, z_{jL'_i}\right) \text{ trans-shipments to Stage } j$$

for all $j \in \{1, \ldots, J\}$ and $L'_j \equiv L_j - 1$. Here $\vec{O}_t$ is the observed part of the demand over periods $\{t, \ldots, t + N - 1\}$, z_{js} is the shipment dispatched from Stage $j - 1$ to Stage j at the beginning of period $t - s$. The echelon net inventory includes shipments placed a lead time earlier and received at the beginning of the period. Under this convention, $\vec{z}_{jt}$ is irrelevant whenever $L_j \in \{0, 1\}$. We define

$$\hat{x}_{jt} = \text{echelon net inventory at Stage } j$$
$$x_{jt} = \text{echelon inventory position at Stage } j$$
$$x_{jt}^{L_j} = \text{modified echelon inventory position at Stage } j$$

the echelon *net* inventory $\hat{x}_{jt}$ to be the inventory on hand at the beginning of period t at stages $\{j, j + 1, \ldots, J\}$ plus the inventory in transit between these stages minus the backorder at Stage J. The echelon inventory *position* at Stage j is the echelon net inventory at stage j plus inventories in transit to Stage j. Notice that $\hat{x}_{Jt} = I_{Jt} - B_t$, $x_{jt} = \hat{x}_{jt} + \sum_{l=1}^{L'_j} z_{jl}$ and $\hat{x}_{jt} = x_{j+1t} + I_{jt}$ for $j \in \{1, \ldots, J - 1\}$. Finally, in order to reduce the dimension of the state space we define, for each stage, the *modified* echelon inventory position as the echelon inventory position minus the observed part of the protection period demand corresponding to that stage, i.e.,

$$x_{jt}^{L_j} = x_{jt} - O_t^{L_j} \quad \text{for } j \in \{1, \ldots, J\},$$

where $O_t^{L_j} \equiv \sum_{s=t}^{t+L_j} O_{t,s}$. For convenience we also define $U_t^{L_j} = \sum_{s=t}^{t+L_j} U_{t,s}$. For each period t the system has the following cost parameters:

$$p_t = \text{penalty cost (at Stage } J) \text{ per unit,}$$
$$h'_{jt} = \text{local inventory holding cost at Stage } j \text{ per unit,}$$
$$h_{jt} = \text{echelon holding cost at Stage } j \text{ per unit}$$
$$= h'_{jt} - h'_{j-1,t} \text{ for } j \in \{2, \ldots, J\} \text{ and } h_{1t} = h'_{1t},$$
$$c_{jt} = \text{shipment cost per unit in period } t.$$

The holding and penalty costs for a period are based on the inventory levels at the end of the period. We assume that the echelon holding costs are strictly positive.

The period t holding and penalty cost is given by

$$\sum_{j=1}^{J-1} h'_{jt}[\hat{x}_{j,t+1} - \hat{x}_{j+1,t+1}] + h'_{Jt}[\hat{x}_{J,t+1}]^+ + p_t[\hat{x}_{J,t+1}]^-$$

$$= \sum_{j=1}^{J} h_{jt}\hat{x}_{j,t+1} + [p_t + h'_{Jt}][\hat{x}_{J,t+1}]^-$$

where $[x]^+ = \max(x, 0)$ and $[x]^- = \max(-x, 0)$. Notice that shipments in transit to Stage $j + 1$ are charged at Stage j's holding cost rate. At the end of period t the updates are given by

$$\vec{O}_{t+1} = (O_{t+1,t+1}, \ldots, O_{t+1,t+N})$$
$$\vec{z}_{j,t+1} = (z_j, z_{j1}, \ldots, z_{jL'_j-1})$$
$$\hat{x}_{j,t+1} = \hat{x}_{jt} + z_{jL'_j} - o_{t,t} - D_{t,t}$$
$$x_{j,t+1} = x_{jt} + z_j - o_{t,t} - D_{t,t}$$
$$x_{j,t+1}^{L_j} = x_{jt}^{L_j} + z_j - \sum_{s=t}^{t+L_j+1} D_{t,s} - O_{t,t+L_j+1}$$

For finite horizon problems, we assume a linear terminal condition of the form $-\sum_{j=1}^{J} c_j x_j$. The economic interpretation is that of a salvage value if x_j is positive, and an acquisition cost if x_j is negative. Notice that the c_j's are actually echelon costs so the terminal condition makes economic sense. To our knowledge, all other papers in the literature charge zero terminal costs. Linear terminal costs are more realistic and allows us to show that myopic policies are optimal for finite horizon problems with stationary costs and demand distributions, a result that fails to hold under zero terminal costs.

To simplify the exposition of the results we assume $J = 2$. At the end of this section we discuss briefly the extention for more than two stages. The problem in this case is to manage the two stage series inventory system for periods $\{t, \ldots, T\}$. As in single location problem, we use dynamic programming to solve the multi-locations in series. The last order for Stage 1 is dispatched at the beginning of period T, and arrives to Stage 1 at the beginning of period $T + L_1$. The last shipment to Stage 2, initiated at the beginning of period $T + L_1$, arrives at Stage 2 at the beginning of period $T + L_1 + L_2$. Thus, for $t > T + L_1$ there will not be any inventory on order between the supplier and the first stage. Likewise, all the shipments would have arrived to the second stage by the beginning of period $T + L_1 + L_2$. We assume that cost continues to accrue up to period $T + L_1 + L_2$. All holding and penalty costs after time $T + L_1 + L_2 + 1$ are assumed to be zero. Also, $\vec{O}_{T+L_1+L_2+1} = 0$. This means that we take advance information up to the period $T + L_1 + L_2$ only. The state space will be initially defined by $(\hat{x}_{1t}, \vec{z}_{1t}, \hat{x}_{2t}, \vec{z}_{2t}, \vec{O}_t)$. As in the single location case we denote random vectors by lower case when their realization is known, e.g. $\vec{O}_t = \vec{o}_t$ when $\vec{O}_t$ is known.

The optimal cost-to-go starting from period t is given by,

$$\hat{J}_t(\hat{x}_{1t}, \vec{z}_{1t}, \hat{x}_{2t}, \vec{z}_{2t}, \vec{O}_t)$$
$$= \min_{(z_1, z_2) \in \mathcal{A}'} \{ c_{1t} z_1 + c_{2t} z_2 + h_{1t} E \hat{x}_{1,t+1} + E g_t(\hat{x}_{2,t+1})$$
$$+ \alpha E \hat{J}_{t+1}(\hat{x}_{1,t+1}, \vec{z}_{1,t+1}, \hat{x}_{2,t+1}, \vec{z}_{2,t+1}, \vec{O}_{t+1}) \} \qquad (5.12)$$

where $\hat{J}_{T+L_1+L_2+1}(\hat{x}_1, \cdot, \hat{x}_2, \cdot, \cdot) \equiv -c_1 \hat{x}_1 - c_2 \hat{x}_2$, $g_t(x) = h_{2t}[x] + (p_t + h'_{2t})[x]^-$, and $\mathcal{A}' = \{(z_1, z_2) \in Z^2 : z_1 \geq 0, z_2 \geq 0, \text{ and } x_{2t} + z_2 \leq \hat{x}_{1t}\}$. Notice that shipment constraint $x_{2t} + z_2 \leq \hat{x}_{1t}$ is equivalent to $z_2 \leq I_{1t}$, meaning that shipments to Stage 2 are bounded by the inventory on hand at Stage 1. Notice also that g_t is a convex function and that $\lim_{|x| \to \infty} g_t(x) = \infty$.

The dynamic program can be decomposed into two simpler problems. First, by virtue of $x_{2t} = \hat{x}_{2t} + \sum_{l=1}^{L'_2} z_{2l}$, and standard cost accounting manipulations, the state space reduces to $(\hat{x}_{1t}, \vec{z}_{1t}, x_{2t}, \vec{o}_t)$, so we don't need to keep track of the vector $\vec{z}_{2t}$ of shipments from Stage 1 to Stage 2. This yields a dynamic program with cost-to-go $J_t(\hat{x}_{1t}, \vec{z}_{1t}, x_{2t}, \vec{o}_t)$. Second, this program can be decomposed into two simpler dynamic programs:

$$J_t(\hat{x}_{1t}, \vec{z}_{1t}, x_{2t}, \vec{o}_t) = \tilde{V}_t^1(\hat{x}_{1t}, \vec{z}_{1t}, \vec{o}_t) + \tilde{V}_t^2(x_{2t}, \vec{o}_t).$$

We will not define these intermediate dynamic programs here since we can further reduce the dimension of each of these programs by using the *modified* inventory position concept.[5]

Theorem 5.8 *The dynamic program for the series system decomposes into two simpler dynamic programs given by equation (5.13) and (5.14) (defined below), which can be interpreted as single location problems under advance demand information.*

The dynamic program for stage two is given by

$$V_t^2\left(x_{2t}^{L_2}, o_t^2\right) = -c_{2t}x_{2t}^{L_2} + \min_{x_{2t}^{L_2} \leq y} \left\{H_t\left(y, o_t^2\right)\right\} \tag{5.13}$$

where $H_t(y, o_t^2) = c_{2t}y + G_t(y) + \alpha E V_{t+1}^2(x_{2,t+1}^{L_2}, O_{t+1}^2)$, this function is convex and goes to infinity as $|y|$ tends to infinity, $V_{T+L_1+L_2+1}^2(x^{L_2}, \cdot) \equiv -c_2 x^{L_2}$ and $o_t^2 \equiv (o_{t,t+L_2+1}, \dots, o_{t,t+N-1})$. Let $y_{2t}(o_t^2)$ denote the smallest minimizer of $H_t(\cdot, o_t^2)$.

Observe that the above dynamic program is similar to single location problem for which the optimal policy is a state dependent base stock policy. Under this policy the manager orders up to $y_{2t}(o_t^2)$ if the modified inventory position is below this level to achieve the minimum of function $H_t(\cdot, o_t^2)$. In a two stage series system, however, the ordering decision will be constrained by the available inventory at the first stage. Hence, if the first stage turns out to be a bottle neck, it should bare the consequences. Let $IP_t(x, o_t^2) = H_t(\min\{x, y_{2t}(o_t^2)\}, o_t^2) - H_t(y_{2t}(o_t^2), o_t^2)$. This implicit cost function appears in the dynamic program for stage one. Let

$$V_t^1\left(x_{1t}^{L_1}, o_t^1\right) = -c_{1t}x_{1t}^{L_1} + \min_{y \geq x_{1t}^{L_1}} \left\{c_{1t}y + C_t\left(y, o_t^1\right) + \alpha E V_{t+1}^1\left(x_{1,t+1}^{L_1}, O_{t+1}^1\right)\right\}$$

$$\tag{5.14}$$

where $V_{T+L_1+1}^1(x_1^{L_1}, \cdot) \equiv -c_1 x_1^{L_1}$, $o_t^1 \equiv (o_{t,t+L_1+1}, \dots, o_{t,t+N-1})$,

$$C_t\left(y, o_t^1\right) = \alpha^{L_1} E \hat{C}_{t+L_1}\left(y - U_t^{L_1}, O_{t+L_1}^1\right), \quad \text{and}$$
$$\hat{C}_t\left(y, o_t^1\right) = h_{1t}y + \alpha E I P_{t+1}\left(y, O_{t+1}^1\right).$$

Let $y_{1t}(o_t^1)$ denote the smallest minimizer of the function inside the $\{c_{1t}y + C_t(y, o_t^1) + \alpha E V_{t+1}^1(x_{1,t+1}^{L_1}, O_{t+1})\}$. To summarize, the problem of finding an optimal policy for a series system under advance demand information reduces to solving two simpler, single stage, dynamic programs. The state space for these programs is $1 + (N - L_j - 1)^+$ for $j = 1, 2$.

Theorem 5.9 *An echelon state dependent base stock policy is optimal for a two stage system in series. In particular, an optimal base stock level for stage j at time t is given by $y_{jt}(o_t^j)$ for $j = 1, 2$.*

The argument to establish the optimality for $J > 2$ stages requires a recursive application of the decomposition argument until system is decomposed into single stage problems. Then we use the modified inventory position to further reduce the state space for each of the single stage problems.

3.3.2   Myopic policies.   In this section, we construct myopic policies for the two-stage serial system. We assume that costs are stationary hence we drop the t from cost parameters, for example $h_{1t} = h_1$ for all t. We further assume that D_t is a stationary vector.

Let y_2^m be the smallest minimizer of

$$\mathcal{L}_2(y) = (1 - \alpha)c_2 y + G(y).$$

Notice that $\mathcal{L}_2$ is convex and $\lim_{|y| \to \infty} \mathcal{L}_2(y) = \infty$, so y_2^m is finite. We refer to the policy that orders up to the base stock level y_2^m as the myopic policy for Stage 2.

Next, we define the myopic implicit penalty cost function and related cost functions

$$IP^m(x) = \mathcal{L}_2\left(\min\{y_2^m, x\}\right) - \mathcal{L}_2\left(y_2^m\right)$$
$$C^m(x) = \alpha^{L_1} E\left[h_1\left(x - U_t^{L_1}\right) + \alpha IP^m\left(x - U_t^{L_1}\right)\right]$$
$$\mathcal{L}_1(y) = (1 - \alpha)c_1 y + C^m(y).$$

Let y_1^m be the smallest minimizer of $\mathcal{L}_1$. Notice that $\mathcal{L}_1$ is convex and $\lim_{|y| \to \infty} \mathcal{L}_1(y) = \infty$, so y_1^m is also finite. We refer to the policy that orders up to the base stock level y_1^m as the myopic policy for Stage 1. Finally, we refer to the policy that orders up to y_1^m for Stage 1 and up to y_2^m for Stage 2 as the myopic policy.

Theorem 5.10 *Under stationary demand and cost parameters the myopic policy is optimal for finite horizon problems.*

3.3.3   Numerical study.   We assume for our numerical study that $L_1 = L_2 = 1$ and that $N = 3$. Recall that the dimension of the state space is for Stage j is $1 + (N - L_j - 1)^+$. Consequently, the state space is two dimensional for both stages. We will assume that $D_{t,t+i}$ is Poisson with parameter $\lambda_{ti} = w_i \Lambda_t$, where $w_i \geq 0$, and $\sum_{i=0}^{3} w_i = 1$. Hence Λ_t

Table 5.2.   Optimal echelon base stock levels ($\Lambda_t = 3$)

No.	(w_0, w_1, w_2, w_3)	(y_{1t}, y_{2t})	No.	(w_0, w_1, w_2, w_3)	(y_{1t}, y_{2t})
1	(1.0, 0.0, 0.0, 0.0)	(15,8)	4	(0.8, 0.2, 0.0, 0.0)	(15,7)
2	(0.4, 0.3, 0.2, 0.1)	(11,4)	5	(0.8, 0.0, 0.2, 0.0)	(14,7)
3	(0.1, 0.2, 0.3, 0.4)	(9,2)	6	(0.8, 0.0, 0.0, 0.2)	(13,7)

$h_1 = 1, h_2 = 3, c_1 = 10, c_2 = 30, p = 19$

is average number of customers placing orders at time t, of which, on
average, λ_{ti} place their orders to be delivered i periods later. By changing the weights, w_i, we can model the degree to which customers place
orders in advance of their needs. Notice also that o_t^1 and o_t^2 are scalars.

The first experiment in Table 5.2 corresponds to customers requiring immediate delivery. The next two experiments represent the case
where the manager obtains more advance demand information perhaps
by inducing customers to place orders for future periods. The echelon base stock levels for both stages decrease as more of the demand
is known in advance. Experiment No. 4–6 yield similar observations.
Figure 5.2, exhibits the optimal echelon base stock levels through each
period for a stationary demand process where the mean is $\Lambda_t = 2$ for

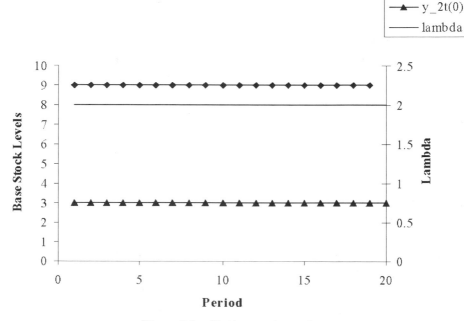

Figure 5.2.   Stationary demand case

all $t = \{1, \ldots, 20\}$. We observe that the myopic policies are stationary and optimal. Similarly, Figure 5.3 exhibits the optimal echelon base stock levels for a ramp-up demand process. We observe the optimality of myopic base stock levels which are non-decreasing, see also Table 5.3. This result can also be established analytically. Thus, advance demand information beyond the protection period has no operational value for both the stationary and the ramp up demand process.

3.4.   One-warehouse multi-retailer systems

In this section, we study a periodic-review distribution system consisting of a central depot and J retailers under advance demand information

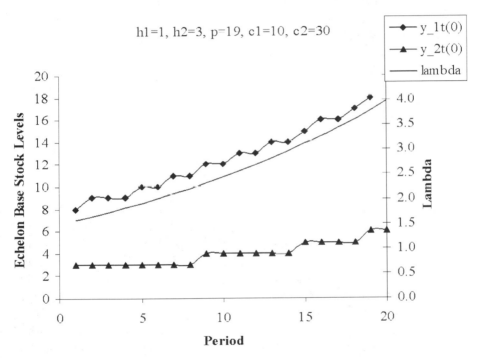

Figure 5.3. Ramp-up demand case

Table 5.3. Optimal echelon base stock levels

	Stationary Demand Case							Ramp-Up Demand Case						
$D_{t-1,t+1}$	0	1	2	3	4	5	y_i^m	0	1	2	·3	4	5	y_i^m
$y_{1t}(D_{t-1,t+1})$	9	9	9	9	9	9	9	8	8	8	8	8	8	8
$y_{2t}(D_{t-1,t+1})$	3	3	3	3	3	3	3	3	3	3	3	3	3	3

Note that $O_t^1 = O_t^2 = D_{t-1,t+1}$ since $L_1 = L_2 = 1$

and centralized control. We assume first that the depot is a coordination and re-packaging center so it does not hold any inventory. The uncertain demand is satisfied through the retailers and each retailer replenishes their inventory through a central depot. The depot satisfies its own requirement from an outside supplier with ample stock. Orders placed by the depot arrive after an exogenously specified fixed lead time L. Shipments to the retailers arrive after an exogenously specified fixed lead time l. Customers are satisfied from on hand inventory at the retailers and unsatisfied orders are backlogged. In section 3.4.3, we discuss briefly how to extend the results for systems where the central depot is allowed to carry inventories.

If holding inventory at the warehouse is not allowed, then the decision maker has to decide (i) whether or not to place an order from an outside supplier, and (ii) how to allocate the incoming order to the retailers to minimize the expected holding and shortage cost over a finite horizon. If central inventory is allowed, the decision maker must decide (i) whether or not to order, (ii) how much to withdraw from the warehouse and (iii) how to allocate the withdrawn amount to the retailers. We remark that the structure, if any, of optimal policies for distribution systems is unknown even in the absence of advance demand information. We, therefore, use a relaxation method to simplify the problem and use the solution to the relaxed problem to develop a heuristic that incorporates advance demand information. Our numerical study suggests that the heuristic performs well.

The distribution system described above can also be interpreted as a multi-item production/distribution system with a common intermediate product. In this interpretation, the depot represents the differentiation point. During the first phase of the production period, L, a common batch is produced. At the end of this period the manager has to decide how much of each item to produce from the batch that has just arrived. This interpretation forms the basis of postponement strategies, see for example Lee et al. [29], Aviv and Federgruen [2]. Through a numerical example we illustrate that advance demand information increases the benefit gained through postponement strategies and that the benefit obtained through advance demand information may far exceed the benefit of postponement strategies.

3.4.1   Warehouse as coordination center.   During time period t at each retailer j we observe the demand vector

$$d_t^j = \left(d_{t,t}^j, \ldots, d_{t,t+N_j}^j\right)$$

where $d^j_{t,s}$ represents the demand for period s observed during period t at retailer j. Customers place order at retailer $j \in \{1, \dots, J\}$ for future periods $s \in \{t, \dots, t+N_j\}$ where $N_j < \infty$ is the length of the *information horizon* for each retailer.

Demand at each retailer is satisfied from on-hand inventory, if any. Unsatisfied demand is backlogged. At the beginning of each period t, the manager reviews the on hand inventory, the pipeline inventory, back-orders and the advance demand information at each retailer. He/she decides whether or not to place an order, say $w_t \geq 0$, from an outside supplier. The orders arrive after a positive fixed lead time, $L \geq 0$. Simultaneously, the orders placed in period $t - L$ arrive to depot at the beginning of period t. The decision maker also decides how to allocate this batch to retailers, say $z^j_t \geq 0$. Clearly, we want total shipments to the retailers equal to the incoming orders to the depot, $\sum^J_{j=1} z^j_t = w_{t-L}$. Shipments to retailers arrive after a positive lead time, $l \geq 0$. A linear shipping cost of $c_{0t} w_t + \sum^J_{j=1} c^j_t z^j_t$ is charged to period t where c^j_t is the variable shipping cost rate for retailer $j \in 1, \dots, J$ and c_{0t} is the variable ordering cost rate from the outside supplier. Demands and costs are not necessarily stationary. The aim is to minimize the expected cost of managing this system over a finite horizon.

At the beginning of each time period t, the demand to be realized at retailer j during a future period $s \in \{t, \dots, t + N_j - 1\}$ is equal to the sum of the part that is *observed* and known to the decision maker $o^j_{t,s} = \sum^{t-1}_{r=s-N_j} d^j_{r,s}$, and the part that is *unobserved* and unknown to the decision maker $u^j_{t,s} = \sum^s_{r=t} d^j_{r,s}$. Hence, at the beginning of period t, the decision maker knows I^j_t on hand inventory, B^j_t backorders, $o^j_{t,s}$ cumulative *observed* part of the demand for periods $s \in \{t, t+1, \dots, t + N_j - 1\}$ for each retailer $j = 1, \dots, J$.

After reviewing this information and receiving the order w_{t-L}, the inventory manager decides on (i) whether or not to place an order from outside supplier, $w_t \geq 0$ and (ii) how to allocate the incoming order w_{t-L} among the retailers. Each of these allocations, z^j_t, arrives after an exogenously specified fixed lead time l. Hence the inventory manager should protect the retailers against the unobserved part of the demand that is to prevail during the next $l + 1$ periods, i.e. over periods $\{t, t + 1, \dots, t + l\}$. We refer to these periods as the protection period. Notice that we can divide the protection period demand into two parts: The observed part $O^{j,l}_t = \sum^{t+l}_{s=t} o^j_{t,s}$ and the unobserved part $U^{j,l}_t = \sum^{t+l}_{s=t} u^j_{t,s}$. The expected holding and penalty cost charged to period t is based on the net inventory (inventory on hand minus the backorders) at the end of period $t + l$. Let,

$$x_t^j = \textit{modified} \text{ inventory position } \underline{\text{before}} \text{ shipment decision is made}$$

$$= I_t^j + \sum_{s=t-l}^{t-1} z_{js} - B_t^j - O_t^{j,l} \text{ for all } j \in \{1, \ldots, J\}$$

$$y_t^j = \textit{modified} \text{ inventory position } \underline{\text{after}} \text{ shipment decision is made}$$

$$= x_t^j + z_t^j \text{ for all } j \in \{1, \ldots, J\}$$

We refer to these variables as modified since they both net the observed part of the demand information for the next l periods, hence they differ from the classical definition of inventory position. The net inventory at the end of period $t + l$ at retailer j is given by $x_t^j + z_t^j - U_t^{j,l}$. Thus, the expected holding and penalty cost charged to period t for retailer j is given by $\tilde{G}_t^j(y_t^j) = \alpha^l E g_t^j(y_t^j - U_t^{j,l})$ where α is the discount factor and the expectation is with respect to the unobserved part of the protection period demand.

The state space for the exact dynamic programming formulation of this problem is $J + L + \sum_{j=1}^{J}(N_j - l - 1)^+$ dimensional. It is impractical to deal with such a large state space. Hence, we develop a lower bound approximation by relaxing the constraints $y_t^j \geq x_t^j$ from the feasible action set. This relaxation is equivalent to assuming that excess inventory at one retailer can be transfered to other retailers without any cost. Also advance orders of a customer can be satisfied through the other retailers. Intuition suggests that under this relaxation all retailers collapses into a single retailer, e.g. single location problem with advance demand information. The state space of the relaxed problem will be based on aggregate quantities. Clearly the optimal solution to this relaxation will be a lower bound. We also apply the same accounting device used for single location problem to subsume the pipeline inventory. We define next the aggregate quantities

$$D_{t,s} = \sum_{j=1}^{J} d_{t,s}^j, \quad O_{t,s} = \sum_{j=1}^{J} o_{t,s}^j, \quad X_t^{\Delta} = \sum_{j=1}^{J} x_t^j + \sum_{s=t}^{t+L-1} w_{s-t}.$$

$N = \max\{N_1, \ldots, N_J\}$. This relaxation yields the following dynamic program:

$$V_t(X_t^{\Delta}, O_t) = c_{0t}X_t^{\Delta} + \min_{Y_t^{\Delta} \geq X_t^{\Delta}} \{c_{0t}Y_t^{\Delta} + \alpha^L E R_{t+L}(Y_{t+L})$$

$$+ \alpha E V_{t+1}(X_{t+1}^{\Delta}, O_{t+1})\} \tag{5.15}$$

where $V_{T+1} \equiv 0$, $Y_{t+L} = Y_t^{\Delta} - \sum_{r=t}^{t+L-1} \sum_{s=r}^{r+l+1} D_{r,s} - \sum_{s=t}^{t+L-1} O_{s,s+l+1}$
and the update for $X_{t+1}^{\Delta} = Y_t^{\Delta} - \sum_{s=t}^{t+l+1} D_{t,s} - O_{t,t+l+1}$ and

$$R_t(Y) = \left\{ \min_{y_t} \sum_{j=1}^{J} G_t^j(y_t^j) : s.t. \sum_{j=1}^{J} y_t^j = Y \right\}. \qquad (5.16)$$

We established earlier the optimality of base stock policies for single stage problems under advance demand information. Equation (5.15) has the same structure as the single location problem. Function $R_t(Y)$ is convex and $\lim_{|Y| \to \infty} R_t(Y) = \infty$. It inherits these properties from $G_t^j(y_t^j)$. Hence, for any fixed vector O_t there exists a $y_t^*(O_t)$, which is defined as the smallest minimizer of function $H_t(Y_t^{\Delta}, O_t) = c_{0t} Y_t^{\Delta} + \alpha^L E R_{t+L}(Y_{t+L}) + \alpha E V_{t+1}(X_{t+1}^{\Delta}, O_{t+1})$. Thus, a base stock policy with base stock level $y_t^*(O_t)$ is optimal for the lower bound problem.

We can now propose a heuristic. Base stock policy solves the lower bound approximation. We solve equation (5.15) to decide how much to order from an outside supplier. The aggregate order quantity then is given by $w_t(O_t) \doteq y_t^*(O_t) - X_t^{\Delta}$. Next we have to decide how to allocate the incoming order, w_{t-L}. We propose an allocation based on the solution of the following problem:

$$\min \left\{ \sum_{j=1}^{J} G_t^i(y_t^i) : s.t. \sum_{j=1}^{J} (y_t^j - x_t^j) = w_{t-L}, y_t^j \geq x_t^j, \forall j \right\} \quad (5.17)$$

which is similar to equation (5.16) with an additional constraint $y_t^j \geq x_t^j$. We call this allocation strategy as *myopic* since it minimizes the total expected cost of managing the retailer inventories at the end of period $t + l$ (the period when the allocations are available for customers) and ignores the future impact of this allocation.

In our numerical study we use a greedy algorithm while evaluating the function $R_t(Y)$ and the equation (5.17). Let $^*y_t^j$ be the minimum of $G_t^j(\cdot)$ and $Y_t^* \doteq \sum_{j=1}^{J} {}^*y_t^j$. The solution to the problem $R_t(Y_t^*)$ is trivial since the constraint allows an allocation such that the minimum of $G_t^j(\cdot)$ is attained for all j. We use this as our initial search point. To evaluate $R_t(Y)$ given $R_t(Y_t^*)$ we allocate the difference $Y - Y_t^*$, if positive, one unit at a time to the jth retailer with the smallest current value of first difference, i.e. choose the $\min_j \{G_t^j(y+1) - G_t^j(y)\}$. Otherwise, we reduce the amount allocated (dis-allocate) one unit at a time from the jth retailer with the largest current value of the first difference, i.e. choose the $\max_j \{G_t^j(y) - G_t^j(y-1)\}$. The solution to the myopic allocation, see equation (5.17), is similar. In this case we start with

$y_t^i = x_t^i$ and allocate one unit at a time to the jth retailer with the smallest current value of the first difference until all w_{t-L} is allocated.

3.4.2   Numerical study.   The computational study focuses on identical retailers, whose demand distribution and cost parameters are equal and stationary. We compare the solution of the lower bound (LB) problem to the solution of the proposed heuristic (UB) for different instances of the model. We report the difference as percentage error, $\epsilon\% = (UB - LB)/LB$, which is a measure of the sub-optimality for the proposed heuristic.

To solve the lower bound problem given in equation (5.15), we use a backward induction algorithm as in the single location problem and obtain the base stock level $y_t(O_t)$. We model the components of the demand vector as Poisson random variables, specifically $d_{t,t+n}^j$ is Poisson with mean λ_n^j. The mean λ_n^j represents the average number of customers who place their orders during time period t to be delivered n periods later. We simulate the system to estimate the cost under the proposed heuristic. We run several replications of each instance to have small sampling errors and sufficiently narrow confidence intervals.

Over 84 simulations, the maximum error was 3.35%, the minimum error was 0.5% and the average was 1.48%. Our numerical study also indicates that optimality gap is insensitive with respect to the number of retailers and the lead times, see Özer [33] for more numerical examples.

One benefit of advance demand information is the resulting decline in inventory levels and inventory related costs. In Table 5.4, the percentage decrease in cost due to advance demand information between the first experiment (in which none of the customers place orders in advance) and the fifth experiment (in which all customers place orders 3 periods in advance) is 13.15%. Notice also that the system maintains lower inventory levels as we incorporate advance demand information. This

Table 5.4.   Cost of LB and UB for two retailers with $L = 1, l = 2$

$h = 1, p = 19, c = 10$		*Lower Bound*		*Simulation*	$\frac{UB-LB}{LB}\%$
No.	$(\lambda_0,\lambda_1,\lambda_2,\lambda_3,\lambda_4)$	$y_t(0)$	Cost	Cost	Gap
1	$(2,0,0,0,0)$	25	5218.43	5273.58 ± 13.04	1.06%
2	$(0,2,0,0,0)$	20	5122.59	5201.65 ± 13.46	1.54%
3	$(0,0,2,0,0)$	14	5023.45	5110.63 ± 14.49	1.74%
4	$(0,0,0,2,0)$	8	4862.93	5017.27 ± 15.82	3.17%
5	$(0,0,0,0,2)$	0	4612.00	4646.28 ± 16.05	0.74%

$\pm$ is based on 95% confidence interval

suggests that advance demand information enables a fundamental shift in production philosophy from build-to-stock to build-to-order.

This distribution system can also be interpreted as a multi-item production system with a common intermediate product. The depot in this case represents the differentiation point. During the first L periods a common batch is produced. At the end of this period the manager has to decide how much of each item to produce and it takes l periods to produce each of these units. In our numerical analysis we also address the impact of advance demand information to postponement strategies. We do this by fixing the information horizon and the total production time to 3 periods $N = 3$, $L + l = 3$. Each of the columns in Table 5.5 represent a different differentiation point. As we move from the first column (in which we have immediate differentiation) to the third (in which we do not differentiate till the last period), we observe a reduction in both the base stock levels and the cost of managing this system. Each of the rows represent a different advance demand information scenario. In terms of reduction in inventory levels and inventory related costs, the impact of advance demand information is more significant than the postponement strategies, which are more difficult to implement. The last column of Table 5.5 shows that cost reduction due to postponement strategies is higher as we incorporate more advance demand information. Advance demand information enhances the outcome of a postponement strategy.

3.4.3   Warehouse as stocking point.   If the warehouse is allowed to hold inventories then similar relaxation approach discussed previously yields a two stage serial system under advance demand information. In previous sections we incorporated advance demand information and established the optimality of state dependent echelon base stock policy for multi-echelons in series. We also provided an algorithm to calculate the echelon state dependent base stock levels for each stage.

Table 5.5.   Advance demand information versus postponement

$(\lambda_0, \lambda_1, \lambda_2, \lambda_3)$	$y_t(0)$			Cost of LB, V_0			
	$L=0$ $l=3$	$L=1$ $l=2$	$L=2$ $l=1$	$L=0$ $l=3$	$L=1$ $l=2$	$L=2$ $l=1$	%
$(1,0,0,0)$	14	13	13	100.00	99.03	98.02	2.02%
$(0,1,0,0)$	10	10	10	97.36	96.30	95.01	2.47%
$(0,0,1,0)$	8	7	7	94.37	93.10	91.25	3.42%
$(0,0,0,1)$	4	4	4	90.91	88.47	87.76	3.59%

$N = 3, h = 1, c = 10, p = 9$ and Index 100.00 corresponds to 2809.94

The solution to this relaxed problem yields an optimal base stock level for the first stage, which can be used to decide how much and when to order from outside supplier. Likewise, base stock level for the second stage can be used to decide on how much to withdraw from the warehouse. We would continue to allocate based on the solution of equation (5.17). We conjecture that a numerical study will illustrate that this heuristic is also close-to-optimal.

4.   Directions for Future Research

Research on the optimal use of demand information in supply chains is in its infancy and there are many fertile opportunities for research in this area. For single location models, research opportunities exist whenever the inventory manager has an option whose realization is based on the demand stream. We mentioned the models by Scarf and by Gallego as two examples of this idea. Additional research opportunities include the option of purchasing through futures markets, and for finite capacity manufacturers the option of selling capacity in futures markets. These ideas can also be extended to multi-echelon settings.

For multi-location models using current demand information under distributed decision making, there are research opportunities even for the serial system with two echelons. For discrete time models, it is frequently assumed that the retailer can instantaneously procure from a different source when the supplier is out of stock. Relaxing this assumption may lead to different results regarding the willingness of the retailer to share demand information with the supplier.

With regard to advance demand information, the most obvious opportunity is in distributed decision making. The simplest model would include a supplier and a retailer where the retailer obtains advance demand information. Should the retailer share this information with the supplier? If so, under what conditions? If not, should the supplier buy this information from the retailer? If so, under what conditions? Additional research opportunities exist for serial systems with more than two echelons, assembly systems, and distribution systems with multiple retailers.

Notes

1. Hariharan and Zipkin [26] coined the term "demand leadtime." A customer who places an order l units of time ahead of his needs is said to have demand leadtime l.

2. A function f is K-concave, if $-f$ is K-convex

3. Under a (Q, r) policy the retailer places an order of size Q whenever the inventory position drops to r. For Poisson demands, (Q, r) policies are equivalent to (s, S) policies with $s = r$ and $S = r + Q$.

4. A function $f : \mathcal{R} \times \mathcal{R}^n \to \mathcal{R}$ is said to have decreasing differences in (x, θ) if it satisfies the following inequality. $f(x_1, \theta) - f(x_2, \theta) \leq f(x_1, \theta') - f(x_2, \theta')$ for all $x_1 \geq x_2$ and $\theta \geq \theta'$.

5. Inventory position that nets the known requirements.

References

[1] APPELL, B., B. GRESSENS, AND C. BROUSSEAU. (2000) The Value Propositions of Dynamic Pricing in Business-to-Business E-Commerce. (http://www.CRMproject.com/crm/wp/appell.html)

[2] AVIV, Y. AND A. FEDERGRUEN. (1999) Capacitated Multi-Item Inventory Systems with Random and Seasonally Fluctuating Demands: Implication for Postponement Strategies. To Appear in *Management Science.*

[3] AXSÄTER, S. (1993) Using the Deterministic EOQ Formula in Stochastic Inventory Control. *Management Science,* 831–834.

[4] BOURLAND, K.E., S.G. POWELL, AND D.F. PYKE. (1996) Exploiting Timely Demand Information to Reduce Inventories. *European Journal of Operations Research,* Vol. 92, 239–253.

[5] BROWN, G., T.M. CORCORAN, AND R.M. LLOYD. (1971) Inventory Models with Forecasting and Dependent Demand. *Management Science,* Vol. 17, 498–499.

[6] CHEN, F. (1998) Echelon Reorder Points, Installation Reorder Points, and Value of Centralized Demand Information. *Management Science,* Vol. 44, No. 12, part 2 of 2, s221–s234.

[7] CHEN, F. (1999) Market Segmentation, Advanced Demand Information, and Supply Chain Performance. To appear in *MS&OM.*

[8] CHEN, F. AND J. SONG. (1999) Optimal Policies for Multi-echelon Inventory Problems with Markov Modulated Demand. Working Paper, Columbia Graduate School of Business.

[9] CHEN, F. AND Y.S. ZHENG. (1994) Lower Bounds for Multi-Echelon Stochastic Inventory Systems. *Management Science,* Vol. 40, 1426–1443.

[10] CHEN, F., Z. DREZNER, J.K. RYAN, AND D. SIMCHI-LEVI. (2000) Quantifying the Bullwhip Effect: The Impact of Forecasting, Leadtime and Information. *Management Science,* Vol. 46, 436–443.

[11] CLARK, A. AND H. SCARF. (1960) Optimal Policies for a Multi-Echelon Inventory Problem. *Management Sciences,* Vol. 6, 475–490.

[12] FEDERGRUEN, A. (1993) Centralized Planning Models for Multi-Echelon Inventory Systems under Uncertainty. Chapter 3 in *Handbook in Operations Research and Management Science,* Vol. 4.

[13] FEDERGRUEN, A. AND P. ZIPKIN. (1984A) Computational Issues in an Infinite Horizon, Multi-Echelon Inventory Model. *Operations Research,* Vol. 32, 818–836.

[14] FEDERGRUEN, A. AND P. ZIPKIN. (1984B) Approximations of Dynamic, Multi location Production and Inventory Problems. *Management Science,* Vol. 30, 69–84.

[15] FRIEDMAN, M. (1953) Essays in the Theory of Positive Economics. University of Chicago Press, 1953, p. 15.

[16] GALLEGO, G., Y. HUANG, K. KATIRCIOGLU, AND Y.T. LEUNG. (2000) When to Share Demand Information in a Simple Supply Chain? Submitted to *Management Science.*

[17] GALLEGO, G. AND Ö. ÖZER. (1999) Integrating Replenishment Decisions With Advance Demand Information. Submitted to *Management Science.*

[18] GALLEGO, G. AND Ö. ÖZER. (2000) Optimal Replenishment Policies for Multi-Echelon Inventory Problems under Advance Demand Information. Submitted to *Operations Research.*

[19] GALLEGO, G. AND B. TOKTAY. (1999) All-or-Nothing Ordering under a Capacity Constraint and Forecasts of Stationary Demand. Working Paper, Columbia University, New York, NY.

[20] GALLEGO, G. AND G. VAN RYZIN. (1997) A Multi-Product Dynamic Pricing Model with Applications to Network Yield Management. *Operations Research,* Vol. 45, 24–41.

[21] GAVIRNENI, S., R. KAPUSCINSKI, AND S. TAYUR. (1999) Value of Information in Capacitated Supply Chains. *Management Science,* Vol. 45, 16–24.

[22] GRAVES, S.C., H.C. MEAL, S. DASU, AND Y. QUI. (1986) Two Stage Production Planning In a Dynamic Environment, in Multi Stage Production Planning and Inventory Control, S. Axsäter, C. Schneeweiss and E. Silver (Eds.), Lecture Notes In *Economics and Mathematical Systems,* Springer-Verlag, Berlin, Vol. 266, 9–43.

[23] GÜLLÜ, R. (1996) On the Value of Information in Dynamic Production/Inventory Problems Under Forecast Evolution. *Naval Research Logistics,* Vol. 43, 289–303.

[24] GÜLLÜ, R. (1998) Optimal Production/Inventory Policies Under Forecasts and Limited Production Capacity. Working Paper, Middle East Technical University.

[25] KATIRCIOGLU, K. AND D. ATKINS. (1998) New Optimal Policies for A Unit Demand Inventory Problem. Working paper, IBM Research Division, Yorktown Heights, NY.

[26] HARIHARAN, R. AND P. ZIPKIN. (1995) Customer Order Information, Lead Times, and Inventories. *Management Science,* Vol. 41, 1599–1607.

[27] HAUSMAN, W.H. (1969) Sequential Decision Problems: A model to Exploit Existing Forecasts. *Management Science,* Vol. 16, B93–B111.

[28] HEAT, D. AND P. JACKSON. (1994) Modeling the Evolution of Demand Forecasts with Application to Safety Stock Analysis in Production/Distribution Systems. *IIE Transactions,* Vol. 26, 17–30.

[29] LEE, H.L., C. BILLINGTON, AND B. CARTER. (1993) Hewlett-Packard Gains Control of Inventory and Service Through Design for Localization. *Interfaces,* Vol. 23, 4 July–August 1–11.

[30] LEE, H.L., K. SO, AND C. TANG. (2000) Information Sharing in a Two-Level Supply Chain. *Management Science,* Vol. 46, 626–643.

[31] MOINZADEH, K. (1999) An Improved Ordering Policy for Continuous Review Inventory Systems with Arbitrary Inter-Demand Time Distributions. Working paper, School of Business. University of Seattle, WA 98195.

[32] MOINZADEH, K. AND S. NAHMIAS. (1988) A Continuous Review Model for an Inventory System with Two Supply Modes. *Management Science,* Vol. 34, 761–773.

[33] ÖZER, Ö. (2000) Replenishment Strategies for Distribution Systems under Advance Demand Information. Working Paper, Stanford University, Department of MS&E.

[34] ÖZER, Ö. (2000) Supply Chain Management With Advance Demand Information. Ph.D. Dissertation, Columbia University.

[35] PORTEUS, E.L. (1990) Stochastic Inventory Theory. Chapter 12 in *Handbooks in Operations Research and Management Science,* Vol. 2, 605–652.

[36] ROSLING, K. (1989) Optimal Inventory Policies for Assembly Systems Under Random Demands. *Operations Research,* Vol. 37, 565–579.

[37] SCARF, H. (1960) The Optimality of (s, S) Policies in the Dynamic Inventory Problem. in: K.A. Arrow, S. Karlin, and P. Suppes (Eds.), *Mathematical Methods in the Social Sciences,* Stanford University Press, Stanford, CA.

[38] SCARF, H. (1963) A Survey of Analytic Techniques in Inventory Theory. Chapter 7 in: Scarf, H., D.M. Gilford, and M.W. Shelly

(Eds.), in *Multistage Inventory Models and Techniques*, 185–225, Stanford University Press, Stanford, CA.

[39] SCARF, H. (2000) Optimal Inventory Policies when Sales are Discretionary. Working Paper, Yale University.

[40] SCHWARZ, L.B., PETRUZZI, N.C., AND K. WEE. (1998) The Value of Advance-Order Information and the Implication for Managing the Supply Chain: An Information/Control/Buffer Portfolio Perspective. Working paper, Kranert Graduate School of Management, Purdue University.

[41] SETHI, S.P., H. YAN, AND H. ZHANG. (2000) Peeling Layers of an Onion: An Inventory Model with Multiple Delivery Modes and Forecast Updates. Working Paper, University of Texas at Dallas, Richardson, Texas.

[42] TOKTAY, L.B. AND L.W. WEIN. (1999) Analysis of a Forecasting-Production-Inventory System with Stationary Demand. Working Paper, Sloan School of Management, MIT.

[43] VEINOTT, A. (1966) The Status of Mathematical Inventory Theory. *Management Science,* Vol. 12, 745–777.

[44] WOODWARD, ROBERT. (2000) Maestro: Greenspan's Fed and the American Boom. *Simon and Schuster,* New York.

[45] ZHENG, Y.S. (1992) On Properties of Stochastic Inventory Systems *Management Science,* Vol. 38, 87–103.

[46] ZIPKIN, P. (1982) Exact and Approximate Cost Functions for Product Aggregates. *Management Science,* Vol. 28, 1002–1012.

Chapter 6

SUPPLY CHAIN INFORMATION SHARING IN A COMPETITIVE ENVIRONMENT

Lode Li
Yale School of Management, New Haven, CT 06520
and CEIBS, Shanghai, China
lode.li@yale.edu

Hongtao Zhang
School of Business and Management
Hong Kong University of Science and Technology
Hong Kong
imhzhang@ust.hk

1. Introduction

The subject of information sharing has regained the interests of both academics and practitioners. Today a huge amount of information is being interchanged between manufacturers and retailers, between retailers and consumers, between companies and investors, and also among the parties in the same level of a vertical chain. How to measure the gains and losses of such activities to the parties involved is an important issue in supply chain management. This chapter studies the incentives for firms to share information vertically in the presence of horizontal competition. We do so in a setting with one manufacturer and many competing retailers where each retailer possesses some private information about the downstream market demand or about its own cost. Horizontal competition in the downstream brings about new effects of information sharing. In general, vertical information sharing, e.g., transmission of point-of-sales data between a retailer and a manufacturer, has two effects, the *"direct effect"* on the payoffs between the parties engaged in information sharing, and the *"indirect effect"* of information sharing on other competing firms. For example, knowing that the manufacturer receives some information from a retailer, other retailers may respond to the fact by changing their strategies, and such reaction may cause additional gains *or* losses to the parties directly engaged in information sharing.

Information, unlike a physical good, may slip beyond the boundaries of the firms between which it is exchanged. This is because a firm who has acquired information would act upon it, and its actions, if observable, could make the otherwise confidential information transparent. Some researchers have voiced their concern about the confidentiality of the shared information, e.g., Lee and Whang [21] point out several hurdles that face information sharing in a supply chain including the issue of confidentiality. They use the example of a supplier who supplies a critical part to two manufacturers who compete in the final product market. Either manufacturer would not share information (like sales data) with the supplier unless it is guaranteed that the information is not leaked to the other manufacturer. However, in a supply chain setting, the shared information may become known to the other manufacturer, not through an intentional act of leakage by a supplier, but rather through the observable behavior of the supplier who reacts to the information.

In fact, vertical information disclosures often result in horizontal information "leakage". For example, when a retailer shares its information with a supplier, even though the information transmission is itself confidential, other retailers may be able to infer such information from the actions of the supplier and hence change their strategies for the game of horizontal competition. A more extreme case is when a firm voluntarily discloses its accounting information to the investors and then the information is made public and becomes known to the competitors. We dub this kind of indirect effect of vertical information sharing the *"leakage effect"*. To study the incentives for vertical information disclosure in the presence of horizontal competition, one must incorporate the leakage effect and possibly other indirect effects. It is well-known in a general conflict situation that obtaining additional information does *not* necessarily make the informed player better off and that giving away private information to other players does *not* necessarily make the original information holder worse off. Therefore, the impact of leakage effect is not always straightforward.

Consider a two-echelon supply chain that consists of a single manufacturer and many retailers. For convenience, we refer to the manufacturer as 'she' and a retailer as 'he'. The manufacturer provides products to the retailers who then sells them to final consumers. The retailers compete in the downstream market where the demand is uncertain. Each retailer possesses some information about the market demand. The manufacturer could benefit from the retailers' information which would enable her to make better production and pricing decisions. This chapter presents a summary of the findings in Li [23], Li and Zhang [25], and Zhang [37], which focus on the following issues in vertical information exchange:

- How does vertical information sharing impact the competition among retailers? Should the confidentiality of shared information be a concern? How does the leakage effect come about?

- What are the additional gains or losses from information exchange due to the indirect effect? Does information sharing increase the profit of each party?

- How much is the value of downstream information to the manufacturer and in what ways will she benefit from information sharing? Does the downstream competition increase or decrease such value?

- What are the incentives for information sharing and when is information sharing achievable on a voluntary basis? When information can be traded between supply chain parties, when does such trade take place and how should prices for information exchange be determined?

- Does information collusion raise anti–trust concerns? Does it benefit consumers or the society as a whole? (Since the firms operate in a macroeconomic environment, public policies do matter.)

The rest of the chapter is structured as follows. In Section 2 we discuss modeling assumptions and illustrate the direct effect through a simple example. Section 3 details the model for the case of homogeneous goods, a make-to-order manufacturer and many identical retailers. Section 4 considers a make-to-stock manufacturer who may produce before observing the actual orders at lower costs. Section 5 focuses on the case of duopolistic retailers who sell differentiated goods and are engaged in either a Cournot or Bertrand competition. In Section 6 we study the sharing of production cost information in the context of a single product and many identical retailers. Section 7 concludes and points to future research directions. Bibliographical notes then follow.

2.   The Models

We consider a two-echelon supply chain that consists of one upstream firm, the manufacturer, and n downstream firms, the retailers. Denote the manufacturer by M and let $N = \{1, 2, \ldots, n\}$ be the set of retailers.

The manufacturer is either make-to-order at a higher marginal cost due to expediting production or make-to-stock at a lower marginal cost due to a longer allowable production lead time. The downstream firms are engaged in either a Cournot (quantity) or Bertrand (price) competition. The products the retailers sell are either identical or differentiated and in the latter case can be either substitutes or complements. We consider two types of demand functions for the downstream market:

1. All the retailers sell an identical good and the inverse demand function for the downstream market is given by

$$p = a + \theta - \sum_{i \in N} q_i \qquad (6.1)$$

where $\sum_{i \in N} q_i$ is the total sales level. If each retailer i picks a level of sales q_i, then p given above is the prevailing price in the downstream market.

2. There are exactly two retailers who sell differentiated goods and the inverse demand functions for the downstream market are given by

$$p_i = a + \theta - q_i - \gamma q_j \quad \text{for } j \neq i,\ i = 1, 2, \qquad (6.2)$$

where γ is a constant with $|\gamma| < 1$ and is positive if the goods are substitutes or negative if they are complements; p_i and q_i are the market price of and demand for, respectively, retailer i's good.

That is, the demand structure is linear and symmetric with respect to the retailers. The intercept of the linear demand function, which represents the potential market size, is $a + \theta$ where a is a constant and θ is a random variable with mean zero and variance σ^2. Before making his quantity or price decision, each retailer i observes a signal Y_i about θ. The joint probability distribution of $(\theta, Y_1, \ldots, Y_n)$ is common knowledge.

The manufacturer provides the retailers with their goods at a constant and equal marginal cost c if making to order. We assume $c < a$. The retailers incur a constant and equal marginal retailing cost. Unless otherwise specified, we assume that the retailing costs are zero without loss of generality. The following is the sequence of events and decisions:

1. Each retailer decides whether to disclose his information and the manufacturer decides whether to acquire such information.

2. Each retailer i observes a signal Y_i and the manufacturer observes shared signals according to the information sharing arrangements made earlier.

3 Based on the available information, the manufacture sets the whole-sale price P. (The manufacturer makes both price (P) and initial production quantity (Q) decisions if make-to-stock is allowed.)

4 Upon receiving the price P, the retailers choose sales levels q_i in a Cournot competition or selling prices p_i in a Bertrand competition.

5 The manufacturer produces to meet the downstream demands q_i that are either chosen by the retailers in Cournot competition or result from their chosen prices p_i in Bertrand competition.

This is a three-stage game. First, retailers decide whether to share their private demand information with the manufacturer. Second, the manufacturer sets the price P. Finally, the retailers choose their sales levels or selling prices. We assume that the information agreements reached in the first stage are known to all parties henceforth and that the information transmission is truthful. It should be emphasized that a retailer must decide whether or not to reveal his private information to the manufacturer *before* he learns the signal Y_i, implying that the manufacturer cannot infer Y_i when making her price decision if retailer i does not reveal his information. It would be a very different model if we allow the retailers to decide whether or not to share *after* they have received the signals. In this latter case, retailers would behave strategically with the actual value of their signals Y_i.

To analyze the game, we first solve the third-stage subgame (quantity or price decisions by the retailers) and then the second-stage subgame (price decision or both price and initial production quantity decisions by the manufacturer) for any given information sharing arrangement from the first stage. In the third stage, a Cournot/Bertrand Bayesian subgame is played. Retailer i receives a signal Y_i. He then formulates a conjecture about the behavior of the manufacturer and the other retailers. This conjecture, together with the joint distribution of $(\theta, Y_1, \ldots, Y_n)$, determines the expected profit of any action retailer i may take, and it should be correct in a Nash equilibrium solution. Retailer i chooses an action that maximizes his expected profit. In the second stage, the manufacturer uses the available information, a result in the first stage of the game, to estimate her expected profit for any price P (and any initial production quantity Q if making to stock) in anticipation of the equilibrium outcome of the third-stage subgame and chooses a value of P (and a value of Q) that maximizes her expected profit. Finally, we analyze the first-stage game in anticipation of the equilibrium behavior in the last two stages of the game.

To facilitate analysis, some assumptions on the information structure are necessary.

(A1) $E[Y_i \mid \theta] = \theta$ for all i.

(A2) $E[\theta \mid Y_1, \ldots, Y_n] = \alpha_0 + \sum_{i \in N} \alpha_i Y_i$ where α_i are constants. And $Y_1, \ldots, Y_n$ are independent, conditional on θ.

(A3) $Y_1, \ldots, Y_n$ are identically distributed.

The first assumption says that each Y_i is an unbiased estimator of θ. The second assumption is general enough to include a variety of interesting prior-posterior distribution pairs such as normal-normal, gamma-Poisson, and beta-binomial. One may raise the concern that since the normal distribution is unbounded on the negative side, the intercept $a + \theta$ of the demand function may take on negative values. However, from a practical standpoint, if a is large relative to the standard deviation σ of θ, unbounded probability distributions such as the normal provide adequate approximation. The last assumption says that firms' private signals are symmetric in probability distribution.

By Lemma 1 of Li [22], the above assumptions imply that, for any given set $K \subseteq N$, $|K| = k$, and any $i \in N \backslash K$,

$$E[\theta \mid Y_j, j \in K] = E[Y_i \mid Y_j, j \in K] = \frac{1}{k+s} \sum_{j \in K} Y_j, \qquad (6.3)$$

where

$$s \equiv \frac{E[\mathit{Var}[Y_i \mid \theta]]}{\mathit{Var}[\theta]}. \qquad (6.4)$$

That is, $\sum_{j \in K} Y_j$ is sufficient to predict θ and other signals Y_i given $(Y_j)_{j \in K}$. One can think of s as a measure of the imprecision of the signal about θ relative to the degree of variation of θ. It can be verified that

$$E[(Y_i)^2] = (1+s)\sigma^2, \quad \text{and} \quad E[Y_i Y_j] = \sigma^2 \quad \text{for } i \neq j. \qquad (6.5)$$

Before concluding this section, we illustrate the direct effect using a model with a single retailer who receives a perfect signal about θ, i.e., $Y_i \equiv \theta$ or $s = 0$. For any wholesale price P set by the manufacturer, the retailer's profit as a function of q is $q(a + \theta - q - P)$. Thus his optimal order quantity is $q^* = (a + \theta - P)/2$ and the optimal profit is $(a + \theta - P)^2/4$. The manufacturer's profit is in turn given by $(P - c)q^* = (P - c)(a + \theta - P)/2$. If the retailer does not reveal the value of θ, the manufacturer's expected profit equals to $(P - c)(a - P)/2$. She maximizes her profit by setting $P^* = (a + c)/2$ and receives an expected profit of $(a - c)^2/8$. The retailer's maximum profit, given $P = P^*$, is $[(a - c)/2 + \theta]^2/4$ and his expected profit is $(a - c)^2/16 + \sigma^2/4$.

On the other hand, if the retailer reveals the value of θ to the manufacturer, the latter would set her wholesale price to $P^* = (a + c + \theta)/2$ and enjoys a profit of $(a - c + \theta)^2/8$. So the manufacturer's expected profit equals $(a - c)^2/8 + \sigma^2/8$. The retailer's maximum profit given $P = P^*$ is $[(a - c)/2 + \theta/2]^2/4$ and his expected profit equals $(a - c)^2/16 + \sigma^2/16$.

Therefore, information sharing increases the manufacturer's but decreases the retailer's profit. That is, the manufacturer seeks more economic rent with better information through pricing. Furthermore, information sharing reduces the total profit of the firms in the supply chain from $3(a - c)^2/16 + \sigma^2/4$ to $3(a - c)^2/16 + 3\sigma^2/16$. In this case, the retailer has no incentive to share his information on a voluntary bases, nor can the manufacturer offer a price high enough that the retailer would accept to sell his information, unless the supply chain decisions could be somehow coordinated.

3. Cournot Retailers and a Homogeneous Product

Throughout this section we assume the manufacturer makes to order and there are n retailers in the downstream who are engaged in a Cournot competition selling an identical good and the demand function is given by (6.1). To solve the third-stage and the second-stage subgames, we represent all possible outcomes in the first stage of the game as follows. Let K, $|K| = k$, be the set of retailers who decide to share their information, $(Y_j)_{j \in K}$, with the manufacturer, who decides to acquire such information. Because the symmetric nature of the game, the sets $K \subseteq N$, $k = 0, 1, \ldots, n$, represent all the possible information sharing arrangements made in the first stage.

The equilibrium behavior of retailers in the third stage is closely related to the Bayesian subgame equilibrium in the Cournot oligopoly described in Li [22], which studies incentives of Cournot oligopolists to share private information with their competitors in a two-stage game. We summarize the related results from that paper next.

3.1. Information sharing in cournot oligopoly

Consider an oligopoly with n identical firms producing a homogeneous product at a constant marginal cost w. The inverse demand function is given by (6.1) where θ is a random variable with zero mean and a known variance. Before deciding its output quantity, each firm i observes a signal Y_i about θ and the signals satisfy (A1–A3). Suppose that, before the firms learn their signals, a set K, $|K| = k$, of firms are committed to release their signals to a common pool to be made "available" to all firms by an "outside agency". The information that firm i can use for an output decision consists of its private signal Y_i and the disclosed information $(Y_j)_{j \in K}$. The expected profit for firm i, given its information, is

$$\left(a + E[\theta \,|\, Y_i, (Y_j)_{j \in K}] - q_i - \sum_{j \neq i} E[q_j \,|\, Y_i, (Y_j)_{j \in K}] - w \right) q_i.$$

The equilibrium quantity decisions must satisfy the first-order condition,

$$2q_i^* = a - w + E[\theta \,|\, Y_i, (Y_j)_{j \in K}] - E\left[\sum_{j \neq i} q_j^* \,|\, Y_i, (Y_j)_{j \in K} \right], \quad \text{for all } i.$$

By Proposition 1 of Li [22], there is a unique equilibrium to this Bayesian subgame and the equilibrium strategies are as follows. For $i \in K$, a firm who releases Y_i to the outside agency,

$$q_i^* = \frac{1}{n+1} \left(a - w + A_1^k \sum_{j \in K} Y_j \right), \tag{6.6}$$

and for $i \in N \backslash K$, a firm who does not release his information,

$$q_i^* = \frac{1}{n+1} \left(a - w + B_1^k \sum_{j \in K} Y_j + B_2^k Y_i \right), \tag{6.7}$$

where

$$A_1^k \equiv \frac{1}{k+s}, \tag{6.8}$$

$$B_1^k \equiv \frac{k+2s}{(k+s)(n+k+1+2s)}, \tag{6.9}$$

$$B_2^k \equiv \frac{n+1}{n+k+1+2s}, \tag{6.10}$$

and s is defined in (6.4). We see that the firms' equilibrium decisions is dependent on the pooled signals $(Y_j)_{j \in K}$ only through their summation $\sum_{j \in K} Y_j$.

Using (6.6) and (6.7), we can derive the expected profit of each firm given any set K of disclosing firms. Denote by $\Pi^S(k)$ the expected profit of a firm if it is one of the k firms that release their private information and by $\Pi^N(k)$ the expected profit of a firm if it does not release its information. By Proposition 3 of Li [22], $\Pi^N(k) > \Pi^S(k+1)$, i.e., a firm is always worse off by disclosing its information and therefore no information disclosure is the unique equilibrium.

We now return to the supply chain context and study the equilibrium behavior of retailers in the third-stage subgame.

3.2.   Equilibrium behavior of retailers

Using the shared information $(Y_j)_{j \in K}$, the manufacturer's best estimate of θ is $E[\theta \,|\, (Y_j)_{j \in K}]$ given by (6.3). We assume that her wholesale price decision P is a strictly monotone function of this best estimate, or equivalently, $P = P(\sum_{j \in K} Y_j)$. (This assumption can be weakened greatly when there are exactly $n = 2$ retailers. All we then need is that P as a function of $(Y_j)_{j \in K}$ is strictly monotone in each Y_j. See Section 5 for details.) Then we argue that although transmission of the shared information $(Y_j)_{j \in K}$ is confidential and known only to the manufacturer, a sufficient amount of information will be known to all parties in the supply chain when the manufacturer use the shared information in her price decision. The reason is as follows. Knowing that the manufacturer knows $(Y_j)_{j \in K}$, each retailer expects the price set by the manufacturer to be a function of $\sum_{j \in K} Y_j$, and he can then deduce the sum $\sum_{j \in K} Y_j$ from P. This sum, being a sufficient statistic, is as good as the signals themselves. In this chapter, we only investigate equilibrium behavior in the subspace of outcomes in which the retailers conjecture that the manufacturer's price is a monotone function of the sum of the shared signals. We shall show that this retailers' conjecture is correct in equilibrium.

In the third stage of the game, (Y_i, P) is the information retailer i has when making his quantity decision. Given the aforementioned retailers' conjecture of P, (Y_i, P) is equivalent to $(Y_i, \sum_{j \in K} Y_j)$. The expected profit for retailer i, given his information, is

$$E[\pi_i \,|\, Y_i, P] = \left(a + E[\theta \,|\, Y_i, P] - q_i - \sum_{j \neq i} E[q_j \,|\, Y_i, P] - P \right) q_i.$$

This is a game of private information, or a Bayesian subgame. Each retailer's equilibrium strategy is a function of his private information Y_i and the manufacturer's price P. We denote retailer i's equilibrium strategy by $q_i^*(Y_i, P)$. The equilibrium sales quantities must satisfy the first-order condition,

$$2q_i^* = a - P + E[\theta \mid Y_i, P] - E\left[\sum_{j \neq i} q_j^*(Y_j, P) \mid Y_i, P\right] \quad \text{for all } i. \quad (6.11)$$

The present Bayesian subgame differs from the Bayesian subgame of horizontal information sharing in Cournot oligopoly in Section 3.1 in the amount of information that is made available. In the present setting retailer i knows only $\{Y_i, \sum_{j \in K} Y_j\}$ as opposed to the specific values of the signals, $\{Y_i; Y_j, j \in K\}$. However, observing that the equilibrium strategies in the Cournot oligopoly only depend on the sum $\sum_{j \in K} Y_j$, one would guess that (6.6) and (6.7) would give an equilibrium of the present Bayesian subgame if we replace the production cost w by the wholesale price P. It is indeed true. Specifically, for $i \in K$, a retailer who shares Y_i with the manufacturer,

$$q_i^*(Y_i, P) = \frac{1}{n+1}\left(a - P + A_1^k \sum_{j \in K} Y_j\right), \quad (6.12)$$

where A_1^k is given in (6.8); for $i \in N \setminus K$, a retailer who does not disclose his information,

$$q_i^*(Y_i, P) = \frac{1}{n+1}\left(a - P + B_1^k \sum_{j \in K} Y_j + B_2^k Y_i\right), \quad (6.13)$$

where B_1^k and B_2^k are given in (6.9) and (6.10), respectively. The coefficients A_1^k, B_1^k and B_2^k satisfy

$$A_1^k(k+s) = 1, \quad \text{and} \quad B_1^k(k+s) + B_2^k = 1 \quad \text{for } k \geq 0.$$

It can be then shown that for any $i \in K$ and $l \in N \setminus K$,

$$E[q_l^*(Y_l, P) \mid (Y_j)_{j \in K}] = q_i^*(Y_i, P). \quad (6.14)$$

Using these facts, we can verify that (6.12) and (6.13) satisfy the first order condition (6.11).

Proposition 6.1 *Given any information sharing agreements reached in the first stage of the game, i.e., a set $K \subseteq N$, and the manufacturer's price P set in the second stage of the game and the conjecture that P is a monotone function of $\sum_{j \in K} Y_j$, there is a unique equilibrium to the* *third-stage subgame, and the equilibrium strategies are given in (6.12) and (6.13).*

By applying the first-order condition, we can write the expected profit of retailer i in the third-stage subgame as a function of his equilibrium strategy:

$$E[\pi_i^* \mid Y_i, P] = [q_i^*(Y_i, P)]^2. \tag{6.15}$$

3.3. Equilibrium price of the manufacturer

In the second stage of the game, the manufacturer sets P in anticipation of retailers' equilibrium decisions $(q_i^*(Y_i, P))_{i \in N}$ in the third stage. To compute the demand function for the manufacturer, we calculate the total quantity demanded by the retailers for any given P,

$$D_M = \sum_{i \in K} q_i^*(Y_i, P) + \sum_{i \in N \setminus K} q_i^*(Y_i, P). \tag{6.16}$$

Denote by D_M^k the demand for the manufacturer conditional on $(Y_j)_{j \subset K}$. Then by (6.14) and (6.12), the expected value of D_M^k as a function of P equals

$$E[D_M^k] \equiv E[D_M \mid (Y_j)_{j \in K}] = \frac{n}{n+1} \left(a - P + A_1^k \sum_{j \in K} Y_j \right). \tag{6.17}$$

Therefore, the expected profit for the manufacturer given her information $(Y_j)_{j \in K}$ is

$$E[\pi_M \mid (Y_j)_{j \in K}] = \frac{n}{n+1}(P - c) \left(a - P + A_1^k \sum_{j \in K} Y_j \right). \tag{6.18}$$

The equilibrium price in the upstream market that maximizes (6.18) is

$$P^*((Y_j)_{j \in K}) = \frac{a+c}{2} + M_1^k \sum_{j \in K} Y_j, \tag{6.19}$$

where

$$M_1^k = \frac{1}{2} A_1^k.$$

When $K = \emptyset$, then $\sum_{j \in K} Y_j = 0$ and $P^* = (a+c)/2$. Note that P^* is a linear function of $\sum_{j \in K} Y_j$. Thus, the retailers' conjecture in the third stage of the game (that P^* is monotone in $\sum_{j \in K} Y_j$) is fulfilled.

Proposition 6.2 *Given any information sharing agreements reached in the first stage of the game, i.e., a set $K \subseteq N$, the manufacturer's equilibrium price P^* is a linear function of $\sum_{j \in K} Y_j$ given by (6.19).*

One natural question to ask is whether the manufacturer has incentives to use price discrimination against the group of retailers who do not disclose their information. That is, she might want to charge different prices for the two groups, P_1 for the retailers who reveal their information and P_2 for those who conceal their information. Li [23] shows that the single price is optimal from the manufacturer's view point.

Proposition 6.3 *Given any information sharing agreements reached in the first stage of the game, i.e., a set $K \subseteq N$, the manufacturer has no incentive to use price discrimination against the group of retailers who do not disclose their information.*

3.4.   Information sharing

We can now compute the expected profits for any information sharing arrangement K given the equilibrium behaviors of the manufacturer in the second stage and of the retailers in the third stage of the game.

Substituting (6.19) into (6.12) and (6.13), we obtain the equilibrium sales quantity of each retailer: for $i \in K$,

$$q_i^*(Y_i, P^*((Y_j)_{j \in K})) = \frac{1}{n+1}\left(\frac{a-c}{2} + (A_1^k - M_1^k)\sum_{j \in K} Y_j\right), \quad (6.20)$$

and for $i \in N \backslash K$,

$$q_i^*(Y_i, P^*((Y_j)_{j \in K})) = \frac{1}{n+1}\left(\frac{a-c}{2} + (B_1^k - M_1^k)\sum_{j \in K} Y_j + B_2^k Y_i\right). \quad (6.21)$$

Then, using (6.15) and (6.5), we can compute the expected equilibrium profits for the retailers as follows: for $i \in K$,

$$\Pi_R^S(k) = \frac{(a-c)^2}{4(n+1)^2} + \frac{\sigma^2}{(n+1)^2}(A_1^k - M_1^k)^2(k^2 + ks), \quad (6.22)$$

and for $i \in N \backslash K$,

$$\Pi_R^N(k) = \frac{(a-c)^2}{4(n+1)^2} + \frac{\sigma^2}{(n+1)^2}\big[(B_1^k - M_1^k)^2(k^2 + ks)$$
$$+ 2(B_1^k - M_1^k)B_2^k k + (B_2^k)^2(1+s)\big]. \quad (6.23)$$

That is, retailer i is expected to get $\Pi_R^S(k)$ if he is one of the k retailers who share their private information with the manufacturer, and get $\Pi_R^N(k)$ otherwise.

Substituting (6.19) into (6.18) and taking the expectation, we can compute the expected profit for the manufacturer:

$$\Pi_M(k) = \frac{n}{n+1}E[(P^* - c)^2] = \frac{n}{n+1}\left(\frac{(a-c)^2}{4} + \sigma^2(M_1^k)^2(k^2 + ks)\right).$$

$$(6.24)$$

Lemma 6.4 *The following holds:*

1 $\Pi_M(k)$ *is strictly increasing and concave in k.*

2 $\Pi_R^N(k-1) > \Pi_R^S(k)$ *for all $k = 1, \ldots, n$, and $\Pi_R^N(k-1) - \Pi_R^S(k)$ is decreasing in k.*

Proposition 6.5 *In the first stage of the game, for any given information sharing agreements K, the expected profit functions for all firms are given in (6.22), (6.23) and (6.24), which have the properties in Lemma 6.4. These facts lead to the following conclusions:*

1 *The manufacturer is better off by acquiring information from more retailers in all circumstances.*

2 *Each retailer is worse off by disclosing his information to the manufacturer in all circumstances.*

Therefore, no information sharing is the unique equilibrium on a voluntary basis.

The proposition indicates that the retailers have no incentive to share their information with the manufacturer although the manufacturer is better off with more information about the downstream market demand. There are two reasons. First, as illustrated at the end of Section 2, a retailer is worse off by disclosing the demand information to the manufacturer due to the direct effect, i.e., the manufacturer takes the advantage of better information to seek more economic rent, information rent, and that hurts the retailers. Second, a retailer will also be worse off by letting the competing retailers learn his demand information due to the leakage effect, as in the case of Cournot oligopoly and a homogeneous product (see Section 3.1 or Proposition 3 of Li, [22]). Therefore, both direct and leakage effects have a negative impact on the incentives of retailers to share information vertically.

The concavity of $\Pi_M(k)$ means that, as more retailers already disclose their information, the incremental gain to the manufacturer from the next sharing retailer becomes smaller. That $\Pi_R^N(k-1) - \Pi_R^S(k)$ is decreasing in k implies that, as more retailers already disclose their information, the drop in profit for the next retailer who chooses to disclose becomes smaller.

3.5.   Information sharing with payment

Since the manufacturer benefits from information sharing, one natural arrangement is for her to pay the retailers for their private information about demand. We first investigate the impact of information sharing on total supply chain profit. Intuitively, information could get traded only if the total supply chain profit is larger with information sharing. Otherwise, the manufacturer would not have enough gain to compensate the retailers for their loss due to information sharing.

Define the expected total profit of the supply chain as:

$$\Pi(k) \equiv \Pi_M(k) + k\Pi_R^S(k) + (n-k)\Pi_R^N(k). \qquad (6.25)$$

It can be shown that

$$\Pi(n) - \Pi(0) = \frac{\sigma^2 n[n(n+1) + 2s][(n-2)(n+1) - 2s]}{4(n+1)^2(n+s)(n+1+2s)^2}.$$

Lemma 6.6 $\Pi(n) > \Pi(0)$ *if and only if*

$$(n-2)(n+1) > 2s. \qquad (6.26)$$

Condition (6.26) says that the total supply chain profit will be larger through information sharing when the information each retailer has is relatively informative in a statistical sense (s is small) or when there is a sufficiently large number of retailers (n is larger). Condition (6.26) will not hold when $n = 2$, i.e., with only two retailers, the total supply chain profit will shrink when information disseminates. Li [22] and Palfrey [31] show that when the number of firms becomes large and competition intensifies, the equilibrium price with privately held information converges to the price in the pooled-information situation. In our case, when n is large, the impact of the leakage effect will diminish and the firms will behave as if the information is shared.

Consider the following *contract signing game* in the first stage. The manufacturer offers a payment δ for each retailer to share his private information. If a retailer signs the contract, he is bound to disclose his information to the manufacturer. Otherwise, he may choose his best strategy without revealing the information.

In order for the manufacturer to benefit by buying the information, δ must satisfy $\Pi_M(n) - n\delta > \Pi_M(0)$, i.e., the manufacturer is willing to pay no more than $(\Pi_M(n) - \Pi_M(0))/n$ for each piece of information. In order for a retailer to benefit by selling his information when all other retailers do the same, δ must satisfy $\Pi_R^S(n) + \delta > \Pi_R^N(n-1)$, i.e., each retailer is willing to accept a price no less than $\Pi_R^N(n-1) - \Pi_R^S(n)$. Complete information sharing would not take place unless both conditions hold, i.e.,

$$\Pi_R^N(n-1) - \Pi_R^S(n) < \delta < \frac{\Pi_M(n) - \Pi_M(0)}{n}, \qquad (6.27)$$

which implies that for such a δ to exist, we must have

$$\Pi_M(n) - \Pi_M(0) > n\left[\Pi_R^N(n-1) - \Pi_R^S(n)\right]. \qquad (6.28)$$

Condition (6.28) can be shown to be equivalent to

$$(n-1)(n+1) > s. \qquad (6.29)$$

Clearly, values n and s that satisfy (6.26) will satisfy (6.29), or (6.26) is a subset of (6.29). That is, information trade could be an equilibrium outcome although such trade makes the total supply chain profit smaller.

Next, we show that partial information sharing is not an equilibrium outcome. Note from Lemma 6.4 that $\Pi_R^N(k-1) - \Pi_R^S(k)$ is decreasing in k. Here $\Pi_R^N(k-1) - \Pi_R^S(k)$ is the minimum price the kth retailer would accept for selling his information given $k-1$ retailers have done so. This minimum compensation is the highest for the first retailer and becomes smaller for each subsequent retailer as more retailers already share their information. Suppose the manufacturer offers any price higher than $\Pi_R^N(n-1) - \Pi_R^S(n)$, say $\delta = \Pi_R^N(k-1) - \Pi_R^S(k)$, and $k-1$ ($1 < k < n$) retailers decide to sell their information. This cannot be an equilibrium because all remaining $n-k+1$ retailers will also have incentive to sell their information at the price. Therefore, no asymmetric equilibrium is possible in which some retailers trade their information and others do not.

Now, we can characterize the conditions under which complete information sharing is an equilibrium that dominates no information sharing. First, each retailer will sell his information regardless of other retailers' decision if and only if

$$\delta > \Pi_R^N(0) - \Pi_R^S(1).$$

Such a δ exists if and only if

$$\Pi_M(n) - \Pi_M(0) > n\left[\Pi_R^N(0) - \Pi_R^S(1)\right]. \qquad (6.30)$$

It can be verified that values n and s that satisfy (6.30) will satisfy (6.26), or (6.30) is a subset of (6.26). Second, when

$$n\left[\Pi_R^N(n-1) - \Pi_R^S(n)\right] < \Pi_M(n) - \Pi_M(0) \leq n\left[\Pi_R^N(0) - \Pi_R^S(1)\right],$$

$$(6.31)$$

the contract signing game has two possible equilibria: all information is traded or no information is traded. In this case, all retailers prefer the equilibrium of complete information sharing to the equilibrium of no information sharing if and only of

$$\delta > \Pi_R^N(0) - \Pi_R^S(n).$$

Such a δ exists if and only if $\Pi_M(n) - \Pi_M(0) > n[\Pi_R^N(0) - \Pi_R^S(n)]$, or equivalently,

$$\Pi_M(n) + n\Pi_R^S(n) > \Pi_M(0) + n\Pi_R^N(0),$$

and this is the same as (6.26), requiring that complete information sharing results in a higher total supply chain profit.

The above results are summarized in the following proposition.

Proposition 6.7 *In the contract signing game:*

1. *Complete information sharing is an equilibrium with the equilibrium price δ determined from (6.27) if and only if (6.29) holds, i.e., $(n-1)(n+1) > s$.*

2. *There exists a δ such that*

$$\Pi_R^N(0) - \Pi_R^S(n) < \delta < \frac{\Pi_M(n) - \Pi_M(0)}{n}$$

   if and only if (6.26) holds. With this δ, complete information sharing Pareto-dominates the equilibrium of no information sharing.

3. *There exists a δ such that*

$$\Pi_R^N(0) - \Pi_R^S(1) < \delta < \frac{\Pi_M(n) - \Pi_M(0)}{n}$$

   if and only if (6.30) holds. With this δ, complete information sharing is a dominant-strategy equilibrium.

In short, information trade that involves all retailers will be more likely to take place if and only if it increases the total supply chain profit, i.e., (6.26) holds.

3.6. Social benefits

Finally, we consider the impact of vertical information sharing on the total social benefits defined as

$$W \equiv \int_0^D (a + \theta - q)\, dq - cD = (a - c + \theta)D - \frac{1}{2}D^2,$$

where D is the total sales volume given by (6.16). The above equals the sum of the total supply chain profit $(a - c + \theta - D)D$ and the consumer surplus $D^2/2$. Denote the expected social benefits and consumer surplus by

$$W(k) \equiv E[W] \quad \text{and} \quad CS(k) \equiv \frac{1}{2}E\left[D^2\right],$$

respectively. The total sales volume D can be expressed as, from (6.20) and (6.21),

$$D = \frac{1}{n+1}\left\{ \frac{n(a-c)}{2} + \left[kA_1^k + (n-k)B_1^k - nM_1^k\right] \right.$$

$$\left. \times \sum_{j \subset K} Y_j + B_2^k \sum_{i \in N \setminus K} Y_i \right\}.$$

It can be shown that

$$W(n) < W(0), \quad \text{and} \quad CS(n) < CS(0).$$

Proposition 6.8 *Complete information sharing reduces both the expected total social benefits and the expected consumer surplus.*

Consumers will be worse off when all retailers share their demand information with the manufacturer, so much so that the total social benefits will be lower, although the total supply chain profit may be higher. Thus, from a viewpoint of social policy making, information trading as described in Proposition 6.7 only benefits the manufacturer and the retailers but reduces the total social welfare, and hence, should be discouraged. In contrast, the result for information disclosure on a voluntary basis, with a unique Nash equilibrium in which no information is shared (Proposition 6.5), is in line with the social benefits.

4. Make-to-Stock Manufacturer

The last section has assumed that the manufacturer is make-to-order with a negligible production lead time. We now consider the case when it is possible to make to stock. That is, before the retailers decide on their sales quantities q_i, the manufacturer has an opportunity to produce an initial quantity Q at the same time she chooses a wholesale price P. We assume that *the manufacturer is obliged to meet the demand from all retailers*. Because of information asymmetry, the manufacturer may not be able to predict the exact order quantities from the retailers at the time the initial quantity is produced. If the total order quantity, $D = \sum_{i \in N} q_i$, exceeds Q, then an additional cost of e dollars would be incurred for each unit produced beyond this initial production quantity. Recall from the previous section that the marginal cost of making to order is c dollars and so the marginal cost for producing the initial Q units to stock is $c - e$ dollars. We may think that the costs of production given a long lead-time (starting before orders are received) are lower than the costs of fast response to urgent demand (handling shortages after the orders are observed). When $D > Q$, each additional unit must be produced using subcontracting or overtime and thus incurs a cost premium e (expediting cost). If $D < Q$, each unit of the surplus has a salvage value of b dollars where $b < c - e$. Alternatively, Q can be viewed as the capacity set by the manufacturer, e as the opportunity cost associated with each unit of capacity shortage and $c - e - b$ as the cost associated with each unit of idle capacity.

We impose a more restrictive information structure. Specifically, we assume that the retailers' signals, Y_i, $i \in N$, are independent draws from a normal distribution with an unknown mean θ and a known variance v^2. This information structure is a special case of the one used in the last section and hence satisfies assumptions (A1) through (A3). It can be shown (see, e.g., Section 6.3 of DeGroot, [12]) that, for any given set $K \subseteq N$, $|K| = k$, and any $i \in N \backslash K$,

$$Var[\theta \,|\, (Y_j)_{j \in K}] = \frac{v\sigma^2}{v^2 + k\sigma^2} = \frac{s\sigma^2}{s + k}, \qquad (6.32)$$

where $Var[\theta \,|\, (Y_j)_{j \in K}]$ is the variance of the posterior distribution of θ given $(Y_j)_{j \in K}$ and s is given in (6.4). Note that $Var[\theta \,|\, (Y_j)_{j \in K}]$ does not depend on the specific values of $(Y_j)_{j \in K}$.

We now show that the manufacturer's problem of choosing an optimal pair (P, Q) has a strikingly simple structure.

In the third stage of the game, since the manufacturer is obliged to satisfy the downstream demand, the retailers would behave exactly the same as before and their equilibrium order quantities and expected profit are given by (6.12), (6.13) and (6.15). Thus, there is no change in the total order quantity from the retailers,

$$D_M = \sum_{i\in N} q_i^*(Y_i, P) = \sum_{i\in K} q_i^*(Y_i, P) + \sum_{i\in N\setminus K} q_i^*(Y_i, P)$$

$$= \frac{1}{n+1}\left\{ n(a-P) + \left[kA_1^k + (n-k)B_1^k\right]\sum_{j\in K} Y_j + B_2^k \sum_{i\in N\setminus K} Y_i \right\}.$$

$$(6.33)$$

In the second stage of the game, the manufacturer sets the price P and the quantity Q in anticipation of retailers' equilibrium order quantity in (6.33). Denote by D_M^k the demand faced by the manufacturer conditional on $(Y_j)_{j\in K}$. From the normality assumption, we know that D_M^k is also normally distributed. The expected value of D_M^k as a function of P is given by (6.17). To find the variance of D_M^k, we can first show by the normality assumption and (6.32) that

$$Var\left[\sum_{i\in N\setminus K} Y_i \mid (Y_j)_{j\in K} \right] = (n-k)v^2 + (n-k)^2 \, Var[\theta \mid (Y_j)_{j\in K}]$$

$$= \frac{(n-k)(n+s)s\sigma^2}{s+k}.$$

Then, by (6.33) and (6.10), the variance of D_M^k equals

$$\sigma_k^2 \equiv Var\left[D_M^k\right] = Var\left[\frac{1}{n+1}B_2^k \sum_{i\in N\setminus K} Y_i \mid (Y_j)_{j\in K} \right]$$

$$= \frac{(n-k)(n+s)s\sigma^2}{(n+k+1+2s)^2(s+k)}. \qquad (6.34)$$

Thus, the upstream market demand D_M^k conditional on the manufacturer's information has a normal distribution with mean in (6.17) and variance in (6.34). Notice that σ_k^2 is independent of the price P and the specific values of $(Y_j)_{j\in K}$.

Lemma 6.9 σ_k^2 *is decreasing in k and is zero when $k = n$.*

The upstream demand uncertainty to the manufacturer is smaller when more retailers disclose their market intelligence and will be completely resolved when all retailers disclose their information. This agrees with the conventional wisdom that vertical information sharing reduces demand information distortion in the upstream.

The expected profit for the manufacturer, given her information $(Y_j)_{j \in K}$, is

$$E[\pi_M \mid (Y_j)_{j \in K}]$$
$$= E[PD_M - (c-e)Q - c(D_M - Q)^+ + b(Q - D_M)^+ \mid (Y_j)_{j \in K}]$$
$$= E[(P - c + e)D_M - (c - e - b)(Q - D_M)^+ - e(D_M - Q)^+ \mid (Y_j)_{j \in K}]$$
$$= (P - c + e)E[D_M^k] - E[(c - e - b)(Q - D_M^k)^+ + e(D_M^k - Q)^+].$$

For any given P, the manufacturer's quantity decision is a newsvendor problem for which the overage cost is $c - e - b$ and the underage cost is e. The optimal initial production quantity as a function of P thus equals

$$Q^* = E[D_M^k] + z^* \sigma_k = \frac{n}{n+1}\left(a - P + A_1^k \sum_{j \in K} Y_j\right) + z^* \sigma_k \qquad (6.35)$$

where z^* is defined by the tail probability of the standard normal distribution,

$$\Pr(z > z^*) = (c - e - b)/(c - b).$$

Note that z^* is independent of P and other parameters of the distribution of D_M^k. The standard deviation σ_k of D_M^k follows from (6.34),

$$\sigma_k = \sqrt{\frac{(n-k)(n+s)s\sigma^2}{(n+k+1+2s)^2(s+k)}}.$$

Given the optimal initial production quantity Q^*, the expected profit for the manufacturer can be written as

$$E[\pi_M \mid (Y_j)_{j \in K}] = (P - c + e)E[D_M] - G(c - b - e, e)\sigma_k,$$
$$= \frac{n}{n+1}(P - c + e)\left(a - P + A_1^k \sum_{j \in K} Y_j\right)$$
$$- G(c - b - e, e)\sigma_k \qquad (6.36)$$

where

$$G(c - b - e, e) = E[(c - b - e)(z^* - z)^+ + e(z - z^*)^+],$$

and z is the standard normal random variable. The function $G(c - b - e, e)$ can be thought of as the expected inventory related costs per unit of standard deviation of demand. It is easy to show that $G(c - b - e, e)$ is increasing in c and decreasing in b, and that $G(c - b - e, e) = (c - b - e)z^* + (c - b)L(z^*)$ where $L(z^*) = E[(z - z^*)^+]$ is the standard normal distribution loss function.

The manufacturer's expected profit in (6.36) is expressed nicely as the difference of two terms. The first term depends on the wholesale price and specific values of the shared information while the second term depends on neither. In fact, the first term is the same as the expected profit in the make-to-order case (6.18) except that the marginal production cost is now $c - e$ instead of c. The second term is the inventory related costs to the manufacturer for not being able to precisely predict orders from the retailers. Since σ_k is decreasing in k, information sharing gives the manufacturer benefits that she can use the information to reduce surpluses and shortages. This is the classical operations management feature captured in our model that information is a substitute for inventory.

To find the manufacturer's optimal price in the second stage we only need to maximize the first term in (6.36). The equilibrium price in the upstream market is given by

$$P^*((Y_j)_{j \in K}) = \frac{a + c - e}{2} + M_1^k \sum_{j \in K} Y_j,$$

It is the same as that for the make-to-order case (6.19) except for the first constant term where the manufacturer's marginal cost is now $c - e$.

In the first stage of the game, for any information sharing arrangement K, the expected profits to the retailers are given by

$$\Pi_R^S(k) = \frac{(a - c + e)^2}{4(n + 1)^2} + \frac{\sigma^2}{(n + 1)^2}(A_1^k - M_1^k)^2(k^2 + ks) \quad \text{for } i \in K,$$

$$(6.37)$$

$$\Pi_R^N(k) = \frac{(a - c + e)^2}{4(n + 1)^2} + \frac{\sigma^2}{(n + 1)^2}[(B_1^k - M_1^k)^2(k^2 + ks)$$
$$+ 2(B_1^k - M_1^k)B_2^k k + (B_2^k)^2(1 + s)] \quad \text{for } i \in N \backslash K, \quad (6.38)$$

and they are the same as in the make-to-order case, (6.22) and (6.23), with c replaced by $c - e$. Note that the retailers' incentives for information sharing do not depend on the common constant term $(a - c + e)^2/4(n + 1)^2$. The expected profit to the manufacturer is

$$\Pi_M(k) = \frac{n}{n + 1}\left(\frac{(a - c + e)^2}{4} + \sigma^2(M_1^k)^2(k^2 + ks)\right) - G(c - b - e, e)\sigma_k.$$

$$(6.39)$$

Compared with the profit function in (6.24) for the make-to-order case, there is now an additional inventory cost term, $G(c - b - e, e)\sigma_k$, which is decreasing in k. Hence the manufacturer receives extra benefits from more information sharing.

It should be clear that the results for information sharing on a voluntary basis in Lemma 6.4 and Proposition 6.5 still holds:

Proposition 6.10 *In the first stage of the game, for any given information sharing agreements K, the expected profit functions for all firms are given in (6.37), (6.38) and (6.39), which have the following properties:*

1 $\Pi_M(k)$ is strictly increasing in k.

2 $\Pi_R^N(k - 1) > \Pi_R^S(k)$ for all $k = 1, \ldots, n$, and $\Pi_R^N(k - 1) - \Pi_R^S(k)$ is decreasing in k.

These facts lead to the following conclusions:

1 The manufacturer is better off by acquiring information from more retailers in all circumstances.

2 Each retailer is worse off by disclosing his information to the manufacturer in all circumstances.

Therefore, no information sharing is the unique equilibrium on a voluntary basis.

Let $\Pi(k)$ be the total supply chain profit defined by (6.25). From (6.37), (6.38) and (6.39),

$$\Pi(n) = \frac{(a - c + e)^2 n(n + 2)}{4(n + 1)^2} + \frac{\sigma^2 n^2(n + 2)}{4(n + 1)^2(n + s)},$$

and

$$\Pi(0) = \frac{(a - c + e)^2 n(n + 2)}{4(n + 1)^2} + \frac{\sigma^2 n(1 + s)}{(n + 1 + 2s)^2}$$
$$- G(c - b - e, e)\frac{\sigma\sqrt{n(n + s)}}{n + 1 + 2s}.$$

Thus,

$$\Pi(n) - \Pi(0) = \frac{\sigma^2 n[n(n + 1) + 2s][(n - 2)(n + 1) - 2s]}{4(n + 1)^2(n + s)(n + 1 + 2s)^2}$$
$$+ G(c - b - e, e)\frac{\sigma\sqrt{n(n + s)}}{n + 1 + 2s}.$$

The second term is always positive and the first term is positive if $(n - 2)(n + 1) > 2s$. Therefore, $\Pi(n) > \Pi(0)$ if and only if

$$G(c - b - e, e) > \frac{\sigma n[n(n + 1) + 2s][2s - (n - 2)(n + 1)]}{4(n + 1)^2(n + s)(n + 1 + 2s)\sqrt{n(n + s)}} \qquad (6.40)$$

Note that condition (6.26), $(n-2)(n+1) > 2s$, for make-to-order case is a sufficient condition for (6.40) to hold, i.e., the additional savings in inventory related costs enlarge the incremental total supply chain profit due to information sharing. When inventory related cost $G(c-b-e, e)$ is sufficiently high, the incremental total benefits could be positive even if $(n-2)(n+1) < 2s$. For example, if there is only two retailers ($n = 2$), the total supply chain profit always shrinks when information disseminates in the make-to-order case, but this is no longer true as long as the inventory related cost is sufficiently high, i.e.,

$$G(c - b - e, e) > \frac{2\sigma(3 + s)s}{9(2 + s)(3 + 2s)\sqrt{2(2 + s)}}.$$

Thus, in a contract signing game defined as in Section 3, the manufacturer is now willing to pay a higher price δ to each retailer for the demand information due to the additional savings in inventory costs.

Proposition 6.11 *In the contract signing game, there exists an information price, δ, such that complete information sharing is an equilibrium which either is undominated or Pareto-dominates the equilibrium of no information sharing if (6.40) holds, and the equilibrium price δ is determined by*

$$\Pi_R^N(n - 1) - \Pi_R^S(n) \le \delta \le \frac{\Pi_M(n) - \Pi_M(0)}{n}.$$

5. Duopoly Retailers with Differentiated Goods

In this section, we assume there are exactly two retailers in the downstream who sell *differentiated* goods that can be either substitutes or complements. The duopolistic retailers are engaged in either a Cournot (quantity) or Bertrand (price) competition with demand functions given by (6.2). The sequence of events and decisions and the information structure described in Section 2 apply.

There are three possible information sharing arrangements in the first stage of the game: $K = \{1, 2\}$; $K = \emptyset$; or $K = \{i\}$, $i = 1$ or 2. We take advantage of the fact that $n = 2$ and assume a much weaker condition than that in Section 3.2. Specifically, we search for equilibria in the

subspace in which the retailers conjecture that the manufacturer's optimal price P^* is strictly monotone in each Y_i, $i \in K$. (This is a much
weaker condition than the assumption in Section 3.2 that P^* is a strictly
monotone function of the sum of disclosed signals.) Information leakage
occurs under arrangements $K = \{1,2\}$ and $K = \{i\}$.

We assume the manufacturer is make-to-order and consider Cournot
and Bertrand subgames separately.

5.1.   Cournot Bayesian subgames

For a given information sharing arrangement K and a given P, retailer
i's expected profit as a function of his sales quantity q_i equals to

$$E[\pi_i \mid Y_i, P] = (a + E[\theta \mid Y_i, P] - q_i - \gamma E[q_j \mid Y_i, P] - P)q_i$$

for $j \neq i$ and $i = 1, 2$. Setting the derivative of the above with respect
to q_i to zero, the retailers' sales quantities at equilibrium satisfy

$$2q_i^* = a - P + E[\theta \mid Y_i, P] - \gamma E[q_j^* \mid Y_i, P]. \tag{6.41}$$

Retailer i's conditional expected profit at equilibrium in the third-stage
subgame equals

$$E[\pi_i \mid Y_i, P] = (q_i^*)^2. \tag{6.42}$$

Next we analyze the third and second stage subgames for three possible information sharing arrangements in the first stage of the game.
For each arrangement we derive the retailers' equilibrium response, the
manufacturer's optimal price, and the expected profit of each party conditional on the information he/she has.

Both retailers sharing information. The wholesale price P^* reveals
the private information so that both retailers know (Y_1, Y_2) after P^* is
announced. The retailers' estimate of θ is given by (6.3) with $k = 2$.
The retailers' equilibrium sales quantities are

$$q_i^* = \frac{1}{2+\gamma}\left[a - P + \frac{1}{2+s}(Y_1 + Y_2)\right] \quad \text{for } i = 1, 2. \tag{6.43}$$

The manufacturer's expected profit conditional on (Y_1, Y_2) equals

$$E[\pi_M \mid Y_1, Y_2] = (P - c) \cdot E[q_1^* + q_2^* \mid Y_1, Y_2]$$

$$= \frac{2}{2+\gamma}(P - c)\left[a + \frac{1}{2+s}(Y_1 + Y_2) - P\right]$$

and her optimal price is

$$P^* = \frac{a+c}{2} + \frac{1}{2(2+s)}(Y_1 + Y_2). \qquad (6.44)$$

The manufacturer's conditional expected profit for $P = P^*$ equals

$$E[\pi_M^* \mid Y_1, Y_2] = \frac{2}{2+\gamma}\left[\frac{a-c}{2} + \frac{1}{2(2+s)}(Y_1 + Y_2)\right]^2. \qquad (6.45)$$

Substituting (6.44) into (6.43) and using (6.42), we obtain the retailers' conditional expected profit

$$E[\pi_i^* \mid Y_1, Y_2] = \frac{1}{(2+\gamma)^2}\left[\frac{a-c}{2} + \frac{1}{2(2+s)}(Y_1 + Y_2)\right]^2 \quad \text{for } i = 1, 2. \qquad (6.46)$$

No retailer sharing information. Suppose no retailer shares his demand information with the manufacturer. A retailer's estimate of θ, and of the other retailer's private signal, is given by (6.3) with $k - 1$. Then

$$q_i^* = \frac{1}{2+\gamma}\left[a - P + \frac{2+\gamma}{2+\gamma+2s}Y_i\right] \quad \text{for } i = 1, 2. \qquad (6.47)$$

The manufacturer's expected profit equals

$$E[\pi_M] = (P - c) \cdot E[q_1^* + q_2^*] = \frac{2}{2+\gamma}(P - c)(a - P),$$

and her optimal price and maximum expected profit are, respectively,

$$P^* = \frac{a+c}{2}, \qquad (6.48)$$

$$E[\pi_M^*] = \frac{2}{2+\gamma}\left(\frac{a-c}{2}\right)^2. \qquad (6.49)$$

Substituting (6.48) into (6.47) and using (6.42), the retailers' conditional expected profit is

$$E[\pi_i^* \mid Y_i] = \frac{1}{(2+\gamma)^2}\left[\frac{a-c}{2} + \frac{2+\gamma}{2+\gamma+2s}Y_i\right]^2 \quad \text{for } i = 1, 2. \qquad (6.50)$$

One retailer sharing information. Suppose retailer i shares his private signal with the manufacturer but retailer j does not. After P^*

is announce, retailer j learns Y_i through P^*. Retailer j's estimate of θ is given by (6.3) with $k = 2$, while retailer i's estimate of θ, and of Y_j, is given by (6.3) with $k = 1$. We can show that

$$q_i^* = \frac{1}{2+\gamma}\left[a - P + \frac{1}{1+s}Y_i\right], \tag{6.51}$$

$$q_j^* = \frac{1}{2+\gamma}\left[a - P + \frac{2-\gamma+2s}{2(1+s)(2+s)}Y_i + \frac{2+\gamma}{2(2+s)}Y_j\right]. \tag{6.52}$$

It can be shown that $E[q_j^* \mid Y_i] = q_i^*$. The manufacturer's expected profit conditional on Y_i equals

$$E[\pi_M \mid Y_i] = (P-c)\cdot E[q_1^* + q_2^* \mid Y_i] = \frac{2}{2+\gamma}(P-c)\left[a - P + \frac{1}{1+s}Y_i\right],$$

and her optimal price is

$$P^* = \frac{a+c}{2} + \frac{1}{2(1+s)}Y_i. \tag{6.53}$$

The manufacturer's expected profit under P^* equals

$$E[\pi_M^* \mid Y_i] = \frac{2}{2+\gamma}\left[\frac{a-c}{2} + \frac{1}{2(1+s)}Y_i\right]^2. \tag{6.54}$$

Substituting (6.53) into (6.51) and (6.52), and using (6.42), we get the retailers' expected profit,

$$E[\pi_i^* \mid Y_i] = \frac{1}{(2+\gamma)^2}\left[\frac{a-c}{2} + \frac{1}{2(1+s)}Y_i\right]^2, \tag{6.55}$$

$$E[\pi_j^* \mid Y_1, Y_2] = \frac{1}{(2+\gamma)^2}\left[\frac{a-c}{2} + \frac{s-\gamma}{2(1+s)(2+s)}Y_i + \frac{2+\gamma}{2(2+s)}Y_j\right]^2. \tag{6.56}$$

5.2.   Bertrand Bayesian subgames

From (6.2) the demand functions can be written as

$$q_i = \frac{1}{1+\gamma}\left[a + \theta - \frac{1}{1-\gamma}p_i + \frac{\gamma}{1-\gamma}p_j\right] \quad \text{for } j \neq i,\ i = 1,2. \tag{6.57}$$

For a given information sharing arrangement and a given P, a retailer's expected profit as a function of his price p_i equals $(p_i - P)E[q_i \mid Y_i, P]$, i.e.,

$$E[\pi_i \mid Y_i, P]$$
$$= \frac{1}{1+\gamma}(p_i - P)\left[a + E[\theta \mid Y_i, P] - \frac{1}{1-\gamma}p_i + \frac{\gamma}{1-\gamma}E[p_j \mid Y_i, P]\right].$$

Setting the derivative of the above with respect to p_i to zero, the equilibrium retailer prices satisfy

$$\frac{2}{1-\gamma} p_i^* = a + E[\theta \mid Y_i, P] + \frac{1}{1-\gamma} P + \frac{\gamma}{1-\gamma} E[p_j^* \mid Y_i, P]. \quad (6.58)$$

Using (6.58) and (6.57) we obtain retailer i's conditional expected sales at equilibrium prices,

$$E[q_i^* \mid Y_i, P] = \frac{1}{(1+\gamma)(1-\gamma)} (p_i^* - P), \quad (6.59)$$

and the conditional expected profit

$$E[\pi_i \mid Y_i, P] = (p_i^* - P) E[q_i \mid Y_i, P] = \frac{1}{(1+\gamma)(1-\gamma)} (p_i^* - P)^2. \quad (6.60)$$

Next we analyze the third and second stage subgames for three possible information sharing arrangements in the first stage of the game.

Both retailers sharing information. After P^* is announced, it reveals the private information and both retailers know (Y_1, Y_2). In this game of complete information, the retailers' equilibrium prices are

$$p_i^* = \frac{1-\gamma}{2-\gamma} a + \frac{1}{2-\gamma} P + \frac{1-\gamma}{(2-\gamma)(2+s)} (Y_1 + Y_2) \quad \text{for } i = 1, 2. \quad (6.61)$$

By (6.59) a retailer's conditional expected sales at equilibrium prices equals

$$E[q_i^* \mid Y_1, Y_2] = \frac{1}{(1+\gamma)(1-\gamma)} (p_i^* - P)$$

$$= \frac{1}{(1+\gamma)(2-\gamma)} \left[a + \frac{1}{2+s} (Y_1 + Y_2) - P \right], \quad i = 1, 2.$$

The manufacturer's expected profit conditional on (Y_1, Y_2) equals

$$E[\pi_M \mid Y_1, Y_2] = (P - c) \cdot E[q_1 + q_2 \mid Y_1, Y_2]$$

$$= \frac{2}{(1+\gamma)(2-\gamma)} (P - c) \left[a + \frac{1}{2+s} (Y_1 + Y_2) - P \right],$$

and so her optimal price is

$$P^* = \frac{a+c}{2} + \frac{1}{2(2+s)} (Y_1 + Y_2). \quad (6.62)$$

The manufacturer's and the retailers' conditional expected profits for $P = P^*$ are:

$$E[\pi_M^* \mid Y_1, Y_2] = \frac{2}{(1+\gamma)(2-\gamma)}\left[\frac{a-c}{2} + \frac{1}{2(2+s)}(Y_1 + Y_2)\right]^2. \quad (6.63)$$

$$E[\pi_i^* \mid Y_1, Y_2] = \frac{1-\gamma}{(1+\gamma)(2-\gamma)^2}\left[\frac{a-c}{2} + \frac{1}{2(2+s)}(Y_1 + Y_2)\right]^2. \quad (6.64)$$

No retailer sharing information. Suppose no retailer shares his demand information with the manufacturer. We can show that

$$p_i^* = \frac{1-\gamma}{2-\gamma}a + \frac{1}{2-\gamma}P + \frac{1-\gamma}{2-\gamma+2s}Y_i. \quad (6.65)$$

From (6.59), retailer i's expected sales is

$$E[q_i^* \mid Y_i] = \frac{1}{(1+\gamma)(2-\gamma)}\left[a - P + \frac{2-\gamma}{2-\gamma+2s}Y_i\right].$$

Thus, the manufacturer's expected profit equals

$$E[\pi_M] = (P-c) \cdot E[q_1^* + q_2^*] = \frac{2}{(1+\gamma)(2-\gamma)}(P-c)(a-P),$$

and her optimal price and the maximum expected profit are, respectively,

$$P^* = \frac{a+c}{2}, \quad (6.66)$$

$$E[\pi_M^*] = \frac{2}{(1+\gamma)(2-\gamma)}\left(\frac{a-c}{2}\right)^2. \quad (6.67)$$

The retailers' equilibrium expected profit is

$$E[\pi_i^* \mid Y_i] = \frac{1-\gamma}{(1+\gamma)(2-\gamma)^2}\left[\frac{a-c}{2} + \frac{2-\gamma}{2-\gamma+2s}Y_i\right]^2. \quad (6.68)$$

One retailer sharing information. Suppose retailer i shares his private signal with the manufacturer but retailer j does not. Then retailer j learns Y_i from P^*. We can show that

$$p_i^* = \frac{1-\gamma}{2-\gamma}a + \frac{1}{2-\gamma}P + \frac{1-\gamma}{(2-\gamma)(1+s)}Y_i, \quad (6.69)$$

$$p_j^* = \frac{1-\gamma}{2-\gamma}a + \frac{1}{2-\gamma}P + \frac{1-\gamma}{2(2+s)}Y_j + \frac{(1-\gamma)(2+\gamma+2s)}{2(2-\gamma)(1+s)(2+s)}Y_i.$$

$$(6.70)$$

It can be shown that $E[p_j^* \mid Y_i] = p_i^*$. From (6.59) retailers' conditional expected sales are

$$E[q_i^* \mid Y_i] = \frac{1}{(1+\gamma)(2-\gamma)}\left[a - P + \frac{1}{1+s}Y_i\right],$$

$$E[q_j^* \mid Y_i, Y_j] = \frac{1}{(1+\gamma)(2-\gamma)}\left[a - P + \frac{2-\gamma}{2(2+s)}Y_j + \frac{2+\gamma+2s}{2(1+s)(2+s)}Y_i\right].$$

Retailer j's sales estimated by the manufacturer is, using (6.59),

$$E[q_j^* \mid Y_i] = \frac{1}{(1+\gamma)(1-\gamma)}(E[p_j^* \mid Y_i] - P)$$

$$= \frac{1}{(1+\gamma)(1-\gamma)}(p_i^* - P) = E[q_i^* \mid Y_i].$$

The manufacturer's expected profit conditional on her information thus equals

$$E[\pi_M \mid Y_i] = (P - c) \cdot E[q_1 + q_2 \mid Y_i]$$

$$= \frac{2}{(1+\gamma)(2-\gamma)}(P - c)\left[a - P + \frac{1}{1+s}Y_i\right]$$

and so her optimal price is

$$P^* = \frac{a+c}{2} + \frac{1}{2(1+s)}Y_i. \tag{6.71}$$

The manufacturer's and retailers' equilibrium conditional expected profits are

$$E[\pi_M^* \mid Y_i] = \frac{2}{(1+\gamma)(2-\gamma)}\left[\frac{a-c}{2} + \frac{1}{2(1+s)}Y_i\right]^2, \tag{6.72}$$

$$E[\pi_i^* \mid Y_i] = \frac{1-\gamma}{(1+\gamma)(2-\gamma)^2}\left[\frac{a-c}{2} + \frac{1}{2(1+s)}Y_i\right]^2, \tag{6.73}$$

and

$$E[\pi_j^* \mid Y_1, Y_2]$$

$$= \frac{1-\gamma}{(1+\gamma)(2-\gamma)^2}\left[\frac{a-c}{2} + \frac{2-\gamma}{2(2+s)}Y_j + \frac{\gamma+s}{2(1+s)(2+s)}Y_i\right]^2. \tag{6.74}$$

Finally, we note that (6.44), (6.48) and (6.53) are equal to (6.62), (6.66) and (6.71), respectively. That is, the manufacturer's optimal wholesale price, under any given information sharing arrangement, is identical for the two types of competition in the downstream market, Cournot or Bertrand.

5.3.   Equilibrium in the first stage

To find the payoffs for decisions in the first stage, we take expectations, with respect to Y_1 and Y_2, of (6.45, 6.46, 6.49, 6.50, 6.54–6.56), and (6.63, 6.64, 6.67, 6.68, 6.72–6.74). The table below summarizes these *ex ante* expected profits of the three firms in each information sharing arrangement and for each type of downstream competition,

Payoff	Cournot	Bertrand
$\Pi_M(0)$	$\frac{2}{2+\gamma}\left[\frac{1}{4}(a-c)^2 + g_0\sigma^2\right]$	$\frac{2}{(1+\gamma)(2-\gamma)}\left[\frac{1}{4}(a-c)^2 + g_0\sigma^2\right]$
$\Pi_M(1)$	$\frac{2}{2+\gamma}\left[\frac{1}{4}(a-c)^2 + g_1\sigma^2\right]$	$\frac{2}{(1+\gamma)(2-\gamma)}\left[\frac{1}{4}(a-c)^2 + g_1\sigma^2\right]$
$\Pi_M(2)$	$\frac{2}{2+\gamma}\left[\frac{1}{4}(a-c)^2 + g_2\sigma^2\right]$	$\frac{2}{(1+\gamma)(2-\gamma)}\left[\frac{1}{4}(a-c)^2 + g_2\sigma^2\right]$
$\Pi_R^N(0)$	$\frac{1}{(2+\gamma)^2}\left[\frac{1}{4}(a-c)^2 + f_0(\gamma)\sigma^2\right]$	$\frac{1-\gamma}{(1+\gamma)(2-\gamma)^2}\left[\frac{1}{4}(a-c)^2 + f_0(-\gamma)\sigma^2\right]$
$\Pi_R^S(1)$	$\frac{1}{(2+\gamma)^2}\left[\frac{1}{4}(a-c)^2 + f_{1S}\sigma^2\right]$	$\frac{1-\gamma}{(1+\gamma)(2-\gamma)^2}\left[\frac{1}{4}(a-c)^2 + f_{1S}\sigma^2\right]$
$\Pi_R^N(1)$	$\frac{1}{(2+\gamma)^2}\left[\frac{1}{4}(a-c)^2 + f_{1N}(\gamma)\sigma^2\right]$	$\frac{1-\gamma}{(1+\gamma)(2-\gamma)^2}\left[\frac{1}{4}(a-c)^2 + f_{1N}(-\gamma)\sigma^2\right]$
$\Pi_R^S(2)$	$\frac{1}{(2+\gamma)^2}\left[\frac{1}{4}(a-c)^2 + f_2\sigma^2\right]$	$\frac{1-\gamma}{(1+\gamma)(2-\gamma)^2}\left[\frac{1}{4}(a-c)^2 + f_2\sigma^2\right]$

where

$$g_0 = 0, \qquad g_1 = \frac{1}{4(1+s)}, \qquad g_2 = \frac{1}{2(2+s)},$$

$$f_0(z) = \frac{(2+z)^2(1+s)}{(2+z+2s)^2}, \qquad f_{1S} = \frac{1}{4(1+s)},$$

$$f_{1N}(z) = \frac{2+(2+z)^2 s + s}{4(1+s)(2+s)}, \qquad f_2 = \frac{1}{2(2+s)}.$$

Note the symmetry between Cournot and Bertrand outcomes. Cournot competition with product substitutes (complements) is like Bertrand competition with product complements (substitutes). The duality in structure between Cournot and Bertrand competitions is noted by Vives [35] in his duopoly model (horizontal competition without the manufacturer). Interestingly the duality is preserved in our supply chain setting. This allows us to apply the following lemma to either competitive situation.

Lemma 6.12 *(a)* $g_2 > g_1 > g_0$, *(b)* $f_{1S} < f_0(z)$ *for* $|z| < 1$, *(c)* $f_2 < f_{1N}(z)$ *for* $|z| < 1$.

The lemma leads to the following result.

Proposition 6.13 *The expected profits of the manufacturer and the retailers in both Cournot or Bertrand competition and under different information sharing arrangements have the following properties:*

$$\Pi_M(2) > \Pi_M(1) > \Pi_M(0),$$
$$\Pi_R^S(1) < \Pi_R^N(0) \quad and \quad \Pi_R^S(2) < \Pi_R^N(1).$$

These facts lead to the following conclusions: The manufacturer is always better off by acquiring demand information from more retailers; each retailer is always worse off by disclosing his demand information to the manufacturer. Therefore, no information sharing is the unique equilibrium.

The proposition indicates that the retailers have no incentive to share their demand information with the manufacturer whereas the manufacturer is better off with more information about the downstream market demand. This is not entirely unexpected if retailers' goods are substitutes in Cournot competition or complements in Bertrand competition, because then both the direct and leakage effects are negative on the retailers' profit. However, even when the leakage effect is positive, i.e., when goods are complements in Cournot competition or substitutes in Bertrand competition, the overall effect is still negative. In other words, the negative direct effect dominates the positive leakage effect.

5.4.   Information sharing with payment

Can information sharing be achieved through a contract signing game? We show in Section 3 that this is not possible for Cournot competition with two retailers and identical goods. We now examine the feasibility of this arrangement for Cournot competition with product complements or Bertrand competition with product substitutes, because the leakage effect is positive in both cases (Vives [35]).

It can be shown that $\Pi_R^N(0) - \Pi_R^S(1) > \Pi_R^N(1) - \Pi_R^S(2)$. Also from the results in Proposition 6.13, the game has a similar structure as the contract signing game considered in Section 3, and the results in Proposition 6.7 apply. That is, there exists a payment δ for each retailer so that complete information sharing is an equilibrium if

$$\Pi_M(2) - \Pi_M(0) - 2\big[\Pi_R^N(1) - \Pi_R^S(2)\big] > 0.$$

The equilibrium is a dominant-strategy equilibrium if

$$\Pi_M(2) - \Pi_M(0) - 2[\Pi_R^N(0) - \Pi_R^S(1)] > 0. \qquad (6.75)$$

And there exists a δ so that complete information sharing Pareto-dominates the equilibrium of no information sharing if the total supply chain profit is larger,

$$\Pi_M(2) + 2\Pi_R^S(2) > \Pi_M(0) + 2\Pi_R^N(0).$$

Of the three conditions above, (6.75) is the strongest. To see when it holds, we consider the two types of competition separately.

Cournot retailers with product complements. It can be verified that

$$\Pi_M(2) - \Pi_M(0) - 2[\Pi_R^N(0) - \Pi_R^S(1)]$$

$$= \frac{C_\gamma(s)\sigma^2}{2(2+\gamma)^2(1+s)(2+s)(2+\gamma+2s)^2}$$

where

$$C_\gamma(s) = 4(-\gamma^2 - 2\gamma + 1)s^3 - (8\gamma^2 + 20\gamma)s^2$$
$$+ (2\gamma^3 + \gamma^2 - 12\gamma - 12)s + (2\gamma^3 + 6\gamma^2 - 8).$$

We note that $C_\gamma(s)$ is a polynomial in s of degree 3 and, for $\gamma < 0$, the coefficient of s^3 is positive. Thus $C_\gamma(s)$ is positive for sufficiently large s. Let $s_\gamma^C \equiv \inf\{s : C_\gamma(x) > 0 \text{ for all } x > s\}$, then (6.75) holds for $s > s_\gamma^C$. Figure 6.1 shows that s_γ^C is increasing in γ and $s_{\gamma=0}^C = 2$.

Bertrand retailers with product substitutes. It can be verified that

$$\Pi_M(2) - \Pi_M(0) - 2[\Pi_R^N(0) - \Pi_R^S(1)]$$

$$= \frac{B_\gamma(s)\sigma^2}{2(1+\gamma)(2-\gamma)^2(1+s)(2+s)(2-\gamma+2s)^2}$$

where

$$B_\gamma(s) = 4(\gamma^3 - 5\gamma^2 + 5\gamma + 1)s^3 + (16\gamma^3 - 68\gamma^2 + 68\gamma)s^2$$
$$+ (17\gamma^3 - 67\gamma^2 + 72\gamma - 12)s + 2(2\gamma - 1)(2 - \gamma)^2.$$

We note that $B_\gamma(s)$ is a polynomial in s of degree 3 and, for $\gamma > 0$, the coefficient of s^3 is positive. Thus $B_\gamma(s)$ is positive for sufficiently large s.

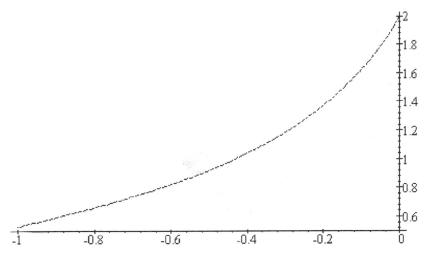

Figure 6.1. s_γ^C as a function of γ for $\gamma \in (-1, 0]$

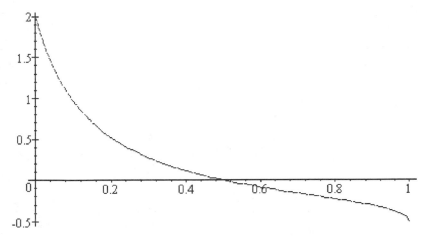

Figure 6.2. s_γ^B as a function of γ for $\gamma \in [0, 1)$

Let $s_\gamma^B \equiv \inf\{s : B_\gamma(x) > 0 \text{ for all } x > s\}$, then (6.75) holds for $s > s_\gamma^B$. Figure 6.2 shows that s_γ^B is decreasing in γ, $s_{\gamma=0}^B = 2$ and $s_\gamma^B < 0$ for $\gamma > 0.5$.

Proposition 6.14 *Suppose the retailers are in a Cournot (Bertrand) competition with product complements (substitutes) and $s > s_\gamma^C$ (s_γ^B). Then there exists a $\delta > 0$ such that to share their information for a* *payment δ from the manufacturer is a dominant strategy for both retailers in the contract signing game.*

By Figures 1 and 2, condition $s > s_\gamma^C$ (s_γ^B) is met when s is large, or when $|\gamma|$ is large so that s_γ^C (s_γ^B) is small. A large s means that the demand signals of the retailers are statistically less accurate. With a large $|\gamma|$, the retailers' goods are nearly perfect complements (in Cournot competition) or nearly perfect substitutes (in Bertrand competition), and this implies, in light of Vives [35], that the leakage effect is more beneficial to the retailers.

An interesting special case is the Bertrand competition with $\gamma > 0.5$ and $s = 0$. With $s = 0$ both retailers have perfect information about the demand function and there is no leakage effect. But since $B_\gamma(0) = 2(2\gamma - 1)(2 - \gamma)^2 > 0$, condition (6.75) holds and information sharing can be achieved through side payment. In this case *the direct effect of information sharing increases the total supply chain profit.* This differs sharply from the simple analysis at the end of Section 2 with only one retailer where the direct effect reduces the total supply chain profit. This is because downstream competition interacts with and alters the incentives for vertical information sharing.

The duopoly model can be extended to the case of make-to-stock manufacturer when the demand uncertainty and the signals form a normal-normal conjugate pair. Again it can be shown that the problem of optimizing over (P, Q) has a simple, separable structure.

6.   Information about Cost Uncertainty

We now consider the case with cost uncertainty and we do so in the setting of a make-to-order manufacturer and n retailers in the downstream who are engaged in a Cournot competition selling an identical good. There is no demand uncertainty, i.e., $\theta \equiv 0$ in the demand function (6.1). We assume that each retailer's marginal cost C_i is a random variable with the same mean and a common variance $\sigma^2 \equiv Var[C_i]$. Without loss of generality, we assume that $E[C_i]$ is normalized to zero. Before making his sales quantity decision, each retailer observes his own marginal cost C_i and must decide whether to share this information with the manufacturer. We assume:

(A4) $E[C_i \,|\, C_{-i}] = \beta_i^i + \sum_{j \neq i} \beta_j^i C_j$ where $C_{-i} = (C_1, \ldots, C_{i-1}, C_{i+1}, \ldots C_n)$ and $\beta_j^i \geq 0$ for all i, j.

(A5) $C_1, \ldots, C_n$ are identically distributed.

In (A4), $\beta_j^i \geq 0$ implies that C_i's are positively (nonnegatively) correlated — a natural case to consider. Under these assumptions, for any given set $K \subseteq N$ and any $i \in N\backslash K$,

$$E[C_i \,|\, C_j, j \in K] = \frac{1}{k+s} \sum_{j \in K} C_j, \qquad (6.76)$$

where

$$s \equiv \frac{1-\rho}{\rho} \quad \text{with} \quad \rho \equiv \frac{Cov[C_i C_j]}{\sigma^2}. \qquad (6.77)$$

That is, $\sum_{j \in K} C_j$ is sufficient to predict other marginal costs C_i given $(C_j)_{j \in K}$.

6.1.   Equilibrium behavior of retailers

Let K, $|K| = k$, be the set of retailers who decide to share their information, $(C_j)_{j \in K}$. The manufacturer's best estimate of C_i for $i \in N\backslash K$ is given by (6.76). Similar to the case of demand uncertainty in Section 3, we only investigate equilibrium behavior in the subspace of outcomes in which the retailers conjecture that the price set by the manufacturer is a strictly monotone function of $\sum_{j \in K} C_j$.

In the third stage of the game, (C_i, P) is the information retailer i has when making his quantity decision, which is equivalent to $(C_i, \sum_{j \in K} C_j)$. The expected profit for retailer i, given his information, is

$$E[\pi_i \,|\, C_i, P] = \left(a - q_i - \sum_{j \neq i} E[q_j \,|\, C_i, P] - C_i - P \right) q_i.$$

His equilibrium strategy $q_i^*(C_i, P)$ must satisfy the first-order condition,

$$2q_i^* = a - C_i - P - E\left[\sum_{j \neq i} q_j^*(C_j, P) \,|\, C_i, P \right]. \qquad (6.78)$$

Proposition 6 of Li [22] shows that there is a unique equilibrium solution to the game if $(C_j)_{j \in K}$ is public information. In fact, that equilibrium solution also solves the last stage of our game in which only $\sum_{j \in K} C_j$ is publicly known. The equilibrium strategies are,

$$q_i^*(C_i, P) = \frac{1}{n+1}\left(a - P + A_1^k \sum_{j \in K} C_j + A_2^k C_i\right) \quad \text{for } i \in K, \quad (6.79)$$

$$q_i^*(C_i, P) = \frac{1}{n+1}\left(a - P + B_1^k \sum_{j \in K} C_j + B_2^k C_i\right) \quad \text{for } i \in N \backslash K. \quad (6.80)$$

where

$$A_1^k \equiv \frac{n+s}{k+s},$$

$$A_2^k \equiv -(n+1),$$

$$B_1^k \equiv \frac{(n+s)(k+2s)}{(k+s)(n+k+1+2s)},$$

$$B_2^k \equiv -\frac{(n+1)(k+1+s)}{n+k+1+2s},$$

and s is defined in (6.77). When $K = \emptyset$, $\sum_{j \in K} C_j = 0$ and each q_i^* only depends on the retailer's own signal. The coefficients A_1^k, A_2^k, B_1^k and B_2^k have the following property:

$$(k+1)A_1^k + A_2^k + (n-k)\left(B_1^k + \frac{1}{k+s}B_2^k\right) = 0. \quad (6.81)$$

Proposition 6.15 *Given any information sharing agreements reached in the first stage of the game, i.e., a set $K \subseteq N$, and the manufacturer's price P set in the second stage of the game, there is a unique equilibrium to the third-stage subgame, and the equilibrium strategies are given in (6.79) and (6.80).*

6.2.   Equilibrium price of the manufacturer

Using (6.79), (6.80) and (6.81), we can express the upstream market demand conditional on the manufacturer's information as

$$E\left[\sum_{i \in N} q_i^*(C_i, P) \mid (C_j)_{j \in K}\right] = \frac{1}{n+1}\left(n(a - P) - A_1^k \sum_{j \in K} C_j\right).$$

In the second stage of the game, the expected profit for the manufacturer, given her information $(C_j)_{j \in K}$, is

$$E[\pi_M \mid (C_j)_{j \in K}] = \frac{n}{n+1}(P - c)\left(a - P - \frac{A_1^k}{n} \sum_{j \in K} C_j\right), \quad (6.82)$$

and so the equilibrium price in the upstream market is

$$P^*((C_j)_{j \in K}) = \frac{a+c}{2} - M_1^k \sum_{j \in K} C_j, \qquad (6.83)$$

where

$$M_1^k \equiv \frac{A_1^k}{2n}.$$

6.3. Information sharing

Substituting (6.83) into (6.79) and (6.80), we obtain the equilibrium sales quantity of each retailer,

$$q_i^*(C_i, P((C_j)_{j \in K})) = \frac{1}{n+1} \left(\frac{a-c}{2} + (A_1^k + M_1^k) \sum_{j \in K} C_j + A_2^k C_i \right)$$

for $i \in K$, and

$$q_i^*(C_i, P((C_j)_{j \in K})) = \frac{1}{n+1} \left(\frac{a-c}{2} + (B_1^k + M_1^k) \sum_{j \in K} C_j + B_2^k C_i \right)$$

for $i \in N \backslash K$. Then, we can compute the expected equilibrium profits for the retailers,

$$\Pi_R^S(k) = \frac{(a-c)^2}{4(n+1)^2} + \frac{\rho \sigma^2}{(n+1)^2} \left[(A_1^k + M_1^k)^2 (k^2 + ks) \right.$$
$$\left. + 2(A_1^k + M_1^k) A_2^k (k+s) + (A_2^k)^2 (1+s) \right]$$

for $i \in K$, and

$$\Pi_R^N(k) = \frac{(a-c)^2}{4(n+1)^2} + \frac{\rho \sigma^2}{(n+1)^2} \left[(B_1^k + M_1^k)^2 (k^2 + ks) \right.$$
$$\left. + 2(B_1^k + M_1^k) B_2^k k + (B_2^k)^2 (1+s) \right]$$

for $i \in N \backslash K$. Substituting (6.83) into (6.82) and taking the expectation, we obtain the expected profit for the manufacturer,

$$\Pi_M(k) = \frac{n}{n+1} \left(\frac{(a-c)^2}{4} + \rho \sigma^2 (M_1^k)^2 (k^2 + ks) \right).$$

Proposition 6.16 *The expected profit functions have the following properties:*

1. *$\Pi_M(k)$ is strictly increasing in k.*

2. *$\Pi_R^S(n) > \Pi_R^N(n-1)$. And $\Pi_R^S(k) > \Pi_R^N(k-1)$ for all $k = 1, \ldots, n$ if*

$$\rho < \frac{2(n^2 - n - 1)}{2n^2 - n - 1}. \tag{6.84}$$

These facts lead to the following conclusions:

1. *The manufacturer is better off by acquiring information from more retailers in all circumstances.*

2. *Each retailer is better off by sharing his information if all other retailers also share their information. And each retailer is better off by sharing his information in all circumstances if condition (6.84) holds.*

Therefore, complete information sharing is an equilibrium and it is the unique equilibrium if condition (6.84) is satisfied. When (6.84) is not satisfied, the only other equilibrium is the one in which no retailer shares his information.

It can be shown the direct effect still discourages the retailers from sharing their cost information with the manufacturer. However, retailers benefit from the information "leakage" and the benefit outweighs the harm of direct effect. Hence, complete information sharing is a Nash equilibrium. Information dissemination is a unique equilibrium only when the correlation between retailers' marginal costs is relatively small or the number of retailers is relatively large. In this case, information sharing can be achieved on a voluntary basis and no side payments are needed.

6.4.   Social benefits

We now examine the impact of vertical information sharing on the expected total supply chain cost, total supply chain profit, consumer surplus and total social benefits, defined as

$$C(k) \equiv E\left[\sum_{i \in N}(C_i + c)q_i^*\right],$$

$$\Pi(k) \equiv E[(a - D)D] - C(k),$$

$$CS(k) \equiv \frac{1}{2}E\left[D^2\right],$$

$$W(k) \equiv \Pi(k) + CS(k),$$

respectively, where $D \equiv \sum_{i \in N} q_i^*$ is the total sales quantity. We can show that

$$\Pi(n) > \Pi(0) \quad \text{and} \quad CS(n) < CS(0).$$

We can further show that $C(n) < C(0)$ if and only if (6.84) holds, and that the same condition also implies $W(n) > W(0)$ for $n > 2$.

Proposition 6.17 *Complete information sharing increases the expected total supply chain profit although it reduces the expected consumer surplus. Complete information sharing reduces the expected total supply chain cost and increases the expected social benefits if condition (6.84) is satisfied.*

Since the total supply chain profit always increases when information is shared, it is possible to find a contract scheme like the one in Proposition 6.7 in which the manufacturer pays each retailer for his information sharing arrangement so that all firms are better off with complete information sharing. This way, even when condition (6.84) is violated, information sharing still occurs, perhaps at the expense of social benefits. On the other hand, when condition (6.84) holds, i.e., when the correlation between retailers' marginal costs is relatively small or the number of retailers is relatively large, information trading is also socially preferable.

7.   Conclusions and Future Research

The key message delivered in this article is that information sharing in a supply chain should not be studied in isolation, namely, restricted to the gains and losses to the parties between which the information is exchanged. The other firms which are not directly involved (e.g., competitors) may react to the information sharing activity, and such reaction may change the strategic interaction and cause additional gains or losses to the parties involved. The shared information may be leaked because other retailers may be able to infer the manufacturer's information from her observable actions. So the manufacturer in this setting plays the role of an outside agency who transmits information back to other competing retailers.

The interaction between vertical information sharing and horizontal competition alters the nature of both. First, vertical information transmission between a downstream retailer and the upstream firm may cause other retailers to change his strategy due to information leakage. Second, horizontal competition may alter the effect of and incentives for vertical information exchange. For example, downstream competition may create an opportunity for information sharing with payments while the same opportunity does not exist in the absence of retailer competition.

We have illustrated these points through various models. In all cases, the effect of vertical information sharing on competition is manifested by information leakage. Sometimes, the leakage effect works in the same direction as the (always negative) direct effect to discourage the downstream firms from passing back their information. At other times, the leakage effect benefits the retailers. When the uncertainty is about downstream costs, the benefit of information leakage may even outweigh the direct effect to encourage the sharing of information.

When vertical information sharing is not possible on a voluntary basis, we identify conditions under which it can be achieved in a noncooperative contract signing game. These conditions are model dependent. For example, when there is an identical good and the downstream competition is Cournot, vertical sharing of demand information will be easier to achieve if the retailers' information is statistically *more* valuable. On the other hand, when there are exactly two retailers and their goods are differentiated, information sharing can be achieved if the retailers' information is statistically *less* valuable. We also show how prices for information trade are determined to achieve a "fair" division of gains from information exchange.

Interestingly, vertical information sharing may not always increase the social benefits. This is quite different from the result in the literature of information sharing in oligopoly, which says that horizontal information sharing alone is always socially preferable. In particular, we show that vertical information sharing will reduce the social benefits in the case with demand uncertainty but may increase the social benefits in the case with cost uncertainty. In our settings, when information sharing is voluntary (without side-payments), the equilibrium behavior of the firms is often socially preferable.

Our models also capture the gain from vertical information sharing to the manufacturer due to reduced inventory and shortage costs. This benefit provides increased incentives for firms in a supply chain to share information and makes information sharing activities more socially desirable.

The approach we have used should be applicable to more general situations. Interesting extensions include competition in the upstream market (multiple suppliers), forward information sharing from suppliers to retailers, endogenizing the leakage effect in voluntary disclosure of accounting information, etc.

In this article, we adopt a strict non-cooperative game approach to study incentives for vertical information sharing. This is an important first step for further investigation of how to coordinate other supply chain decisions together with information sharing arrangements. Essentially, the results in this article show what the consequences would be if the effort of coordination and cooperation fails, the *status quo* in any cooperative game. For example, we may look into information sharing through a revenue sharing contract by which the manufacture can choose a price P to induce system-optimal behavior of the retailers. A related issue is how to induce truthful reporting of demand or cost signals.

A weakness of our model is assumption (A2) of linear conditional expectations on which the quoted oligopoly results on information exchange rest. This assumption limits the improvement of forecast accuracy from additional signals. Malueg and Tsutsui [28] shows that, in the linear-conditional-expectations duopoly framework, at most half of the forecast error can be eliminated through information exchange. It will be interesting to examine the leakage effect under other distributional assumptions.

Another limitation of our model is that the uncertainty is only about the intercept of the demand function. Malueg and Tsutsui [27] studies the duopoly information exchange when the slope of the demand function is unknown and obtain very different results from those for the case of intercept uncertainty. One may certainly study vertical information exchange along this line.

8.   Bibliographical Notes

Many recent papers study the direct effect of vertical information sharing where the vertical parties involved are insulated from horizontal competition. This line of research includes Bourland et al. [4], Chen [10], Gavirneni et al. [18], Lee et al. [19], Aviv and Federgruen [2], and Cachon and Fisher [7]. The reported benefits of information sharing include improved ordering function, better inventory allocation, etc. Such benefits are due in part to the fact that information sharing mitigates the information distortions along the vertical linkages and results in lower inventory and/or shortage costs (see Lee et al. [20]). However, the majority of the papers use a serial system that is isolated from horizontal competition.

Even in the papers that do use a system with one supplier and multiple retailers (e.g., Cachon and Fisher [7]), or one production center and multiple markets (Anand and Mendelson [1]), the demand for each retailer or in each market is assumed to be independent — each retailer is provided with an exclusive territory — and hence there is no competition among the retailers. Cachon and Zipkin [9] study the inventory competition through stocking policies in a supply chain but their study is limited to the single-retailer situation. Cachon [6] considers the stocking competition in a two-echelon supply chain with one supplier and many retailers where each retailer faces an independent Poisson demand. In his model, a retailer's cost only depends on her own reorder point and the reorder point of the supplier and hence there is no direct competition among the retailers.

The above papers only capture the "direct effect" of vertical information sharing but ignore the impact of such activities on horizontal competition. It appears that the role of horizontal competition has not received sufficient attention from supply chain researchers and certainly has not been carefully studied in the context of vertical information exchange.

There is a related literature in economics on the horizontal effect of information sharing in oligopoly. Pioneered by Novshek and Sonnenschein [29] and followed by Clarke [5], Vives [35], Gal-Or [14, 15], Li [22], Shapiro [34] and Raith [33], this body of research studies whether a firm has incentives to share its private information with its competing firms in an oligopoly, i.e., horizontal information sharing among competitors. These papers examine how disclosure of the private information of a firm to its horizontal competitors would affect the profitability of all firms involved. In the stereotypical model with demand uncertainty, for example, an oligopoly of firms produce products that are either substitutes or complements, facing linear demand functions with the same, but a priori unknown, intercept. Before deciding its output quantity or sales price, each individual firm privately observes a noisy signal of the true value of the intercept. Whether the firms have incentives to voluntarily disclose their private demand information depends on model specifics (Cournot or Bertrand competition; substitute or complementary products). We quote two examples here. Li [22] establishes, for the case of Cournot oligopoly with one identical product and many retailers, that disclosing the private demand information has a negative impact on a firm's profit. Vives [35] establishes the following result: in a Cournot duopoly, if the firms' goods are substitutes, disclosing the demand information to the other party hurts a firm's profit, but if the firms' goods are complements, disclosing the demand information improves a firm's profit; in a Bertrand duopoly, the situation is reversed.

One ought to be very careful when trying to draw conclusions about the leakage effect of vertical information sharing on the basis of the above studies alone. These studies, in the absence of interaction with vertical parties, can only serve as a benchmark for the leakage effect.

There are other papers in the supply chain literature on the (competitive) interaction between firms. Most of these papers first show that competition hurts the supply chain performance and then propose coordination mechanisms that pull the supply chain performance closer to the overall optimum. Cachon and Lariviere [8] examine the value of demand forecast from the downstream in helping the upstream firm make better capacity decisions in the context where the credibility of the forecast is a concern. They study contracts that allow the supply chain to align incentives and share demand forecasts credibly. Corbett and Tang [11] study the designing of supply contracts when the retailer's internal marginal cost is unknown to the supplier. They find supply contracts that would maximize the supplier's profit and examine the interaction between the type of contract and the supplier's knowledge about the buyer's cost structure.

The vertical contracting model of Gal-Or [16] incorporates both the demand uncertainty and cost uncertainty in a single-manufacturer single-retailer situation. She finds that, in general, neither the franchise fee pricing (wholesale price plus fixed franchise fee for the right to sell the product) nor retail price maintenance (supplier forces a particular retail price as part of the contract) can achieve the vertically-integrated solution under information asymmetry. Gal-Or [17] also studies the situation of two duopoly suppliers, each having an exclusive retailer, with a model that essentially consists of two competing supply chains.

Padmanabhan and Png [30] consider the impact of two factors — retail competition and demand uncertainty — on a manufacturer's decision whether to accept returns. In their model the manufacturer behaves like a Stackelberg leader and there are either multiple retailers with no demand uncertainty or a single retailer with demand uncertainty. They assume that at all times the manufacturer and the retailer have equal information (so there are no asymmetries of information in the channel). They show that returns policies can increase the retail competition and that there is a trade-off between the benefit to the supplier due to more intensive retail competition and the costs to the supplier of excess stocking.

Also related to our work is the considerable research that deals with simultaneously choosing an optimal stocking quantity and a price in the newsvendor setting with price sensitive random demand. Whitin [36] was the first to formulate a newsvendor model with price effects. Porteus (1990) and Pertuzzi and Data [32] are good sources of review for this problem. Federgruen and Heching [13] consider the dynamic version of this problem and show that a base stock list price policy is optimal. Recently Li and Atkins [26] address the issues of capacity planning and consumer pricing in the face of demand uncertainty in a decentralized supply chain.

Acknowledgments

This work is supported by the Research Grants Council of Hong Kong (Project no. HKUST6019/99H), Yale School of Management Faculty Research Fund and CEIBS Faculty Research Fund.

References

[1] ANAND, K.S. AND H. MENDELSON. (1997) Information and Organization for Horizontal Multimarket Coordination. *Management Science*, Vol. 43, 1609–1627.

[2] AVIV, Y. AND A. FEDERGRUEN. (1998) The Operational Benefits of Information Sharing and Vendor Managed Inventory (VMI) Programs, Washington University.

[3] BASAR, T. AND Y.C. HO. (1973) Information Properties of the Nash Solutions of Two Stochastic Nonzero-Sum Games, *Journal of Economic Theory*, Vol. 7, 370–387.

[4] BOURLAND, K., S. POWELL, AND D. PYKE. (1996) Exploiting Timely Demand Information to Reduce Inventories, *European Journal of Operational Research*, Vol. 92, 239–253.

[5] CLARKE, R. (1983) Collusion and Incentives for Information Sharing. *Bell Journal of Economics*, Vol. 14, 383–394.

[6] CACHON, G. (1999) Stock Wars: Inventory Competition in a Two-Echelon Supply Chain with Multiple Retailers. Duke University.

[7] CACHON, G. AND M. FISHER. (2000) Supply Chain Inventory Management and the Value of Shared Information. *Management Science*, Vol. 46, No. 8, 936–953.

[8] CACHON, G. AND M. LARIVIERE. (1999) Contracting to Assure Supply: How to Share Demand Forecasts in s Supply Chain. Duke University and The University of Pennsylvania.

[9] CACHON, G. AND P. ZIPKIN. (1999) Competitive and Cooperative Inventory Policies in a Two-Stage Supply Chain. *Management Science*, Vol. 45, No. 7, 1032–1048.

[10] CHEN, F. (1998) Echelon Reorder Points, Installation Reorder Points, and the Value of Centralized Demand Information. *Management Science*, Vol. 44, S221–S234.

[11] CORBETT, C. AND C. TANG. (1999) *Designing Supply Contracts: Contract Type and Information Asymmetry. Quantitative Models for Supply Chain Management*, edited by S. Tayur, R. Ganeshan and M. Magazine, Kluwer Academic Publishers.

[12] DEGROOT, M. (1986) *Probability and Statistics*, 2nd edition, Addison-Wesley Publishing Company.

[13] FEDERGRUEN, A. AND A. HECHING. (1999) Combining Pricing and Inventory Control under Uncertainty. *Operations Research*, Vol. 47, No. 3, 454–475.

[14] GAL-OR, E. (1985) Information Sharing in Oligopoly. *Econometrica*, Vol. 53, 329–343.

[15] GAL-OR, E. (1986) Information Transmission: Cournot and Bertrand Equilibria. *Review of Economic Studies*, Vol. 53, 85–92.

[16] GAL-OR, E. (1991A) Vertical Restraints with Incomplete Information. *The Journal of Industrial Economics*, Vol. 39 (September), 503–516.

[17] GAL-OR, E. (1991B) Duopolistic Vertical Restraints. *European Economic Review*, Vol. 35, 1237–1253.

[18] GAVIRNENI, S., R. KAPUSCINSKI, AND S. TAYUR. (1999) Value of Information in Capacitated Supply Chains. *Management Science*, Vol. 45, 16–24.

[19] LEE, H., K. SO, AND C. TANG. (2000) The Value of Information Sharing in a Two-Level Supply Chain. *Management Science*, Vol. 46, No. 5, 626–643.

[20] LEE, H., P. PADMANABHAN, AND S. WHANG. (1997) Information Distortion in a Supply Chain: The Bullwhip Effect, *Management Science*, Vol. 43, 546–558.

[21] LEE, H.L. AND S. WHANG. (2000) Information Sharing in a Supply Chain. *International Journal of Technology Management*, Vol. 20, 373–387.

[22] LI, L. (1985) Cournot Oligopoly with Information Sharing. *Rand Journal of Economics*, Vol. 16, 521–536.

[23] LI, L. (1999) Information Sharing in a Supply Chain with Horizontal Competition. Yale School of Management.

[24] LI, L., R.D. MCKELVEY, AND T. PAGE. (1987) Optimal Research for Cournot Oligopolists, *Journal of Economic Theory*, Vol. 42, 140–166.

[25] LI, L. AND H. ZHANG. (2000) Information Sharing in a Supply Chain with Horizontal Competition: The Case of a Make-to-Stock Supplier. Yale School of Management and Hong Kong University of Science and Technology.

[26] LI, Q. AND D. ATKINS. (2000) Competition and Coordination in a Capacity/Price Setting Supply Chain. The University of British Columbia.

[27] MALUEG, D. AND S. TSUTUI. (1996) Duopoly Information Exchange: The Case of Unknown Slope. *International Journal of Industrial Organization*, Vol. 14, 119–136.

[28] MALUEG, D. AND S. TSUTUI. (1998) Distributional Assumptions in the Theory of Oligopoly Information Exchange. *International Journal of Industrial Organization*, Vol. 16, 785–797.

[29] NOVSHEK, W. AND H. SONNENSCHEIN. (1982) Fulfilled Expectations Cournot Duopoly with information Acquisition and Release. *Bell Journal of Economics*, Vol. 13, 214–218.

[30] PADMANABHAN, V. AND I. PNG. (1997) Manufacturer's Returns Policies and Retail Competition. *Marketing Science*, Vol. 16, No. 1, 81–94.

[31] PALFREY, T. (1985) Uncertainty Resolution, Private Information aggregation, and Cournot Competitive Limit, *Review of Economic Studies*, Vol. 52, 69–84.

[32] PETRUZZI, N. AND M. DATA. (1999) Pricing and the Newsvendor Problem: A Review with Extensions. *Operations Research*, Vol. 47, No. 2, 183–194.

[33] PORTEUS, E.L. (1990) Stochastic Inventory Theory. *Stochastic Models,* edited by D.P. Heyman and M.J. Sobel, North-Holland.

[34] RAITH, M. (1996) A General Model of Information Sharing in Oligopoly. *Journal of Economic Theory*, Vol. 71, 260–288.

[35] SHAPIRO, C. (1986) Exchange of Cost Information in Oligopoly. *Review of Economic Studies*, Vol. 53, 433–446.

[36] VIVES, X. (1984) Duopoly Information Equilibrium: Cournot and Bertrand. *Journal of Economic Theory*, Vol. 34, 71–94.

[37] WHITIN, T. (1955) Inventory Control and Price Theory. *Management Science*, Vol. 2, 61–18.

[38] ZHANG, H. (2000) Vertical Information Exchange in a Supply Chain with Duopoly Retailers. Hong Kong University of Science and Technology.

Chapter 7

PLANNING AND SCHEDULING IN AN ASSEMBLE-TO-ORDER ENVIRONMENT: SPICER-OFF-HIGHWAY PRODUCTS DIVISION

Nico J. Vandaele

University of Antwerp, Prinsstraat 13, 2000 Antwerp, Belgium

nico.vandaele@ufsia.ac.be

Marc R. Lambrecht

Katholieke Universiteit Leuven, Naamsestraat 69, 3000 Leuven, Belgium

marc.lambrecht@econ.kuleuven.ac.be

1.   Introduction

Spicer Off-Highway Products Division (a division of Dana Corporation) offers their customers a full range of powershift transmissions and torque convertors. The structure of the manufacturing process is typically a hybrid of make-to-stock and make/assemble-to-order. The main problem is that the production lead time for the make parts is substantial due to the large variety of components and the complexity of the production process. Consequently, a large number of manufacturing operations are performed in a highly uncertain environment inevitably resulting in mismatches at the time of the actual customer commitment.

In this chapter we report on our modeling effort to improve the management of such structures. We modeled the manufacturing system as a queueing model and used the model to analyze and evaluate improvement schemes (layout changes, product-mix decisions, lot-sizing decisions and lead time estimations). Next, we developed a (deterministic) finite scheduler to improve the detailed scheduling of the shop. Our methodology successfully links stochastic and deterministic aspects of the planning process and it combines analysis, planning and detailed scheduling. We first introduce the business case, then we give an overview of the ACLIPS approach, along the lines of the software, next we develop the model and finally we report on the benefits and the impact of our approach.

2.   The Business Case: Spicer-Off-Highway

2.1.   The company

Spicer Off-Highway Products Division (SOHPD), located in Bruges (Belgium) is responsible for the worldwide development and manufacturing of powershift transmissions, torque convertors and electronic control systems within Dana Corporation. Dana Corporation is one of the world's largest independent suppliers to vehicle manufacturers and related aftermarkets. The company operates some 330 major production facilities in 32 countries and employs more than 86,000 people. The customers of SOHPD are manufacturers of off-highway vehicles and are located all around the world. The products are applied in a wide range of vehicles such as telescopic boom handlers, asphalt machines, compacters, mining machines, cranes, railway maintenance vehicles, forestry machinery, etc. SOHPD employs 680 people and realized $150 million sales in 1999. We refer to Vandaele, Lambrecht, De Schuyter and Cremmery [27] for more details; this reference and the video of the 1999 Franz Edelman finalists also offer interesting complimentary material for several issues covered in this chapter.

2.2.   The production process

The process of producing powershift transmissions is roughly a four-step process. First are rough machining operations on raw steel parts (mainly forgings and bars). This takes place in the cold (soft) steel shop. Second, the steel parts are hardened through a heat-treatment. The third step consists again of machining operations but now on the hardened parts. This takes place in the hard-steel shop. The fourth step is a final assembly operation in which the rotating steel parts are built into housings. The housings are produced in the casting department. Our study deals with the planning of the rotating steel parts and more specifically the planning of the cold-steel shop. The orders for the cold steel shop are based on the final assembly schedule, which is derived from effectively booked customer orders. The facility layout in Figure 7.1 illustrates the production process.

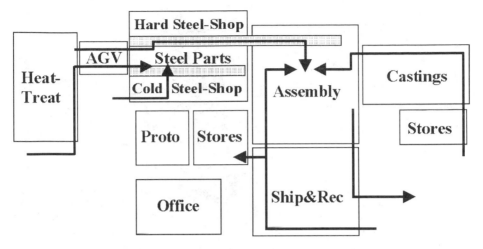

Figure 7.1. The four step production process of powershift transmissions: cold steel shop, heat-treatment, hard steel shop, and final assembly

In Figure 7.2 we give a detailed view of the cold and hard-steel shop. Material handling is done by AGV's and an AS/RS system. The AGV system moves the parts from the cold-steel shop to heat-treatment and brings them back to the hard-steel shop. The AS/RS system stores all the steel parts' work-in-process. The pallets containing the steel parts move automatically to the input slot of the machining centre when that particular operation is activated or released. Machine operators make use of a 'queue' list (called the queue manager) and determine in a myopic way the sequence of the jobs to be done. They can activate the AS/RS system, so that parts are moved automatically to the machine and subsequently moved to a temporary stock location in the AS/RS system, from which they move again to another machine centre until all operations are performed. The shop orders are generated by an ERP (Enterprise Resource Planning) system, as will be discussed later on in this chapter.

2.3.   The planning process

With respect to the customer, SOHPD follows an assemble-to-order strategy. It results in a final assembly schedule (FAS), and the material requirements routines translate the FAS into requirements for the hard-steel shop and the heat-treat department. The planned order releases of the heat-treat department in turn determines the production requirements for the cold-steel shop, stating both quantities and

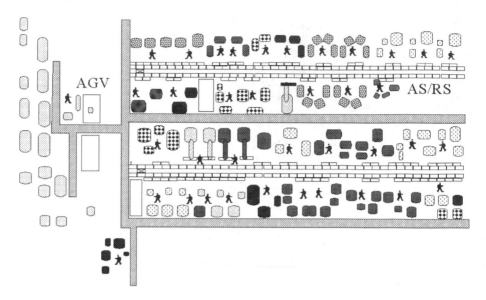

Figure 7.2. Detailed view of the hard (top part) and cold (lower part) -steel shop, the AGV (Automative Guided Vehicle) and the AS/RS system (Automatic Storage and Retrieval System). All machines are arranged around two high-stacker cranes. At the start of the project, machines were typically arranged in a job shop layout. Later on in the project management decided to adopt cellular manufacturing

due dates. The cold-steel shop and the heat-treat department typically operate in a make-to-stock environment. After heat-treat the plant switches to an assemble-to-order mode. In Figure 7.3 we summarize the lead time structure.

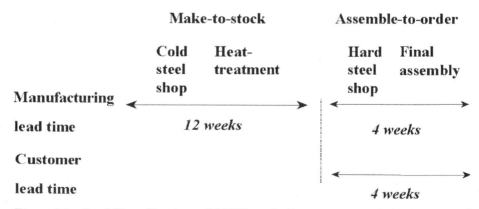

Figure 7.3. Lead Time Structure: SOHPD typically operates in a hybrid system of make-to-stock and assemble-to-order

The decoupling point is after heat-treatment, before this process the components are build to stock, whereas the end product is assembled to order. At the start of our project (1993), the total manufacturing lead time equalled 16 weeks, including the time needed to assemble the end product which was equal to 4 weeks on average. This lead time structure is extremely important for a good understanding of the complexity of our business case. The very first manufacturing operations have to be initiated up to 3 months before a final customer commitment is made. In other words, there is an considerable high level of uncertainty. Add to that the phenomenon of product variety and it is easy to understand the complexity of the manufactruing environment. SOHPD assembles hundreds of different end products, 80% of these end items have an annual volume of less than 50 units; the mode correpsonds to an annual volume of 5 units. See Figure 7.4 for more detail.

This automatically results in an enormous variety of make parts, mostly produced in small volumes. Long lead times and a large product variety result in substantial mismatches. The costs of these mismatches are considerable: both excess stock and shortages of components, frequent rescheduling, inefficiencies in production, missed due dates, etc.

To reconcile the long lead times with the need to respond to orders on time, we have to rely on forccasted demand. Most customers of SOHPD communicate information about future purchases on a rolling horizon basis. SOHPD's management team kept track of all these forecasts and

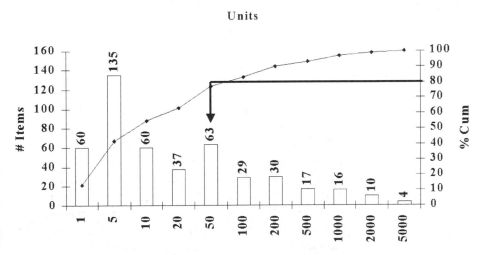

Figure 7.4. The product variety at SOHPD

Average Deviation

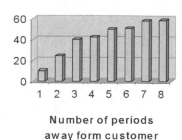

**Number of periods
away form customer**

Figure 7.5.   Forecast accuracy

its updates for the period 1993–1998. Based on the data we were able to compute the average deviation of a forecast provided by the customer compared to their actual purchase quantity and this as a function of the number of periods away from final customer commitment; e.g. forecasts made 3 months in advance of the real commitment deviate on average by 40% compared to the true quantity purchased. Forecasts made 2 months in advance deviate by 24% and the forecasts made 1 month in advance only deviate by 10%. Figure 7.5 summarizes the data.

Rolling horizon updating of forecasts is in many businesses a standard procedure. It is clear that the manufacturing lead time of the 'make-to-stock' part of the hybrid production structure (at the start of the project in 1993, this equalled 3 months on average) is the main determinant for the forecast accuracy. If we can reduce the lead time to one or two months, the impact on parts inventories will obviously be substantial, being beneficial to all the problems mentioned before.

2.4.   The key problem

The above description of the busines case clearly shows that a lead time reduction program for the make-to-stock part of the business is at order here. We therefore focussed our attention on the cold steel shop and the heat treat department. We need a tool that is able to address all lead time related parameters such as lot sizes, the capacity structure, degree of commonality, layout, outages, order acceptance, forecast accuracy, product design, etc. A queueing network approach was suggested. We also wanted to complement the queueing model with a finite capacity scheduling system to remedy the scheduling problem of the cold steel shop. All of this resulted in the development of ACLIPS: A Capacity and Lead Time Integrated Procedure for Scheduling.

3. The ACLIPS Approach

ACLIPS is a methodology in which we link separate applications into an integrated planning and scheduling system. It takes the realities of a typical job shop environment into account: utilisation, stochasticity, variability, part variety, heterogeneity, complexity, managerial decision making, etc. Products follow a particular routing throughout the shop and orders are characterised by a due date and a quantity. For shop efficiencies, orders are grouped into manufacturing orders, for which the lead time characteristics can be obtained. This information is used to obtain optimal production lots minimizing overall product lead time. Subsequently, after lead time off-setting, release dates are calculated. For given release dates and due dates, a detailed scheduling procedure is invoked to produce robust production schedules.

In this section we first describe the overall ACLIPS approach and second we detail the various steps of the procedure based on a discussion of the graphical user-interface of ACLIPS.

3.1. The overall ACLIPS approach

Basically, ACLIPS consists of three important decisions. The first decision is a lot sizing decision. Individual orders for the same product are grouped into manufacturing orders. These manufacturing lot sizes are the outcome of a queueing model. The whole production system is modeled as a queueing network, in which all operations (and its parameters) and arrival streams are stochastically represented. The outcome of this exercise is a list of lot sizes per product which minimize the overall aggregate expected lead time. Although there are many multi-product, multi-machine queueing networks described in the literature (e.g. Leung and Suri [25], Bitran and Tirupati [8], Whitt [38], [37]), we developed our own network approach which is described in detail, including examples, in Lambrecht, Ivens and Vandaele [26]. The main idea of the queueing network approach is the convex relationship between lot sizes and lead times (see e.g. Karmarkar [17], Lambrecht, Chen and Vandaele [23], and Lambrecht and Vandaele [24]). This phenomenon is particularly exploited as we determine the manufacturing lot sizes which minimize the overall aggregrate lead time of the entire shop. By using queueing models we quantify the product lead times, especially the waiting times at each resource. Therefore input data are necessary concerning shop orders (quantity and timing of order arrivals) and production parameters (setup and processing times, routings). The queueing network provides target lot sizes which give an indication of how orders have to be grouped into manufacturing orders. Recall that the orders are based on MRP calculations, which themselves are derived from forecasted orders for the end-products which is materialzed in the Final Assembly Schedule. As stated before, the importance of accurate forecasts can hardly be underestimated. We group orders in such a way that we approach the target lot sizes as close as possible. Given the time varying nature of the customer demands (the ungrouped net MRP requirements), the lot size (number of units in a manufacturing order) may actually differ from manufacturing order to manufacturing order, but on the average we aim for lot sizes minimizing the expected lead time and expected work-in-process inventories.

The second major decision is the determination of the release dates of the manufacturing orders. The release date is set equal to the due date minus the lead time estimate of the manufacturing order (a grouping of MRP requirements). The estimate of the lead time is equal to the expected lead time plus a safety lead time. The safety lead time depends on the desired customer service. The lead time estimate is such that we expect to satisfy orders on time, in P% of the cases, based on knowledge about the lead time distribution. Therefore, an estimate of the variance of the lead time together with a postulated distribution is mandatory. One of the major managerial challenges is the determination of the amount of safety time to be incorporated. A substantial amount of safety time will increase customer service but at the price of long lead times and thus choking in-process inventories and demanding for forecasts of more remote future periods. If there is not enough safety time, customer service will suffer while scheduling (see 4.2) will become more cumbersome. Again the important relationship with forecasts is predeterminant: the forecast deviation depends heavily on the length of the forecast horizon.

The third major decision concerns the sequencing policy. The release dates obtained together with the given due dates form time windows within which all operations can be scheduled in detail. Note that the length of the scheduling window is equal to the average lead time plus the safety time. Whereas any good scheduling engine would suffice, we opted for the shifting bottleneck procedure (Adams, Balas and Zawack [1]) for various reasons, one being its excellent and efficient performance as described by Ivens and Lambrecht [16]. Of course the shifting bottleneck procedure has to be adapted so that it can be used to sequence the operations for our general job shop environment including assembly operations, release dates, due dates, overlapping operations, multiple resources (machines and labour force), setup times, calendars and many other real life features. The ESBP (Extended Shifting Bottleneck

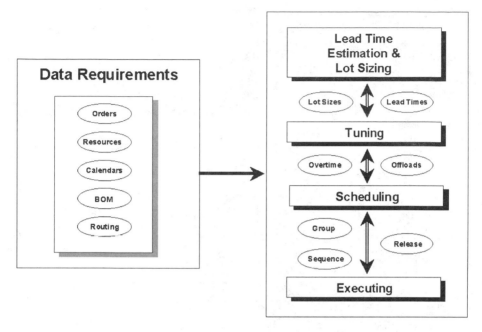

Figure 7.6. The four phase hierarchical approach of ACLIPS

procedure) is described in section 4.2. Now it can be understood that the scheduling window (and thus safety time) determines how many manufacturing orders will meet their due date.

This methodology, based on three major decisions (lot sizing, the release date setting and the detailed production sequence) is embedded in a four phase, operational implementation scheme as summarized in Figure 7.6.

The four phases (and the associated data requirements) are described in the next section.

3.2. The ACLIPS software

3.2.1 The business data. In order to obtain a realistic model of the job shop we need data describing the details of the shop. Order information, as can be seen in the '**Orders**' window (Figure 7.7), consists of the requirements for the various parts manufactured in the shop. In the '**Order Detail**' window, the order identification number (order code '**AUQ9540**'), together with a list of parts ordered, is provided. Very important information is laid down in the due date and the quantity required. From these data, the average order quantities and order arrival processes (in terms of the average and the variance of

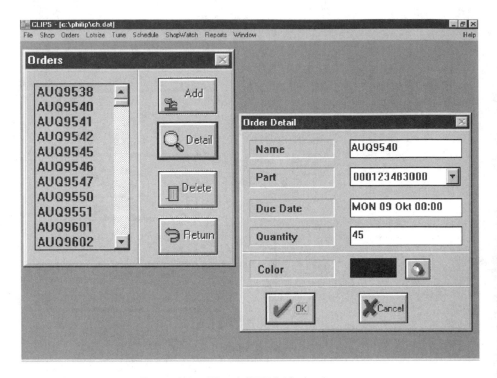

Figure 7.7. The ACLIPS 'Orders' screen

order interarrival times) can be derived. Taking all orders for all parts together, this consitutes the total load for the cold steel shop for the period studied. Typically, within SOHPD, this period covers some 6 months ahead.

The shop consists of various resources, accessible through the '**Resources**' window in Figure 7.8. Besides a small search engine, it shows a list of the available resources (e.g. '**264 - DEBUR CAE**') allowing access to a '**Resource Details**' window. Along descriptions and the like, important capacity related data is provided here: the number of units available, the resource setup characteristics and the possibility to link the resource to calendars.

Calendars are extremely important, because the availability of the production capacity of the shop is determined by the 'active calendar'. In ACLIPS, a resource is subject to the overall shop calendar(s) (describing shifts, collective holidays, weekends, planned breaks, etc.), the resource calendar(s) (specifying additional shift information such as overtime, overhaul downtime, resource specific unavailabilities, etc.) and the unit calendar(s) (for unit specific time tables. The 'active calendar' for a particular unit of a resource type is obtained by overlaying all the applicable

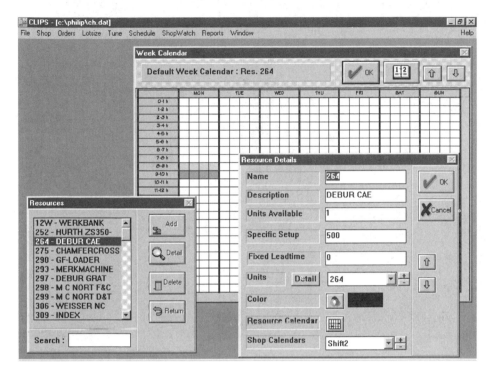

Figure 7.8. The ACLIPS 'Resources' window

calendars (shop, resource and unit) and taking the intersection to obtain the effective availability of the unit. For instance, in Figure 7.8, the '**Default Week Calendar for resource 264**' is given. As an example, the '**Week Calendar**' in Figure 7.8 shows that production is allowed on monday from 8.00 until 9.00 am, production is not allowed from 9.00 until 10.00 am while all other time slots for the week are open for allocation of overtime.

In Figure 7.9 the '**Shop Calendars**' window is given showing a typical one shift calendar: five days (monday through friday) from 5.30 am. up to 13.30 pm. Note the quarter break from 9.30 to 9.45.

In the '**Parts**' window, the necessary data concerning the parts are given (see Figure 7.10). A list of part numbers allows access to the '**Part Details**' window, which gives a list of all the '**Processes**' necessary to process the part. In this way, the sequence of the processes constitute the routing of the part. If a particular process needs other parts, these are listed in the '**Components**' list box. Components are in fact the next level parts in the Bill Of Material. Note that in our case the number of parts composed of assembled parts is very limited. More important for our case at hand are the data provided by the '**Process**

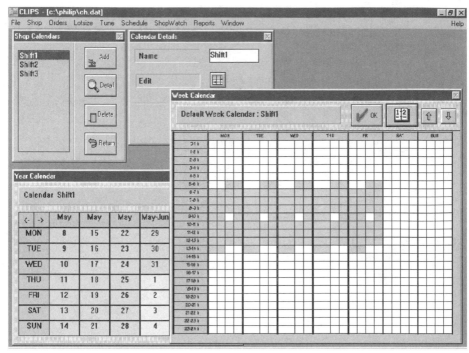

Figure 7.9.   The ACLIPS 'Calendars' window

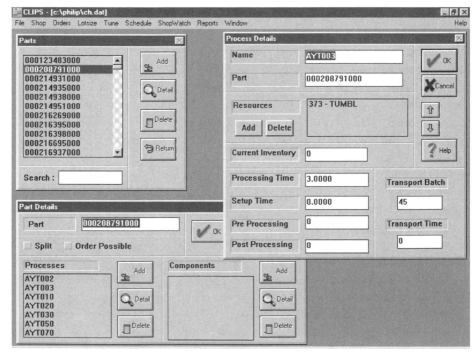

Figure 7.10.   The ACLIPS 'Parts' window

Details'. This pop up window describes the strictly necessary process characteristics such as the unit processing time and setup time. Other information includes the time needed for both 'pre processing' (e.g. loading) and/or 'post processing' (e.g. unloading, cooling), transport time and transport batch. As an example, process '**AYT003**' on part '**000208791000**' is processed on resource '**373 - TUMBL**', with zero setup time and 3 minutes process time per unit; the transport batch equals 45 units.

For planner/user convenience, a visual representation of the routing is desirable. This can be seen in Figure 7.11 with the '**Bill of Processes**' for part number '**000123483000**'. All operations are listed in their logical order, including the strictly necessary data on process and setup times. In Figure 7.11 the seven operations for part number '**00012348300**' are listed from bottom to top. The first process '**AUQ002**', takes 16 minutes per unit without a setup time.

It should be clear that at this point all necessary data are available to activate the lead time estimation and lot sizing routine for which the theoretical details are given in 4.1. In this phase the manufacturing system is transformed into a queueing network (based on estimated order

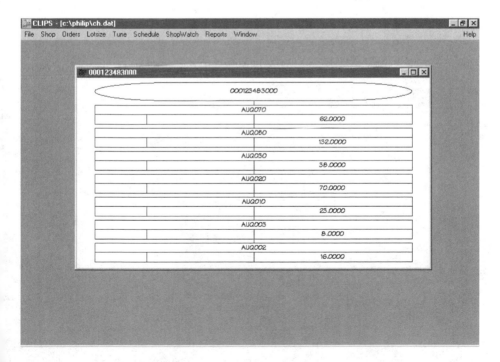

Figure 7.11. The ACLIPS 'Bill Of Processes' window

rates, the various resource types, calendars, routings, etc.). This phase results in lot sizes and lead time estimations.

3.2.2   The tuning interface.   The second step is a tuning phase. In this phase management intervention is required. Management may consider the lead times as unacceptable. To remedy the situation, management may decide to adjust the capacity structure (e.g. overtime, capacity expansion), to off-load heavily loaded resources, to consider alternative routings, etc. The adjustments may result in a new run of the queueing model. The queueing model is an excellent tool to respond quickly to these requests. We found out that this capability of the model was one of the key success factors of the ACLIPS implementation. The queueing model can be used to evaluate the impact of a cell layout on lead times. Indeed, management decided later on in the project, to implement cellular manufacturing. Analysis of the product mix (unique product, repeat orders, large and small volume orders) is another interesting issue that can be tackled by the queueing model. Therefore analyzing tools are provided such as a visual representation of the order patterns and an overview of the total demand volumes over the reference period. Both give evidence about the arrival process of the product: both timing and quantities can be of interest. In Figure 7.12 a list of the total

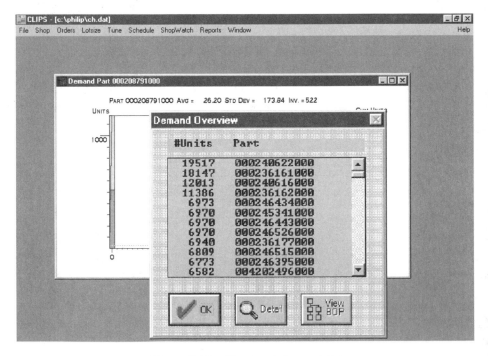

Figure 7.12.   The total demand list

demand quantities can be seen for all products. Highlighted is product **000240622000** for which 19517 units are required. As already stated, these high volumes are very rare at SOHPD. This list serves as a navigator through the products. There are direct links with the product details (see Figure 7.7 and the bill of processes (see Figure 7.11).

In Figure 7.12 also the demand pattern over a reference period is shown for product '**000208791000**'. In this way one can analyse the demand (arrival) pattern including average and variances of order interarrival times and the average and variance of the demand quantities. Also the current inventory on hand can be visualized. Both the demand time series and the cumulative demand can be viewed. This window provides a quick look on the (ir)regularity of the demand pattern, one of the major concerns for controlling lead times.

The tuning phase is an essential intermediate stage before diving into finite scheduling. Phases one and two are typically executed once a month. The actions to be taken here depend upon the practical situation at hand. The ACLIPS software provides some nice graphical tools, with which the user can analyse both the data and the outcomes for the job shop. These include a very efficient search engine to query the database concerning manufacturing orders (lots), orders, parts, processes and resources. A view on the 'Plan Search Filter' is shown in Figure 7.13. The intention is to guide the user through the most logical links in the database. This can save a substantial amount of 'tuning' time. In Figure 7.13 a typical search can be seen: for part number **000123483000** the production lots are listed (**AUQ9510,..., lot1,...**. These lots have to be processed in a number of steps: the processes (**AUQ002,...**) on various resources (**252 HURTH ZS350**).

Another interesting graphical tool allowing the user to analyze the various utilisations of the resources can be seen in Figure 7.14. For a selection of resources, the overall utilisation is shown graphically where the different shade partitions indicate different utilisation modes: processing, setup, down, etc. It speeds up the identification of heavy loaded (overloaded) resources. As utilisation is one of the main determinants of long lead times, high utilisation levels can be countered in a focussed way, shrinking unacceptable lead times quickly. On Figure 7.14 it is clear that resources **334**, **965** and **X80** are overloaded.

3.2.3   The scheduling interface.   The next phase is the scheduling phase, including (a) the grouping of orders into manufacturing orders (approaching the target lot sizes as close as possible); (b) determining the release date for each manufacturing order (which is set equal to the due date minus the lead time estimate); and (c) the detailed sequencing

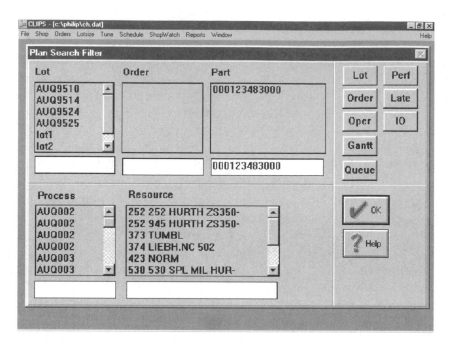

Figure 7.13.   The ACLIPS search engine

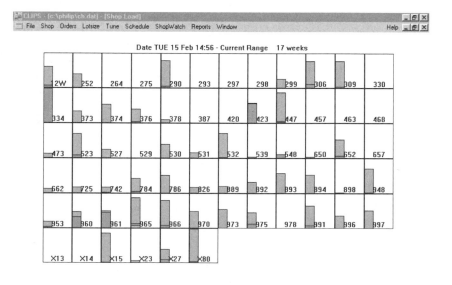

Figure 7.14.   A view on the resource utilisations

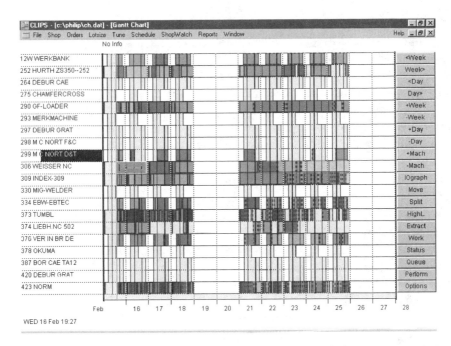

Figure 7.15. A gantt chart of the cold steel shop

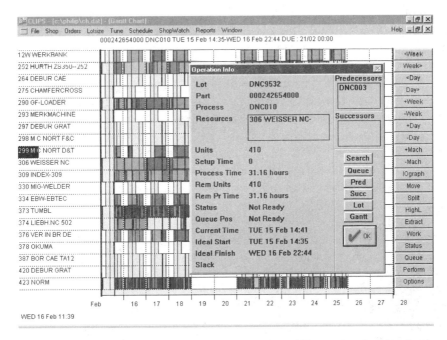

Figure 7.16. An efficient way to query management information by simple mouse moves

of all operations. These calculations are performed upon request and can be visualized in different ways.

The most common visualization of a schedule is a gantt chart. It is a list of resources on which the availability and the allocation of activities (processing, setup, down, unavailable, break, weekend, etc.) is shown on the time axis. As this is the major work document for the scheduler, a whole battery of scrollers is provided. The gantt chart can be scrolled through by day, week, machine, link with the Input/Output graph (see also Figure 7.18), options to modify the schedule manually (move, split, highlight, extract) and links to other useful information. Note that if a modification of the schedule is induced by the human scheduler, ACLIPS immediately updates the schedule and eventually indicates infeasibilities.

If a particular operation is touched, a pop up screen with additional information is shown: operation **DNC010** from lot **DNC9532** for part **000242654000**on resource **306 Weisser NC**; the lot consists of 410 units and runs for 31.16 hours. No setup is necessary. The operation has not yet started, as can be derived from **Remaining units 410** and **Remaining Processing Time**. Given the current time **Tuesday February 15, 14:41**, the process is scheduled to start on **Tuesday February 15** at **14:35** and is scheduled to finish on **Wednesday February 16** at **22:44**. Note that the lot due date is on **February 22**, at **00:00**. The predecessor of this operation is on process **DNC003**; there are no successors. Again some commonly used links are directly provided (as with queue, predecesors, successors, lot and back to gantt).

As to summarize the overall quality of a schedule, some schedule performance indicators can be consulted. For instance, a list of all late manufacturing lots is given in order of lateness. Some well known objective functions and scheduling criteria are provided: percential number of late orders, maximum and average lateness, average tardiness and average lead time, average waiting time and capacity utilization. The latter two give an idea how the schedule exploits the trade-off between lead time and capacity.

3.2.4   The execution interface.   In the final phase, the detailed plans are transferred to the shop floor on a real-time basis. As to visualize the list of manufacturing lots currently waiting for a particular machine, the screen 'Machine Queue' in Figure 7.17 gives all necessary details. All parts available for processing on resource **527 - PFAUT P400 - 528** are listed with all the relevant scheduling information. For part **000233227000** some details about the current as well as about the succeeding operations can be obtained. It is this queue of manufacturing

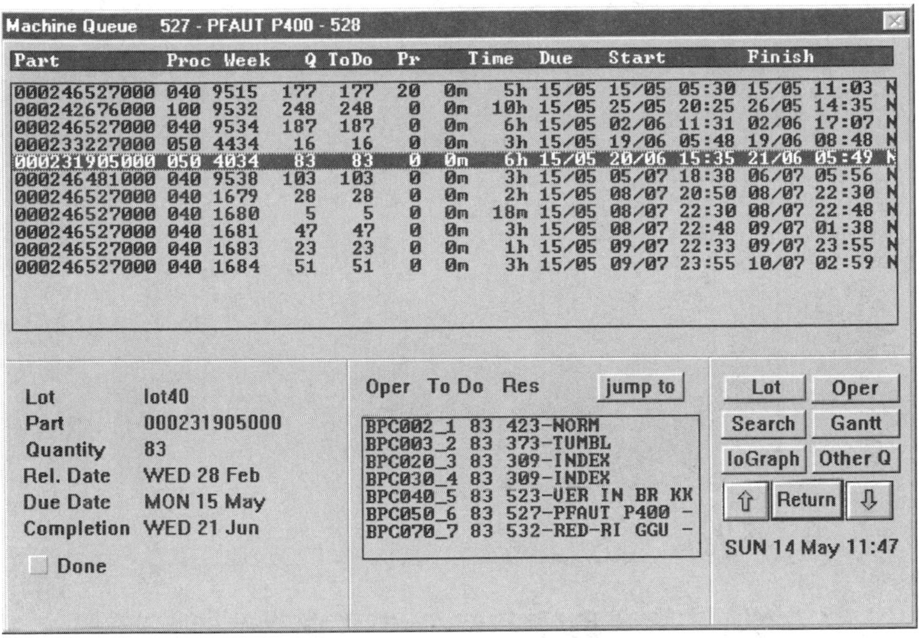

Machine Queue	527 - PFAUT P400 - 528								✕

Part	Proc	Week	Q	ToDo	Pr	Time	Due	Start	Finish	
000246527000	040	9515	177	177	20	0m	5h	15/05	15/05 05:30	15/05 11:03 N
000242676000	100	9532	248	248	0	0m	10h	15/05	25/05 20:25	26/05 14:35 N
000246527000	040	9534	187	187	0	0m	6h	15/05	02/06 11:31	02/06 17:07 N
000233227000	050	4434	16	16	0	0m	3h	15/05	19/06 05:48	19/06 08:48 N
000231905000	050	4034	83	83	0	0m	6h	15/05	20/06 15:35	21/06 05:49 N
000246481000	040	9538	103	103	0	0m	3h	15/05	05/07 18:38	06/07 05:56 N
000246527000	040	1679	28	28	0	0m	2h	15/05	08/07 20:50	08/07 22:30 N
000246527000	040	1680	5	5	0	0m	18m	15/05	08/07 22:30	08/07 22:48 N
000246527000	040	1681	47	47	0	0m	3h	15/05	08/07 22:48	09/07 01:38 N
000246527000	040	1683	23	23	0	0m	1h	15/05	09/07 22:33	09/07 23:55 N
000246527000	040	1684	51	51	0	0m	3h	15/05	09/07 23:55	10/07 02:59 N

			Oper To Do Res	jump to	Lot	Oper
Lot	lot40					
Part	000231905000		BPC002_1 83 423-NORM		Search	Gantt
Quantity	83		BPC003_2 83 373-TUMBL			
			BPC020_3 83 309-INDEX		IoGraph	Other Q
Rel. Date	WED 28 Feb		BPC030_4 83 309-INDEX			
Due Date	MON 15 May		BPC040_5 83 523-UER IN BR KK		⇧ Return ⇩	
Completion	WED 21 Jun		BPC050_6 83 527-PFAUT P400 -			
			BPC070_7 83 532-RED-RI GGU -		SUN 14 May 11:47	
☐ Done						

Figure 7.17. The queue manager for operations sequencing

lots a shop floor manager can adjust dynamically to exploit occasional opportunities or to respond to unforeseen events.

Through electronic data captation, information concerning the execution of the detailed plan, is fed back so that rescheduling can be done. It must be noted that the Gantt chart is automatically updated each 15 minutes. A rerun is of course needed to incorporate the schedule changes and messages form data capatation. An interesting on-line help for monitoring and controlling the shop floor operations is the moving cumulative Input/Output graph as can be seen for resource **12W Werkbank** in Figure 7.18. It provides the scheduler with a fast graphical interpretation of the work performed. If Input and Output are getting out of order, besides investigating for the cause, rescheduling is often necessary.

The nature and frequency of rescheduling heavily depends on the dynamics of the situation and the level of responsiveness required.

The objective of the ACLIPS approach is to obtain an integrated planning and scheduling system. Lead times are estimated through a queueing model taking into account congestion phenomena and the queueing impact of lot sizing (which results in a simultaneous treatment of both capacity and material flow). Standard lead time off-setting (as is e.g.

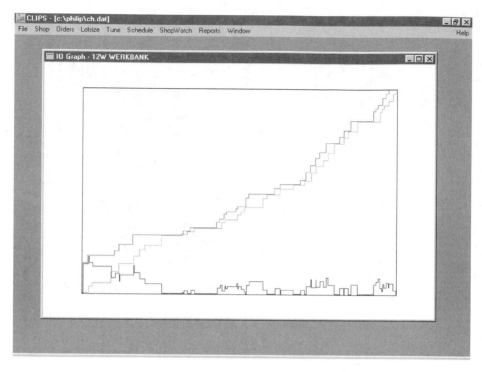

Figure 7.18. The cumulative input-output curves for a resource

done in MRP) is replaced by realistic estimates of order release dates. The lead time estimates include a safety margin so that customer service targets can be specified. The detailed real-time scheduler operates within time windows allowing to deal explicitly with the dynamics of the floor. The tuning phase allows a management intervention to cope with the capacity/inventory(lead time) trade-off.

4.   Methodological Issues

This section describes some important methodological issues. The queueing model for lead time estimation and lot sizing step is discussed in subsection 4.1 while the scheduling model is explained in subsection 4.2. We use a small example to illustrate the algorithms. This section heavily relies on Lambrecht, Ivens and Vandaele [26].

4.1.   Lead time estimation and lot sizing based on an open queueing network

In ACLIPS, the production environment is modeled as a network of queues. Our model clearly builds on well known approximations found in the literature, but our approach adds some additional features making it more suitable for practical applications. The most significant differences are explained in Lambrecht, Ivens and Vandaele [26]. Basically, our expressions for the expected lead time and the variance of the lead time are written as a function of the lot size (including setup times). This alllows us to use an optimization routine to find the optimal lot sizes for all products on all resources simultaneously. Most approximations in the literature only allow to evaluate the expected lead time for a set of predetermined lot sizes.

We now review the most relevant research contributions from the literature. First there is the approach described by Shantikumar and Buzacott [29] and Buzacott and Shantikumar [12], from whom, to the best of our knowledge, no software package is available. Other work can be found in Karmarkar, Kekre and Kekre [19], [18]. They have developed a lot sizing tool, called Q-LOTS, which is used to model a manufacturing cell (see Karmarkar, Kekre and Kekre [20]). The work done by Suri is probably the most closely related to our approach. This research is described for instance in Suri and Diehl [33] and in Suri [30]. Two software implementations are known: the first one is Manuplan which is described in Suri, Diehl and Dean [34] and in Suri, Tomsicek and Derleth [35]. A more recent version called MPX is described in Suri and DeTreville [31], [32]. Other work is done by Bitran and Tirupati [8], [9]. A software implementation called Operations Planner is described in [8]. Whitt [38], [37] described the QNA (Queueing Network Analyzer) approach. All the above mentioned approaches can be considered as queueing network implmenentaions with general applicability. Only Karmarkar describes the lot sizing issue. To the best of our knowledge, they have not published a generalized version of their work, including multiple products and multiple machines.

In our approach, equations (for the expected lead time and the variance of the lead time) are derived capturing the dynamics of the system in an aggregate way. The arrival process for each product is characterized by the expected customer demand and the average and variance of the order interarrival times. The exogenous arrival rate can be estimated from historical data or from demand forecasts or even confirmed orders depending on the availability of data. The other parameters are: the service times (average and variance of both setup and unit processing times) and shop parameters such as routings and calendars. The outcome of the model are expressions for the expected lead time and variance of the lead time as a function of the lot size. Although we rely on approximations, simulation studies (for small examples) turned out that the approximations behave satisfactory. Detailed simulation experiments

are described in Vandaele [36]. The deviations between the approximations and the simulation results are in line with other results available
in the literature. We assume a constant lot size per product over the
entire routing. Next an optimization routine is used to find the lot sizes
that minimize the expected lead time. We call these lot sizes 'target lot
sizes'. A lognormal distribution is postulated to characterize the lead
time distribution. This in turn allows the user to specify a lead time,
satisfying a predetermined customer service (lead time percentile).

Throughout the chapter a small example taken form Vandaele [36]
will be used to numerically illustrate the various steps of our procedure.
The shop, a small metal shop, is shown in Figure 7.19.

The metal shop fabricates two products, P and S, and has three machine (centers) types: a cutter (C), a grinder (G) and a lathe (L). Product P has three stages on its route (on machine C, G and L) and product
S has two stages (on machine L and G). The shop runs three shifts per
day, seven days a week. There is one machine available of each type. The
customer demands for both products are summarized in Table 7.1.

Table 7.1 is interpreted as follows: for product P we expect a order every 144 hours, while the average order size equals 3 units. The processing

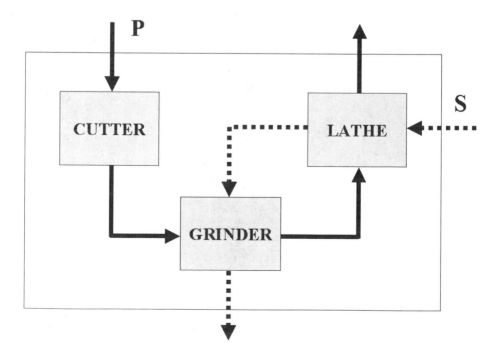

Figure 7.19.   The layout of the small metal shop

Table 7.1.   Demand characteristics for products P and S

	Product P	Product S
Average interarrival time (hours)	144	48
Variance interarrival time	3744	494
Average order quantity (units)	3	2

Table 7.2.   Production characteristics of the metal shop in hours

Product	Operations	Machine	Setup Average	Setup Variance	Processing Average	Processing Variance
Product P	3	cutter	20	0	30	0
		grinder	20	400	10	100
		lathe	24	0	12	0
Product S	2	lathe	16	0	8	0
		grinder	20	400	10	100

and setup times are summarized in Table 7.2 (all times are expressed in hours). From Table 7.2 it can be seen that both the cutter and the lathe have deterministic setup and processing times. The grinder faces exponentially distributed setup and unit processing times.

4.1.1   Model derivation.   We will now discuss the formal treatment of the lead time estimation and lot sizing phase. Assume k to be the product index ($k = 1 \ldots K$), m the machine index ($m = 1 \ldots M$) and o the operation index for product k ($o = 1 \ldots O_k$), where O_k is the number of operations for product k. Each product k is characterized by an average order quantity $\overline{OQ}_k$, an average order interarrival time $\overline{Y}_k$, the variance of the order interarrival time $s^2_{Y_k}$, the squared coefficient of variation (scv) of the order interarrival time $c^2_{Y_k}$ and the arrival rate $\lambda_k = 1/\overline{Y}_k$. For the small metal shop we assume the following characteristics

$$
\begin{array}{llll}
\overline{Y}_P & = 144 & \overline{Y}_S & = 48 \\
s^2_{Y_P} & = 3744 & s^2_{Y_S} & = 494 \\
c^2_{Y_P} & = 13/72 & c^2_{Y_S} & = 3/14 \\
\lambda_P & = 1/144 & \lambda_S & = 1/48 \\
\overline{OQ}_P & = 3 & \overline{OQ}_S & = 2
\end{array}
$$

As far as the production characteristics are concerned, the following is defined for product k and operation o, expressed in hours: T_{ko}, setup time random variable; X_{ko}, unit processing time random variable; $\overline{T}_{ko}$, expected setup time; $\overline{X}_{ko}$, expected unit processing time; μ_{ko}, unit processing rate $(= 1/\overline{X}_{ko})$; $s^2_{T_{ko}}$, variance of the setup time; $s^2_{X_{ko}}$, variance of the unit processing time; $c^2_{T_{ko}}$, scv of the setup time; $c^2_{X_{ko}}$, scv of the unit processing time. In addition, define $\delta_{kom} = 1$ if operation o for product k is on machine m and 0 otherwise. The routing of the metal shop consequently results in: $\delta_{P1C} = 1$, $\delta_{P2G} = 1$, $\delta_{P3L} = 1$, $\delta_{S1L} = 1$, $\delta_{S2G} = 1$ and all other δ_{kom}'s equal 0.

At this point all the input parameters are given. We use a queueing network approach to model the job shop. The job shop is viewed as a network of machines which are linked by the various flows (routings). Each machine is modeled as a multi product lot sizing model with queueing delays. The multiple arrival processes of the k products are superposed into one aggregate arrival process. All characteristics of the aggregate arrival process and the aggregate production process are functions of the lot sizes Q_k. Note that we express Q_k as a multiplier of the average order quantity $\overline{OQ}_k$. For each machine m we have to obtain: l_m, the aggregate batch arrival rate; ca^2_m, the scv of the aggregate batch interarrival time; ca'^2_m, the scv of the external aggregate batch interarrival time; μ_m, the aggregate batch processing rate; cs^2_m, the scv of the aggregate batch processing time; ρ'_m, the adapted traffic intensity.

The aggregate arrival process at machine m is characterized by the average and the scv of the aggregate batch interarrival times. Note that the batch arrival rate of product k at the first machine of its routing equals $\lambda_{b_k} = \lambda_k/Q_k$ which is a result of grouping the order quantities into a manufacturing batch of size $Q_k\overline{OQ}_k$ (expressed in units). The aggregate batch arrival rate of product k at machine m equals $l_{mk} = \sum_{o=1}^{O_k} \lambda_{b_k}\delta_{kom}$. Then the aggregate batch arrival rate at machine m equals $l_m = \sum_{k=1}^{K} \sum_{o=1}^{O_k} \lambda_{b_k}\delta_{kom}$ which includes both the internal and the external batch arrivals at machine m. The external aggregate batch arrival rate at machine m equals $l'_m = \sum_{k=1}^{K} \lambda_{b_k}\delta_{k1m}$. For our numerical example we obtain $\lambda_{b_P} = 1/144Q_P$, $\lambda_{b_S} = 1/48Q_S$ and $l_C = 1/144Q_P$, $l_G = 1/144Q_P + 1/48Q_S$, $l_L = 1/144Q_P + 1/48Q_S$ and $l'_C = 1/144Q_P$, $l'_G = 0$, $l'_L = 1/48Q_S$.

We now turn to the production process at machine m. The aggregate batch processing time on machine m equals

$$1/\mu_m = \sum_{k=1}^{K} \frac{l_{mk}}{l_m} \sum_{o=1}^{O_k} \frac{\lambda_{b_k}\delta_{kom}}{l_{mk}} (\overline{T}_{ko} + Q_k\overline{OQ}_k\overline{X}_{ko})$$

where l_{mk}/l_m is the probability that a randomly picked product in front of machine m is of product type k while $\lambda_{b_k}\delta_{kom}/l_{mk}$ is the probability that a randomly picked operation received by product k on machine m is operation o.

The expression for $1/\mu_m$ is a weighted average over product batch processing times, which are in turn weighted averages of the operations on machine m for the same product.

For the numerical example we obtain

$$\frac{1}{\mu_C} = 20 + 90Q_P$$

$$\frac{1}{\mu_G} = \frac{48Q_S}{48Q_S + 144Q_P}(20 + 30Q_P) + \frac{144Q_P}{48Q_S + 144Q_P}(20 + 20Q_S)$$

$$\frac{1}{\mu_L} = \frac{48Q_S}{48Q_S + 144Q_P}(24 + 36Q_P) + \frac{144Q_P}{48Q_S + 144Q_P}(16 + 16Q_S)$$

Along the same lines, we obtain the scv of the aggregate batch processing time

$$cs_m^2 = \left[\sum_{k=1}^{K} \frac{l_{mk}}{l_m} \sum_{o=1}^{O_k} \frac{\lambda_{b_k} \delta_{kom}}{l_{mk}} [\overline{T}_{ko} + Q_k \overline{OQ}_k \overline{X}_{ko}]^2 \right] \mu_m^2 - 1$$

$$+ \sum_{k=1}^{K} \frac{l_{mk}}{l_m} \sum_{o=1}^{O_k} \frac{\lambda_{b_k} \delta_{kom}}{l_{mk}} \frac{[s_{T_{ko}}^2 + Q_k \overline{OQ}_k s_{X_{ko}}^2]}{[\overline{T}_{ko} + Q_k \overline{OQ}_k \overline{X}_{ko}]^2} \qquad (7.1)$$

Applied for the small metal shop

$$cs_C^2 = \frac{(20 + 90Q_P)^2}{(20 + 90Q_P)^2} - 1 + 0 = 0$$

$$cs_G^2 = \frac{\frac{48Q_S}{48Q_S + 144Q_P}(20 + 30Q_P)^2 + \frac{144Q_P}{48Q_S + 144Q_P}(20 + 20Q_S)^2}{\left[\frac{48Q_S}{48Q_S + 144Q_P}(20 + 30Q_P) + \frac{144Q_P}{48Q_S + 144Q_P}(20 + 20Q_S) \right]^2}$$

$$- 1 + \frac{48Q_S}{48Q_S + 144Q_P} \left[\frac{400 + 300Q_P}{(20 + 30Q_P)^2} \right]$$

$$+ \frac{144Q_P}{48Q_S + 144Q_P} \left[\frac{400 + 200Q_S}{(20 + 20Q_S)_2} \right]$$

$$cs_L^2 = \frac{\frac{48Q_S}{48Q_S + 144Q_P}(24 + 36Q_P)^2 + \frac{144Q_P}{48Q_S + 144Q_P}(16 + 16Q_S)^2}{\left[\frac{48Q_S}{48Q_S + 144Q_P}(24 + 36Q_P) + \frac{144Q_P}{48Q_S + 144Q_P}(16 + 16Q_S) \right]^2}$$

$$- 1 + 0$$

When setup times are included in the machine utilization, we define the adapted traffic intensity ρ', which includes both the utilization due to setups and the utilization due to processing. The utilization without setup is the traditional traffic intensity ρ. Now we can determine the adapted traffic intensity for machine m

$$
\rho'_m = \frac{l_m}{\mu_m} = \sum_{k=1}^{K} \sum_{o=1}^{O_k} \lambda_{b_k} \delta_{kom} [\overline{T}_{ko} + Q_k \overline{OQ}_k \overline{X}_{ko}]
$$

$$
= \sum_{k=1}^{K} \sum_{o=1}^{O_k} \lambda_{b_k} \delta_{kom} \overline{T}_{ko} + \rho \tag{7.2}
$$

and applied for the metal shop

$$
\rho'_C = \frac{20 + 30Q_P}{144Q_P} = \frac{5}{36Q_P} + \frac{5}{24}
$$

$$
\rho'_G = \frac{20 + 30Q_P}{144Q_P} + \frac{20 + 20Q_S}{48Q_S} = \frac{5}{36Q_P} + \frac{5}{12Q_S} + \frac{5}{8}
$$

$$
\rho'_L = \frac{24 + 36Q_P}{144Q_P} + \frac{160 + 16Q_S}{48Q_S} = \frac{1}{6Q_P} + \frac{1}{3Q_S} + \frac{7}{12}
$$

Further define f_{0n}, the proportion of batches from outside and going to machine n, f_{mn}, the proportion of batches leaving machine m and going to machine n, f_{m0}, the proportion of batches leaving machine m and going outside, and $\mathbf{F}$, the transition matrix of f_{mn} ($m, n = 0 \dots M$).

Solving the following set of linear equations yields the M unknowns ca_m^2, $m = 1, \dots, M$:

$$
- \sum_{n=1}^{M} l_n f_{nm}^2 (1 - \rho'_n) ca_n^2 + l_m ca_m^2
$$

$$
= \sum_{n=1}^{M} l_n f_{nm} \left(f_{nm} {\rho'_n}^2 cs_n^2 + 1 - f_{nm} \right) + l'_m {ca'_m}^2 \tag{7.3}
$$

Equations (7.3) are a slightly adapted version (in terms of general exogenous arrivals instead of Poisson arrivals) of the results obtained by Shantikumar and Buzacott [29]. The entrances of the transition matrix

F are obtained as follows:

$$f_{0n} = l'_n \bigg/ \sum_{m=1}^{M} l'_m \tag{7.4}$$

$$f_{mn} = (1/l_m) \sum_{k=1}^{K} \sum_{l=1}^{O_k-1} \lambda_{b_k} \delta_{kom} \delta_{ko+1n} \tag{7.5}$$

$$f_{m0} = (1/l_m) \sum_{k=1}^{K} \lambda_{b_k} \delta_{k\,O_k\,m} \tag{7.6}$$

for $n = 1 \ldots M$ and $m = 1 \ldots M$. Note that in our model, due to the fact that the routings are given, we face deterministic routing. Therefore, the transition matrix **F** can be derived in the way described above.

Returning to the small metal shop we have the following transition matrix **F**:

	0	C	G	L
0	0	$\frac{48Q_S}{48Q_S+144Q_P}$	0	$\frac{144Q_P}{48Q_S+144Q_P}$
C	0	0	1	0
G	$\frac{144Q_P}{48Q_S+144Q_P}$	0	0	$\frac{48Q_S}{48Q_S+144Q_P}$
L	$\frac{48Q_S}{48Q_S+144Q_P}$	0	$\frac{144Q_P}{48Q_S+144Q_P}$	0

To obtain ca'^2_m we use the following approximation: if $\sum_{k=1}^{K} \delta_{k1m} \geq 2$, then

$$ca'^2_m \approx \frac{1}{3} + \frac{2}{3} \sum_{k=1}^{K} \frac{\lambda_{b_k} \delta_{k1m}}{\sum_{k=1}^{K} \lambda_{b_k} \delta_{k1m}} \frac{c^2_{Y_k}}{Q_k} = \frac{1}{3} + \frac{2}{3} \sum_{k=1}^{K} \frac{\lambda_{b_k} \delta_{k1m}}{l'_m} \frac{c^2_{Y_k}}{Q_k}$$

If $\sum_{k=1}^{K} \delta_{k1m} = 1$ then $ca'^2_m = \frac{c^2_{Y_k}}{Q_k}$.

The approximation for ca'^2_m is the sum of a constant and a weighted average of the scv's of all the external batch arrivals at machine m. It is an interpolation between complete deterministic arrivals (where the aggregate scv is approximated by the scv of a uniform distribution $U[0, 2/l'_m]$ and the scv of Poisson arrivals (where all scv's equal one). The latter is the only known exact result in the literature for the superposition of arrival processes. The weights $1/3$ and $2/3$ in the expression for ca'^2_m are a particular instance of a general approximation described by Albin [2], [3].

For our illustrative example we obtain $ca'^2_C = \frac{13}{72}\frac{1}{Q_P}$, $ca'^2_G = 0$, $ca'^2_L = \frac{3}{14}\frac{1}{Q_S}$. Then finally the lead time for product k on machine m for operation o is

$$E(W_{ko}) = \sum_{m=1}^{M} E(Wq_m)\delta_{kom} + \overline{T}_{ko} + Q_k\overline{OQ}_k\overline{X}_{ko}$$

with

$$E(Wq_m) = \frac{\rho'_m{}^2(ca^2_m + cs^2_m)}{2l_m(1-\rho'_m)}\exp\left\{\frac{-2(1-\rho'_m)(1-ca^2_m)^2}{3\rho'_m(ca^2_m + cs^2_m)}\right\} \quad \text{if } ca^2_m \le 1$$

$$E(Wq_m) = \frac{\rho'_m{}^2(ca^2_m + cs^2_m)}{2l_m(1-\rho'_m)} \quad \text{if } ca^2_m > 1$$

This approximation is based on the well-known Kraemer-Lagenbach-Belz [21] approximation, which has been tested widely in the literature (see e.g. Shantikumar and Buzacott [28]).

The aggregated objective function for machine m can be stated as follows

$$E(W_{M_m}) = E(Wq_m) + \sum_{k=1}^{K} \frac{\sum_{o=1}^{O_k}\lambda_k\overline{OQ}_k\delta_{kom}}{\sum_{k=1}^{K}\sum_{o=1}^{O_k}\lambda_k\overline{OQ}_k\delta_{kom}}$$
$$\times \left(\sum_{o=1}^{O_k}\frac{\lambda_k\overline{OQ}_k\delta_{kom}}{\sum_{o=1}^{O_k}\lambda_k\overline{OQ}_k\delta_{kom}}[\overline{T}_{ko} + Q_k\overline{OQ}_k\overline{X}_{ko}]\right)$$

This objective function for machine m is the weighted average over the products visiting machine m, which on their turn are weighted averages over the operations on machine m for product k. Note that weight $\sum_{o=1}^{O_k}\lambda_k\overline{OQ}_k\delta_{kom} / \sum_{k=1}^{K}\sum_{o=1}^{O_k}\lambda_k\overline{OQ}_k\delta_{kom}$ is independent from the manufacturing lot size. It measures the relative importance of product k for machine m. The weights are different from the ones derived by Baker [5], who uses weights which are a function of the manufacturing lot size. Because the manufacturing lot size is a decision variable, these weights can turn to zero for large lot sizes so that the respective term in the objective function can grow to infinity without penalty. This observation led us to remove the decision variable from the weights in the objective function, while the relative importance of the products is still present. In addition, simulation results showed that, as far as the optimal lot sizes are concerned, our objective (with weights which are not a function of the lot size) yield a small deviation from the lot sizes obtained by simulation.

The objective function for the total job shop becomes

$$E(W) = \sum_{m=1}^{M} E(Wq_m) + \sum_{k=1}^{K} \frac{\lambda_k \overline{OQ}_k}{\sum_{k=1}^{K} \lambda_k \overline{OQ}_k} \frac{[Q_k \overline{OQ}_k - 1]\overline{Y}_k}{2\overline{OQ}_k}$$

$$+ \sum_{m=1}^{M} \sum_{k=1}^{K} \frac{\sum_{o=1}^{O_k} \lambda_k \overline{OQ}_k \delta_{kom}}{\sum_{k=1}^{K} \sum_{o=1}^{O_k} \lambda_k \overline{OQ}_k \delta_{kom}}$$

$$\times \left(\sum_{o=1}^{O_k} \frac{\lambda_k \overline{OQ}_k \delta_{kom}}{\sum_{o=1}^{O_k} \lambda_k \overline{OQ}_k \delta_{kom}} [\overline{T}_{ko} + Q_k \overline{OQ}_k \overline{X}_{ko}] \right) \qquad (7.7)$$

The weight $\lambda_k \overline{OQ}_k / \sum_{k=1}^{K} \lambda_k \overline{OQ}_k$ takes care of the relative importance of product k for the total job shop. The second sum of equation (7.7) measures the average waiting time of finished batches until their due date. For the metal shop the objective function (as a function of the lot size multiplier Q_k) for the entire job shop equals

$$E(W) = E(W_{qC}) + E(W_{qG}) + E(W_{qL}) + 20 + 90Q_P$$
$$+ \frac{1}{3}(20 + 30Q_P) + \frac{2}{3}(20 + 20Q_S) + \frac{1}{3}(24 + 36Q_P)$$
$$+ \frac{2}{3}(16 + 16Q_S) + 8(3Q_P - 1) + 8(2Q_S - 1)$$

At this point the formulation of the job shop is complete.

4.1.2 Optimization and decomposition. The minimization problem involves a non-linear objective function and a set of simultaneous, non-linear constraints. A dedicated optimization routine has been developed to solve the problem. Our numerical experience indicates that the optimization routine always converges towards the unique global minimum although we lack a general proof of the convexity of the objective function. Convergence is accomplished by approximate gradient calculations and a backtrack procedure in case of infeasibility (constraint violation). Rounding the lot sizing variable is always done to the next higher integer, because this always guarantees feasibility (rounding to the next smaller integer could cause the utilization exceeding 100% due to a prohibitive number of setups).

The optimal lot sizes for the small metal shop are $Q_P^* \overline{OQ}_P = 4$ and $Q_S^* \overline{OQ}_S = 6$. The decomposition, after the optimization, can be summarized as follows. The optimal lot sizes Q_k^* (or the vector $\mathbf{Q}^*$) for each product are used to calculate the expected lead time of operation o of product k on machine m, $E(W_{ko}) = \sum_{m=1}^{M} E(Wq_m(\mathbf{Q}^*))\delta_{kom} + \overline{T}_{ko} + Q_k^* \overline{OQ}_k \overline{X}_{ko}$. The first term is clearly common for all products using machine m. The total lead time of product k (for the whole routing) is given by

$$
E(W_k) = \frac{Q_k^*(\overline{OQ}_k - 1)\overline{Y}_k}{2\overline{OQ}_k}
$$
$$
+ \sum_{o=1}^{O_k} \left(\sum_{m=1}^{M} E(Wq_m(\mathbf{Q}^*))\delta_{kom} + \overline{T}_{ko} + Q_k^* \overline{OQ}_k \overline{X}_{ko} \right) \quad (7.8)
$$

The numerical outcomes are summarized in Table 7.3. From this table it can be seen that there is a small queue in front of the cutter. On the other hand, both the grinder and the lathe face long waiting times compared to their processing times. This is mainly due to the high adapted traffic intensities. The waiting time for the grinder is even larger. This is due to the stochastic nature of that machine. The operation 'stock' is the average time that a particular order (from a completed manufacturing batch) has to wait until its due date.

The variance of the total lead time of product k is approximated by

$$
V(W_k) = \frac{Q_k^* \overline{OQ}_k - 1}{2\overline{OQ}_k^2} s_{Y_k}^2 + \frac{(Q_k^* \overline{OQ}_k - 1)(Q_k^* \overline{OQ}_k + 1)}{12\overline{OQ}_k^2} \overline{Y}_k^2
$$
$$
+ \sum_{o=1}^{O_k} V(Wq_m(\mathbf{Q}^*))\delta_{kom} + \sum_{o=1}^{O_k} s_{T_{ko}}^2 + \sum_{o=1}^{O_k} Q_k^* \overline{OQ}_k s_{X_{ko}}^2 \quad (7.9)
$$

Table 7.3.   Optimal lot size and lead time for the metal shop

Product	Optimal Lot Size	Operation	Adapted Traffic Intensity (%)	Waiting Time	Setup Time	Processing Time	Lead Time
P	4	cutter	73	7	20	120	147
		grinder	87	109	20	40	169
		lathe	82	42	24	48	114
		stock					72
		total					502
S	6	lathe	82	42	16	48	106
		grinder	87	109	20	60	189
		stock					60
		total					355

The term $V(Wq_m)$ is obtained as follows

$$V(Wq_m) = [E(Wq_m)]^2 cw_m^2$$

$$cw_m^2 = \frac{cd_m^2 + 1 - \sigma_m}{\sigma_m}$$

$$\sigma_m = \rho_m' + (ca_m^2 - 1)\rho_m'(1 - \rho_m')h(\rho_m', ca_m^2, cs_m^2)$$

$$h(\rho_m', ca_m^2, cs_m^2) = \begin{cases} \dfrac{1 + ca_m^2 + \rho_m' cs_m^2}{1 + \rho_m'(cs_m^2 - 1) + \rho_m'^2(4ca_m^2 + cs_m^2)} & ca_m^2 \leq 1 \\[2ex] \dfrac{4\rho_m'}{ca_m^2 + \rho_m'^2(4ca_m^2 + cs_m^2)} & ca_m^2 \geq 1 \end{cases}$$

$$cd_m^2 = 2\rho_m' - 1 + \frac{4(1 - \rho_m')ds_m^3}{3(cs_m^2 + 1)^2}$$

$$ds_m^3 = \begin{cases} \dfrac{3}{4}\left[\dfrac{1}{q_m^2} + \dfrac{1}{(1 - q_m)^2}\right] & cs_m^2 \geq 1 \\[2ex] (2cs_m^2 + 1)(cs_m^2 + 1) & cs_m^2 < 1 \end{cases}$$

$$q_m = \frac{1}{2} + \sqrt{\frac{cs_m^2 - 1}{cs_m^2 + 1}}$$

where cw_m^2, the scv of the batch waiting time, σ_m, the batch probability of delay, $P(Wq_m > 0)$, cd_m^2, scv of the conditional batch waiting time i.e. the batch waiting time, given that the server is busy and $ds_m^3 = E[S_{b_m}^3]/E(S_{b_m})^3$. For our example, the standard deviation of the total lead time is 158 hours for product P and 154 hours for product S which suggests that the lead times are highly variable. If the lognormal distribution is assumed, then the parameters are $\beta_k = \ln(E(W_k)/\sqrt{\frac{V(W_k)}{E(W_k)^2} + 1})$ and $\gamma_k^2 = \ln(\frac{V(W_k)}{E(W_k)^2} + 1)$. The lead times, including safety time, are obtained in the following way. W_{P_k} is the total lead time guaranteeing a service of $P_k\%$. This means that the manufacturer will satisfy this lead time $P_k\%$ of the time for product k. Then

$$W_{P_k} = \exp\{\beta_k + z_{P_k}\gamma_k\} \tag{7.10}$$

where z_{P_k} can be obtained from the standard normal table (P_k is the required percentile for product k). Below, we will call W_{P_k} the planned lead time, because it will be used to fix the planned release date. For our example we obtain (for some values of P_k) the planned lead times as can be seen in Table 7.4.

Table 7.4.   Some lead time percentiles

P_k	70%	80%	90%	95%	99%
Product P	484	621	710	794	980
Product S	338	463	554	644	855

4.2.   Scheduling phase

In the scheduling phase, basically three types of decisions have to be taken. First, booked orders have to be grouped into manufacturing orders, approaching the previously calculated target lot sizes as close as possible. Next, we have to establish a release date for each manufacturing order. Finally all non-completed operations of both newly released and in-process manufacturing orders have to be sequenced on the different machines. We briefly discuss each of these three steps.

4.2.1   Grouping of orders into manufacturing orders.   The problem addressed here is the grouping of C_k orders of product k, characterized by an order quantity OQ_{kc} $(1 \le c \le C_k)$ and a due date DD_{kc} $(1 \le c \le C_k)$, into a number of manufacturing orders $L_{kl}(l = 1, \ldots, S_k)$ of which the number of units ideally approach the previously fixed target lotsize Q_k^*. In Table 7.5 we give the various booked orders, covering roughly a time period of one month. As can be seen, we have 5 orders for product P and 15 orders for product S. Each order is characterized by an order quantity and a due date. Table 7.5 has to be interpreted as follows: one unit of product P has to be delivered at day 22, 5 units at day 28, 3 units at day 37, etc.

Table 7.5.   Booked orders for products P and S

Product P	Order	1	2	3	4	5
	quantities	1	5	3	2	4
	due dates (days)	22	28	37	41	44
Product S	order	1	2	3	4	5
	quantities	1	3	2	3	1
	due dates (days)	17	18	19	22	24
	order	6	7	8	9	10
	quantities	1	3	2	3	1
	due dates (days)	26	27	30	33	34
	order	11	12	13	14	15
	quantities	2	3	3	1	1
	due dates (days)	35	36	39	42	44

For each product k, we first fix the number of setups,

$$S_k = \left\lfloor \frac{1}{Q_k^*} \sum_{c=1}^{C_k} OQ_{kc} \right\rfloor,$$

where $\lfloor x \rfloor$ is the largest integer smaller than or equal to x. This is again
a conservative rounding precluding infeasibility. In our case, S_P equals 3
($\lfloor 15/4 \rfloor$) and S_s equals 5 ($\lfloor 30/6 \rfloor$). The grouping into manufacturing lots
can be done in several ways. It is clear that this problem can be formu-
lated as an integer programming model or, more elegantly, transformed
into a dynamic program. Given the standard nature of this problem
we omit the formulation. It is however important to mention that the
objective function we used minimizes the number of inventory-days. In
Table 7.6 we summarize the results for our illustrative case where QL_{kl}
stands for the lot size of the new manufacturing orders. It should be
clear that as long as the manufacturing lots are not physically released,
the optimal grouping can be recalculated when new information becomes
available. In this way we are able to react fast to changing circumstances,
enabling the planner to fit in quickly new incoming orders. However, if
the changes are drastic, such as a significant increase or decrease in the
size and the number of orders, we opt for re-optimizing the target lot
sizes, so that these changes will be reflected in the lead time estimates.

4.2.2   Release of new manufacturing orders.   For the newly
determined manufacturing order quantities, QL_{kl}, we have to compute
the corresponding expected lead time and the planned lead time
(expected lead time plus safety time). Because each manufacturing order is due at the due date of the first order from this manufacturing
order, we have to remove the terms accounting for the stock time from
the equations (7.8) and (7.9). We therefore restate the expression for the

Table 7.6.   The manufacturing lot sizes

Product	L_{kl}	Grouped Orders	QL_{kl}
P	L_{P1}	1-2	6
	L_{P2}	3-4	5
	L_{P3}	5	4
S	L_{S1}	1-2-3	6
	L_{S2}	4-5-6	5
	L_{S3}	7-8	5
	L_{S4}	9-10-11-12	9
	L_{S5}	13-14-15	5

expected lead time as follows:

$$E(W_k) = \sum_{o=1}^{O_k} \left(\sum_{m=1}^{M} E(Wq_m(\mathbf{Q}^*))\delta_{kom} + \overline{T}_{ko} + QL_{kl}\overline{X}_{ko} \right) \quad (7.11)$$

and for the variance of the lead time

$$V(W_k) = \sum_{o=1}^{O_k} V(Wq_m(\mathbf{Q}^*))\delta_{kom} + \sum_{o=1}^{O_k} s_{T_{ko}}^2 + \sum_{o=1}^{O_k} QL_{kl}s_{X_{ko}}^2 \quad (7.12)$$

Next, the planned lead time (70% service) is obtained using expression (7.10) from section 4.1.2. Subsequently, the planned lead times are deducted from the due dates to obtain the release dates for each manufacturing order. These results are summarized in Table 7.7. The due date for L_{P1} is day 22 (it includes orders 1 and 2) so the due date equals 528 hours from now. The expected lead time is calculated for each manufacturing batch. Across manufacturing batches for the same product, the waiting time and setup time are equal but the batch production times differ due to the different manufacturing quantities. The same is true for the lead time variance so that each manufacturing batch of a given lot size ends up with its own planned lead time (for a batch of 4 units of product P and a batch of 6 units of product S, the planned lead time coincides with the lead time percentiles from Table 7.4). The planned lead time is substracted from the due date and the release date is obtained. Negative release dates mean that the guaranteed service level will not be reached because the batch can only be released at the current moment. Due to the fact that the planned lead time incorporates both waiting time and safety time it is still possible, but less likely, that the manufacturing order is finished before the due date.

Table 7.7.   Release dates of the manufacturing orders

Manufacturing Lot	Due Date (hours)	Expected Lead Time (hours)	Planned Lead Time (hours)	Release Date (hours)
$L_{P1}(6)$	528	534	593	−65
$L_{P2}(5)$	888	482	539	349
$L_{P3}(4)$	1056	430	484	572
$L_{S1}(6)$	408	295	338	70
$L_{S2}(5)$	528	277	318	210
$L_{S3}(5)$	648	277	318	330
$L_{S4}(9)$	792	349	398	394
$L_{S5}(5)$	936	277	318	618

To conclude, this phase has set the time windows (planned lead time between release date and due date) for each manufacturing order. The various operations of each manufacturing order will be sequenced within these time windows. This will be covered in the next phase, the detailed scheduling phase

4.2.3 Detailed scheduling of the operations.

At this stage of our procedure, all non-completed operations of manufacturing orders are scheduled between the release date (or the current moment if the order is overdue) and the due date of the order. Detailed scheduling requires to specify for each operation of each manufacturing order L_{kl} ($k = 1, \ldots K; l = 1, \ldots S_k$) when it has to be performed and by what resource, explicitly taking into account the limited availability of the various resources and many other constraints such as precedence among operations, release dates and due dates. A schedule needs to optimize a predetermined objective. Many production managers strive for due date performance, short lead times and low in-process inventory levels.

The well known job shop scheduling problem is the theoretical abstraction of this problem and has been subject of numerous research efforts. Both optimal and heuristic solution procedures are proposed in the literature. Recent integer programming based models can be found in Balas [6] and Applegate and Cook [4]. Among others, Lageweg, Lenstra and Rinnooy Kan [22], Carlier and Pinson [14], Brücker, Jürisch and Sievers [11] and Brücker, Jürisch and Krämer [10] propose implicit enumeration methods for solving the job shop problem. The problem can be stated as follows. N operations have to be scheduled on M resources. Each operation requires a particular resource. A resource can process only one operation at a time and preemption of processing is not allowed. Precedence constraints among operations may exist (e.g. between operations of the same order). A schedule has to be found so that the makespan is minimal. For our application, we use other objectives as well. The job shop scheduling problem can be formalized as follows

$$
\begin{aligned}
\min \quad & C \\
\text{Subject to} \quad & C \geq t_i + p_i & \forall i \in N \\
& t_j \geq t_i + p_i & \forall (i,j) \in A \\
& t_i \geq t_j + p_j \lor t_j \geq t_i + p_i & \forall (i,j) \in E_m, \forall m \in M \\
& t_i \geq 0 & \forall i \in N
\end{aligned}
$$

with C, the makespan of the schedule, M, the set of all available resources, N, the set of all operations to be scheduled, p_i, the processing time of operation i, t_i, the starting time of operation i, A, the set of all pairs of operations (i, j) for which i has to precede j, E_m, the set of all pairs of operations that require the same resource m. Unfortunately, the job shop scheduling problem is NP-hard in the strong sense. This implies that there is little hope to find optimal solutions to large real-life scheduling problems within reasonable computer time. For practical applications heuristic schedule generation procedures with priority dispatching rules are often used. Well known dispatching rules are, FCFS (First Come First Served), SPT (Shortest Processing Time), EDD (Earliest Due Date), MWR (Most Work Remaining), CR (Critical Ratio), to mention only a few.

Adams, Balas and Zawack [1] introduced the Shifting Bottleneck Procedure (SBP), a new, powerful heuristic for the job shop scheduling problem. Extensions to the SBP, such as Dauzère-Pérès and Lasserre [15] and Balas, Lenstra and Vazacopoulos [7] increase its performance. Experiments by Adams, Balas and Zawack [1], Dauzère-Pérès and Lasserre [15], Ivens and Lambrecht [16] and Balas, Lenstra and Vazacopoulos [7] indicate that SBP offers exceptionally good results compared to other heuristics such as priority dispatching rules. Because of the SBP's good balance between computational complexity and the quality of the generated schedules, we have chosen this procedure as the engine of our detailed scheduling phase.

However, the scope of the theoretical job shop scheduling problem is far too limited to be applicable in practical environments. We therefore extended the SBP so that non-standard features such as release dates, due dates, assembly structures, split structures, overlapping operations, setup times, transportation times, parallel machines and in-process inventory can be modelled (see Ivens and Lambrecht [16]). Recent extensions include the use of resource calendars and the possibility that operations require more than one resource at a time. In addition, other performance criteria could be considered. A brief overview of some extensions is given below in section 4.2.6. We will first discuss the disjunctive graph representation of the scheduling problem and next we will explain the approach of the Extended Shifting Bottleneck Procedure (ESBP) to solve the problem.

4.2.4 The network representation of the job scheduling problem.

Problems solved by the Shifting Bottleneck Procedure are represented by an activity-on-the-node disjunctive graph $DG = (N, A, E)$. $N = N' \cup \{b\} \cup \{e\}$, N' is the set of nodes each representing one operation, and $\{b\}$ and $\{e\}$ are dummy nodes indicating the start and the end of the schedule. Each node $i \in N'$ has a label p_i, the processing time of operation i. A is the set of directed (conjunctive) arcs, representing precedence relations between nodes (i, j) of N. E is the set of undirected (disjunctive) arcs, which represent precedence relations between nodes (i, j) of N' that require the same resource. Initially, the arcs E are undirected. A solution to the problem corresponds with the choice of a direction for each arc in E. Let E' be such a selection, i.e. a set of directed arcs. $DG' = (N, A, E')$ is the graph obtained when replacing the disjunctive arcs in DG by the selection E'. The longest path in DG' from $\{b\}$ to $\{e\}$ corresponds to the makespan of the schedule. A feasible solution requires that DG' is acyclic. The set $E(E')$ can be partitioned into m subsets $E_1, \ldots, E_m$ $(E'_1, \ldots, E'_m)$ where each set E_k represents the disjunctive arcs connecting nodes which require the same resource k. The job shop scheduling problem consists of finding a selection which minimizes the longest path between the starting and ending node.

4.2.5   The shifting bottleneck procedure.   The Shifting Bottleneck Procedure is an iterative procedure which schedules one resource at a time. In each iteration two decisions have to be taken; we have to decide which resource is the bottleneck, and we have to determine the sequence on that resource. Next, at the end of each iteration there is a reoptimization run on all resources scheduled so far. We refer to the paper of Adams, Balas and Zawack [1] for a description of the Shifting Bottlerneck Procedure. In this paper it is shown that the heads and tails can be calculated within linear time complexity. The single machine scheduling problem with release dates and due dates (the subproblem of the SBP) is NP-hard. Nevertheless, large instances can be solved by a branch-and-bound procedure proposed by Carlier [13]. For many practical applications, we found out that computer time needed for the one machine problem constitutes no problem.

4.2.6   The extended shifting bottleneck procedure.   In this section we will briefly describe the ESBP which is more suitable for real-life applications. For an in-depth treatment of these extensions we refer to Ivens and Lambrecht [16]. The extensions are modeled by an Extended Disjunctive Graph (EDG). This representation is similar to the DG, but arcs can have labels to represent general precedence relationships such as Start-Start (SS), Start-Finish (SF), Finish-Start (FS) and Finish-Finish (FF). In practice, it is common to allow overlapping, i.e. units of a batch do not have to wait till the whole batch is finished; instead, products flow from one resource to another in several smaller transfer batches. When preemption of processing is not allowed, overlapping can be modelled by using SS precedence relations (or negative FS). Overlapping may have a considerable impact on lead time performance. Also all sorts of delays, forced waiting times or cooling times can be modeled by FS precedence relationships.

In the standard job shop problem, operations can have only one technological predecessor or successor. Product assemblies and splits can easily be modelled by allowing multiple predecessors and successors.

Customer or manufacturing lots can have a release date and a due date. These can be incorporated in the EDG by assigning FS precedence relations respectively between {b} and the first operation(s) of the lot (in case of release dates) or between the last operation(s) of the lot and {e}. We also allow restrictions on starting times or finishing times of individual operations (e.g. due to temporary unavailability of raw materials).

In the standard job shop scheduling problem there is only one unit available of each resource. For some applications however, it is possible that some resources are available in multiple units (i.e. parallel machines). Thus, in addition to sequencing, an assignment of operations to resources has to be done. The EDG is able to model this complication. In each iteration of the shifting bottleneck procedure a resource type is scheduled, which requires the solution to a single or parallel machine problem, depending on the availability of the resource.

In addition, the SBP has been extended towards other objective functions. In this chapter we use the minimal average lateness criterion.

4.2.7   The metal shop example.   We will now illustrate the Extended Shifting Bottleneck Procedure for the small metal shop example. The eight manufacturing orders from Table 7.6 have to be scheduled. The corresponding extended disjunctive graph is given in Figure 7.20. Each node corresponds with one operation. The numbers above the nodes indicate the processing time of the operation. In order to keep the figure transparent, the operations which have to be scheduled on the same machine share the same shade.

Currently, two previously released manufacturing orders reside on the shop floor. A first manufacturing lot for 5 units of product P has already passed the cutter and has been loaded on the grinder, on which the setup (20 hours work content) is already performed. So we still have to schedule 50 hours processing time for the operation on the grinder and a complete operation on the lathe (84 hours). A second manufacturing order for 6 units of product S has already spend 34 hours on the lathe. In addition to these two in-process orders, there are three manufacturing orders with three operations and five manufacturing orders with two operations. For instance nodes 4, 5 and 6 symbolize manufacturing lot L_{P2}. Node 4 has a processing time of 170 hours (20 hours setup and 5*30 hours processing).

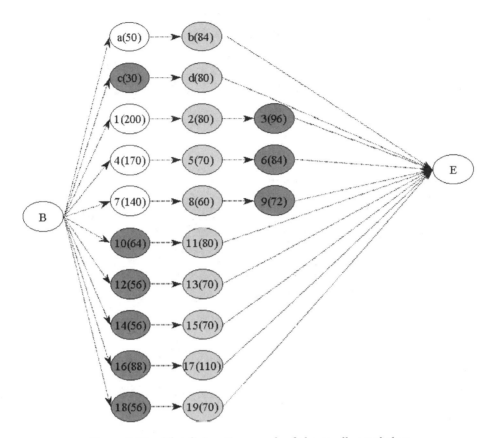

Figure 7.20. The disjunctive graph of the small metal shop

The disjunctive graph contains all the necessary input for the ESBP. The output of the ESBP is visualized in Figure 7.21 as a Gantt-chart. The Gantt-chart shows the two manufacturing orders already in process (the unlabeled blocks) and the eight newly released manufacturing orders. The shaded blocks stand for product P while the white blocks visualize product S. Next, for each manufacturing order the time window (planned lead time) is shown. The release and due date for each order can be seen. Although order P1 could not start on its release date (as it is already late), due to the safety time the order can still be finished on time. From the Gantt chart it can also be seen that the orders are scheduled as early as possible in their respective time windows (cf. Minimal average lateness criterion). Of course the sequencing of the various operations causes some operations to start later than the realease date. This can be seen for manufacturing order S2. These time windows have been designed is such a way that the deterministic schedule will not be useless as soon as disruptions occur; the in-build slack (safety) time will

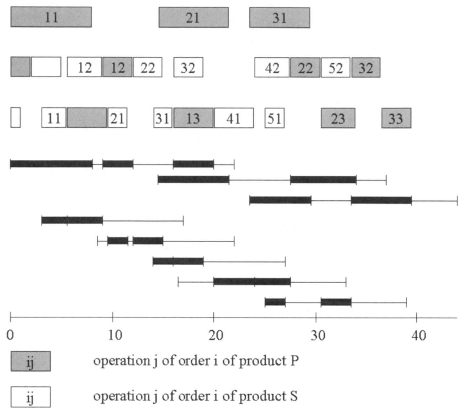

Figure 7.21. The Gantt chart of the detailed schedule

absorb the impact of variability. The in-build slack is clearly visualized in Figure 7.21. A re-optimization may be required in case there are too many changes.

4.3.   The execution phase

In this phase, the detailed schedule will be executed. From the detailed schedule, dispatching and picking lists can be drawn. A data captation system can transmit information concerning work progression back to ACLIPS. From time to time, a recalculation of the detailed schedule will be necessary because of the numerous changes on the shop floor. The frequency of recalculation is of course a function of the dynamics of the shop. A re-optimization of the lot sizing and lead time estimation phase will also be required now and then but, of course, less frequently. This needs some argumentation.

Remind that ACLIPS is a hierarchical approach. The lot sizing and lead time estimation phase is on an aggregate level and focusses more on a long term planning horizon. The information needed for this phase does not change very quickly. It also allows the user to evaluate what-if questions. The grouping of orders, the lead time off-setting and the detailed scheduling on the other hand are short term decisions requiring more frequent recalculations. We have of course to avoid rescheduling each time that a single disruption occurs. That's where we benefit from the safety time included in the time windows.

5.   The Results of ACLIPS

In this section we provide details on the results of the impelmentation of ACLIPS at SOHPD. First we deal with some more scientific results and secondly we introduce the benefits and impact of ACLIPS on SOHPD.

5.1.   Some scientific results

In order to give an idea about the scientific results, we conducted a couple of experiments.

In the first planning experiment we evaluated the existing planning practice. The current planning practice involves the use of heuristically determined lot-sizes (fixed at 1, 2, 4, 8, or 16 weeks of supply) and the use of local, myopic priority rules for scheduling. In the second experiment we set the lot-sizes as obtained through the optimization routine and the scheduling is done by our extended shifting bottleneck procedure. The exact problem size concerned 556 different components, 70 machines, 3,484 operations and about 10,000 orders (cold-steel shop). To make things comparable, we used the ACLIPS model (the lead-time estimator and the finite scheduler) for both experiments but of course they had different parameter settings. The conclusion is that the average lead time (for the cold-steel shop) per order can be reduced by a factor three to four. We figured out that 85 percent of the lead time improvement is due to a better lot sizing policy and for 15 percent due to improved scheduling. [Lambrecht et al. 1998] The above mentioned lead time reduction is based on a computer experiment. Real-life data show that the lead time for final products decreased from the original 16 weeks on average, to lead times in the range of 6 to 8 weeks. A lead time reduction by a factor 2 to 3 is a realistic estimate. Again, the positive effect on the forecasts cannot be underestimated.

Next, we examined the resulting lot sizes and the resulting lead times
in the tuning phase. Here, often unsatisfactory long lead times are de-
tected. To remedy this, the various options of the tuning phase can be
used. For instance, extremely large lead times occur at heavy loaded
machines. Off-loading and the provision of additional capacity (extra
shifts or overtime) can cause a considerable decrease in utilization and
consequently in the average and variance of the lead times. It is crucial
that all serious capacity problems, whether permanent or occasional,
are solved before diving into the detailed scheduling phase. A detailed
analysis with the lot sizing module of ACLIPS revealed that there was
one machine with a utilization level of 97%, resulting in unsatisfactory
average waiting times of 226 hours on that machine. At that time this
machine ran in one shift. Scheduling an additional shift seemed to be the
most plausible solution. If this were not possible, then other capacity expansion measures have to be taken (off-loading, outsourcing, additional
capacity). We increased the capacity of that machine by one shift, and
we ran the lot sizing and lead time estimation routine again. This resulted in a lead time of 16.9 hours per operation. This is an additional
improvement of 23% achieved by adjusting the capacity of only one resource. The computational effort for the optimization is rather high,
but we forced an accuracy of 10^{-10}, which is not neccessary for practical applications. Moreover, the optimization routine is run infrequently.
An evaluation of a given set of lot sizes takes only about 0.3 seconds,
which allows the practitioner to evaluate what-if data modifications
quickly.

In the scheduling phase, we group the various orders and forecasts
(for a one-year time horizon) into new manufacturing orders. After determining a release date for each manufacturing order (with a service
rate of 95%), we ran the Extended Shifting Bottleneck Procedure. To
test our approach we ran the following simulation. We released all new
lots resulting in a huge detailed scheduling problem of about 30,000 operations. For the practical application however, the scheduling problems
are much smaller; since it is not necessary to schedule all manufacturing
orders, we can restrict ourselves to manufacturing orders whose release
date is within an acceptable scheduling period. In Table 7.8 we compare the outcome of our detailed schedule with current practice in the
company. The first column refers to current practice. These results include the old MRP lot sizes, no intelligent way of grouping orders and a
slack-based priority scheduling rule. In the second column we display the
ACLIPS outcome, including the target lot sizes minimizing the average
lead time, the optimal grouping of the orders into manufacturing orders,
the determination of the time windows and a detailed schedule based on

Table 7.8.   The overall performance of the detailed schedule

Performance Measure	Without ACLIPS	With ACLIPS
proportion of late orders	13.20%	10.63%
maximum lateness	19.57 days	3.71 days
average lateness	−29.61 days	−5.00 days
average tardiness	0.60 days	0.14 days
average leadtime	15.72 days	2.31 days

the ESBP (we used the minimization of average lateness as an objective function).

On all performance measures ACLIPS outperfomed current practice. The maximum lateness is drastically reduced. The average (in process) lead time decreased by 85%. The average lateness increases to −5 days, which indicates that the lead time estimation is much more accurate, resulting in enormous lead time and work in process savings. Still, 10.63% of the orders are late. This is of course a major concern. The main reason is that in certain time periods the shop experiences a heavy workload and too many orders are due in a small time period. In order to make the comparison with current practice meaningful we did not manipulate due dates nor the available capacity. However the tuning phase offers plenty of opportunities to manage the workload better, so that late orders can be avoided. The distribution of the lateness however is seriously improved with the ACLIPS approach.

5.2.   Benefits and impact on SOHPD

In this section we like to report not only on the benefits but also on the impact ACLIPS had on SOHPD and on the community at large. Recall that due to the large variety of parts, the high precision required and the small volumes, most of the steel shop was originally organized as a job shop, with a lot of general purpose machinery grouped in a functional way, in which products follow their own routings through the shop. Before implementation, the job shop suffered from some well recognized problems: rigid lot-sizing, predetermined lead times, the infinite loading practice, uncoordinated material and capacity plans, and a sequencing of jobs based on myopic priority dispatching rules (called the queue-manager). The diagnosis was clear: installing a finite scheduling system alone could never do the job. The real problems are rooted at a higher, more aggregated, decision level; such issues as lot sizing, the capacity structure, order acceptance and release policies are to be settled first. In addition, procedures, parameter settings,

Table 7.9. Quantitative and qualitative benefits from the implementation

	SOHPD	Community at Large
Quantitative	50%–60% Lead time Reduction From 3.5 to 6 Inventory Turns 27.3 Percentage Points Productivity Improvement More Reliable Forecast	41% Workforce Increase 66% Sales Improvement
Qualitative	Development of Supporting Activities Benchmark for new Implementations OR/MS Awareness Training Programs	Leveraged MRP/ERP OR and Competitive Advantage

shop-floor layout had to be reconsidered. Analysis was needed on the capacity structure, on the lead time determinants, the product design, setup times, etc. Sales-order-entry people need to be convinced that they were key players in the whole process, especially due to the forecast issue.

The benefits and impact of our project are summarized in Table 7.9.

The lead-time performance of ACLIPS has been dealt with in detail in section 5.1, where a lead time reduction with a factor two or three was discussed. The main issue is probably the need for less remote forecasts. The positive effects of this have already been discussed. One of the most important secundary effects of this lead time reduction is a significant reduction of internal quality problems while yearly inventory turns have increased from 3.5 to 6. A relentless search to reduce the total manufacturing lead time has led to a complete new layout of the plant, focused on cellular manufacturing. This was possible only after managers made a strategic decision, supported by Activity Based Costing information, to outsource all non-core processes and components. SOHPD also uses a simple but effective productivity performance measure. The output of the company is translated into total numbers of standard hours produced (earned hours) and compared to the total direct hours (present hours). The ratio of standard hours produced over total direct hours is a measure of the direct productivity. This measure accounts for all kinds of outages. The standard times for the operations are regularly adjusted to reflect process improvement. The combined effect is what is called productivity improvement in Table 7.10.

SOHPD 's business succes has had a measurable impact for the community. The measurable impact for the community can be derived from the overall business success of SOHPD. The total number of employees increased by 41 percent over the period 1993–1998. Over the same period, sales per full time equivalent improved by 66 percent. The qualitative

Table 7.10. Direct productivity equals earned hours (standard processing time) over present hours. The standard times are regularly adjusted, consequently the sum of both measures fairly accurately describes productivity improvements

	1993	1994	1995	1996	1997	1998
Direct Productivity	60.6%	65.5%	67.5%	67.8%	70.6%	72.6%
Standard Time Reduction	0.0%	2.9%	2.3%	2.8%	4.2%	3.1%
Productivity Improvement		7.8%	12.1%	15.2%	22.2%	27.3%

impact of our project is important. The ACLIPS model is only a relatively small component in the network of logistic activities that supports the business strategy. The use of the OR model induced several other supporting activities. Let's illustrate this point with a simple example. The data required to run the shifting bottleneck procedure are such that several specific tools had to be developed by the systems department to obtain the data in a structured and systematic way. Each machine operation for example is now identified by its 5 most important setup characteristics ranked in order of changeover time required. This allows us to implement a sequence dependent setup time routine to minimize the setup time loss. Indices to measure product variety were developed as well. SOHPD also developed a monotoring system to manage the projected availability of raw materials against the planned release dates of manufacturing orders. It thereby coordinates its capacity and material plans. The ACLIPS project has fostered a general awareness for scheduling and planning issues throughout the whole organization. Operations management issues are on the agenda for management training and also for the training of the shop floor workers. As a matter of fact, a four hour intensive training session on production planning and scheduling has become an integral part of the training program for all employees. A final qualitative benefit for SOHPD is the fact that the production planning and control system of the plant in Bruges is used as a benchmark for the implementation of a new Y2K compliant planning system for the plant in Statesville, NC.

Finally we believe there are also important benefits for the community. First the insights obtained from the ACLIPS approach are applicable to a wide range of manufacturing companies facing hybrid planning realities. In addition, it remedies some basic flaws in MRP/ERP, mainly lot sizing and the fixed lead time assumption. Second, we have a more general reflection that puts our work in an industrial policy context. European manufacturers operate under severe cost pressures. The high cost of labor and restrictive practices are often cited as major competitive disadvantages resulting in the delocalization of manufacturing activities. The SOHPD case shows that a focus on optimization accompanied by supporting, complementary activities can easily outweigh these disadvantages and shows that a coherent bundle of OR tools creates a competitive advantage.

6.   Conclusion

In this chapter we proposed a general methodology to analyze and to schedule a job shop as part of a hybrid manufacturing environment. A four phase methodology is proposed including a lot sizing and lead time estimation phase, a tuning phase, a scheduling phase and an execution phase. In each phase we use analytic approaches which are suitable for practical applications. The methodology is illustrated with an example and an application is given. The ACLIPS methodology is embedded in a software package. Our experience indicates that our approach has a great potential both in terms of computational effort required and in terms of the quality of the generated schedules.

Acknowledgments

This research was supported by the NFWO/FKFO (National Science Foundation) Belgium, under the project G.0063.98, the TRACTEBEL chair, the BOF fund of UA, the SFO fund of UFSIA and WTCM-Belgium. Special thanks to the SOHPD personnel who made this implementation possible.

References

[1] ADAMS, J., E. BALAS, AND D. ZAWACK. (1988) The shifting bottleneck procedure for job-shop scheduling. *Management Science*, Vol. 34, No. 3, 391–401.

[2] ALBIN, S.L. (1981) *Approximating queues with superposition arrival processes*. PhD thesis, Department for Industrial Engineering and Operations Research, Columbia University.

[3] ALBIN, S.L. (1982) Approximating a point process by a renewal process, II: superposition arrival processes to queues. Technical report, Department of industrial engineering, Rutgers University.

[4] APPLEGATE, D. AND W. COOK. (1991) A computational study of the job-shop scheduling problem. *ORSA Journal of Computing*, Vol. 3, No. 2, 149–156.

[5] BAKER, K.R. (1993) Requirements planning. In S.C. Graves, A.H.G. Rinnooy Kan, and P.H. Zipkin (Eds.), *Logistics of production and inventory*, Chapter 11, pp. 571–627, North Holland.

[6] BALAS, E. (1985) On the facial structure of scheduling polyhedra. *Mathematical Programming Study*, No. 24, 179–218.

[7] BALAS, E., J.K. LENSTRA, AND A. VAZACOPOULOS. (1995) The one-machine problem with delayed precedence constraints and its use in job shop scheduling. *Management Science*, Vol. 41, No. 1, 94–109.

[8] BITRAN, G.R. AND D. TIRUPATI. (1988) Multiproduct queueing networks with deterministic routing: decomposition approach and the notion of interference. *Management Science*, Vol. 34, No. 1, 75–100.

[9] BITRAN, G.R. AND D. TIRUPATI. (1989) Approximations for product departures from a single-server station with batch processing in multi-product queues. *Management Science*, Vol. 35, No. 7, 851–878.

[10] BRÜCKER, P., B. JÜRISCH, AND A. KRÄMER. (1992) The job shop problem and immediate selection. Technical report, Univ. Osnabrück/Fachbereich Mathematik.

[11] BRÜCKER, P., B. JÜRISCH, AND B. SIEVERS. (1991) A fast branch-and-bound algorithm for the job-shop scheduling problem. Technical report, Univ. Osnabrück/Fachbereich Mathematik.

[12] BUZACOTT, J.A. AND J.G. SHANTIKUMAR. (1985) On approximate queueing models of dynamic job shops. *Management Science*, Vol. 31, No. 7, 870–887.

[13] CARLIER, J. (1982) The one-machine scheduling problem. *European Journal of Operational Research*, Vol. 11, 42–47.

[14] CARLIER, J. AND E. PINSON. (1989) An algorithm for solving the job-shop problem. *Management Science*, Vol. 35, No. 2, 164–176.

[15] DAUZÈRE-PÉRES, S. AND J.B. LASSERRE. (1993) A modified shifting bottleneck procedure for job-shop scheduling. *International Journal of Production Research*, Vol. 31, No. 4, 923–932.

[16] IVENS, P.L. AND M.R. LAMBRECHT. (1996) Extending the shifting bottleneck procedure to real-life applications. *European Journal of Operational Research*, Vol. 90, 252–268.

[17] KARMARKAR, U.S. (1987) Lot sizes, lead times and in-process inventories. *Management Science*, Vol. 33, No. 3, 9–423.

[18] KARMARKAR, U.S., S. KEKRE, AND S. KEKRE. (1992) Multi-item batching heuristics for minimization of queueing delays. *European Journal of Operational Research*, Vol. 58, 99–111.

[19] KARMARKAR, U.S., S. KEKRE, AND S. KEKRE. (1985) Lotsizing in multi-item multi-machine job shops. *IIE Transactions*, Vol. 17, No. 3, 290–297.

[20] KARMARKAR, U.S., S. KEKRE, S. KEKRE, AND S. FREEMAN. (1985) Lot-sizing and lead-time performance in a manufacturing cell. *Interfaces*, Vol. 15, No. 2, 1–9.

[21] KRAEMER, W. AND M. LAGENBACH-BELZ. (1976) Approximate formulae for the delay in the queueing system $GI/GI/1$. In *Congressbook*, pages 235–1/8, Melbourne, Eighth International Teletraffic Congress.

[22] LAGEWEG, B.J., J.K. LENSTRA, AND A.H.G. RINNOOY KAN. (1977) Job shop scheduling by implicit enumeration. *Management Science*, Vol. 24, 441–482.

[23] LAMBRECHT, M.R., CHEN SHAOXIANG, AND N.J. VANDAELE. (1996) A lot sizing model with queueing delays: the issue of safety time. *European Journal of Operational Research*, Vol. 89, 269–276.

[24] LAMBRECHT, M.R. AND N.J. VANDAELE. (1996) A general approximation for the single product lot sizing model with queueing delays. *European Journal of Operational Research*, Vol. 95, No. 1, 73–88.

[25] LEUNG, Y. AND R. SURI. (1990) Performance evaluation of discrete manufacturing systems. *IEEE Control Systems Magazine*, Vol. 10, 77–86.

[26] IVENS, P.L., M.R. LAMBRECHT, AND N.J. VANDAELE. (1998) Aclips a capacity and lead time integrated procedure for scheduling. *Management Science*, Vol. 44, No. 11, 1548–1561.

[27] VANDAELE, N.J., M.R. LAMBRECHT, R. CREMMERY, AND N. DE SCHUYTER. (2000) Spicer off-highway products division- brugge improves its lead time and scheduling performance. *Interfaces*, Vol. 30, No. 1, 83–95.

[28] SHANTIKUMAR J.G. AND J.A. BUZACOTT. (1980) On the approximations to the single server queue. *International Journal of Production Research*, Vol. 18, No. 6, 761–773.

[29] SHANTIKUMAR, J.G. AND J.A. BUZACOTT. (1981) Open queueing network models of dynamic job shops. *International Journal of Production Research*, Vol. 19, No. 3, 255–266.

[30] SURI, R. (1989) Lead time reduction through rapid modeling. *Manufacturing Systems*, Vol. 7, 66–68.

[31] SURI, R. AND S. DETREVILLE. (1991) Full speed ahead: A look at rapid modeling technology in operations management. *OR/MS Today*, Vol. 18, 34–42.

[32] SURI, R. AND S. DETREVILLE. (1992) Rapid modeling: The use of queueing models to support time based competitive manufacturing. In *Proceedings of the German/US Conference on Recent Developments in Operations Research*, 34–42, Berlin, Springer.

[33] SURI, R. AND G.W. DIEHL. (1985) Manuplan: a precursor to simulation for complex manufacturing systems. In D. Gantz and G. Blais ans S. Solomon, editors, *Proceedings of the 1985 Winter Simulation Conference*, 411–420. Institute of Electrical and Electronics Engineers.

[34] SURI, R., G.W. DIEHL, AND R. DEAN. (1986) Quick and easy manufacturing systems analysis using manuplan. In D. Gantz and G. Blais ans S. Solomon (Eds.), *Proceedings of the Spring IIE Conference*, 195–205, IIE.

[35] SURI, R., M. TOMSICEK, AND D. DERLETH. (1990) Manufacturing systems modeling using manuplan and simstarter: a tutorial. In D. Gantz and G. Blais ans S. Solomon (Eds.), *Proceedings of the 1990 Winter Simulation Conference*, 168–176. IEEE Computer Society Press.

[36] VANDAELE, N.J. (1996) The impact of lot sizing on queueing delays: multi product, multi machine models. PhD thesis, Department of Applied Economics, Katholieke Universiteit Leuven.

[37] WHITT, W. (1983) Performance of the queueing network analyzer. *The Bell System Technical Journal*, Vol. 62, No. 9, 2817–2843.

[38] WHITT, W. (1983) The queueing network analyzer. *The Bell System Technical Journal*, Vol. 62, No. 9, 2779–2815.

Chapter 8

NETWORK SERVER SUPPLY CHAIN AT HP: A CASE STUDY

Dirk Beyer

Hewlett-Packard Laboratories
Hewlett Packard Company
Palo Alto, CA
dirk_beyer@hpl.hp.com

Julie Ward

Hewlett-Packard Laboratories
Hewlett Packard Company
Palo Alto, CA
julie_ward@hpl.hp.com

1.   Introduction

The confluence of several trends in manufacturing organizations has created many opportunities for supply chain management practitioners to apply their skills. These trends include the increasingly ubiquitous implementation of Advanced Planning Systems, the growing complexity of supply chains, and the recognition among managers of the importance of supply chain costs to profitability.

Advanced Planning Systems, or APS, is the general name given to a class of software solutions for supply chain management. Well-known products in this class include i2's Rhythm® and SAP's Advanced Planner and Optimizer®. Planning tools such as these, which provide a greater degree of visibility and coordination across supply chain entities than was previously available, are rapidly replacing more traditional planning methods.

Though these software solutions are continually expanding in functionality, their core function is to determine coordinated procurement, production and shipment plans that meet demand, adhere to capacity and scheduling constraints, and honor inventory policies. Through expanded visibility of the supply chain, coordination across planning functions, explicit modeling of constraints, automation of complex planning rules, and search methods based on optimization and heuristics, these tools offer many advantages over traditional planning methods.

However, the core challenge in supply chain management, namely that of managing uncertainty, is not addressed by APS. These systems regard all inputs as deterministic, leaving uncertainty to be addressed outside their scope. A planner hedges against demand and supply uncertainty through placement of inventory. Practitioners of supply chain management know that determining effective inventory policies that trade off costs and order fulfillment goals can be extremely difficult for complex supply chains. Rather than determining inventory policies, APS software solutions require them as input. Without a systematic, analytical approach to determine effective inventory policies, planners cannot extract the purported benefits of costly APS implementations. It is this limitation of APS, coupled with their ubiquity, that creates an enormous opportunity for supply chain practitioners. The need for effective inventory management will follow the growth trajectory of the APS business.

Meanwhile, as the demand for inventory management expertise grows, so does the complexity of supply chains. Global differences in taxes, duties and labor costs have created incentives for companies to spread their supply chains out geographically. Furthermore, supply chains increasingly span several organizations, and so operational control may be distributed across supply chain entities rather than centralized. There are often multiple alternate shipment modes available between locations. These structural complexities are compounded with other factors such as short product life cycles to make inventory management difficult. These trends require supply chain practitioners to develop new, sophisticated tools.

As the need for supply chain management expertise grows, receptivity among managers to adopt the needed sophistication in their planning methods is also on the rise. Supply chain costs become a critically important component of the bottom line as competition erodes margins. This trend is particularly apparent in high tech industries, with its endemic short product lifecycles and high demand variability. At the beginning of a product's life, insufficient inventory can compromise market share, whereas excess inventory at the end of the product's life can consume all of a product line's profits. These threats have further increased support among managers for supply chain management activities.

The trends that are creating opportunities for supply chain practitioners come hand in hand with challenges. The growing complexity of supply chains calls for rich inventory models. Most existing models in the literature are stylized to lend insight into a particular supply chain feature and how that feature affects optimal inventory policies. For example, there is extensive literature on single location inventory management with two replenishment modes. There are few models, however, that address the simultaneous presence of multiple confounding factors, such as dual replenishment modes and non-stationary stochastic demand. Practitioners facing complex supply chains must draw upon the general principles and intuition derived from several models to create a useful approach to their multi-faceted problems.

Another difficult aspect of practical inventory management is finding strategies that are practical to implement. At least as important as the optimality or cost-effectiveness of a given policy is that it is compatible with existing information technology, the data supporting it is readily available and reliable, its computational complexity is low, and it is intuitive to planners. In many cases, these requirements create severe restrictions in the types of inventory strategies that can be selected. For example, if in-transit inventory data is unreliable, policies based on inventory position instead of on-hand inventory are difficult to execute with accuracy, and thus may be inappropriate.

To compound the difficulty in selecting an effective of inventory management policy, supply chain practitioners face the additional challenge of keeping pace with constant flux in business conditions and organizations. In this climate, the quest for optimality takes a back seat to timeliness of a solution. When sophisticated analytical tools are required, the successful implementation of these tools relies upon a strong collaboration between the ultimate users of the tool, who understand the business and the data, and the supply chain experts, who build an analytical model of the business. The lifecycle of a project should fall well within the tenure of each participant in his or her position. Moreover, the analytical model should be general enough to survive changes in business conditions. If these criteria are not met, the project's likelihood of being implemented is low.

This chapter describes a supply chain management project that was carried out at Hewlett-Packard Laboratories. This project came about due to the same forces that are creating widespread opportunities for supply chain practitioners. Managers at HP's Network Server Division (NSD) approached the authors in late 1997 to ask for help in managing inventory of a component of network servers. NSD was on the verge of implementing an APS called Red Pepper®. In order to extract benefit from Red Pepper's planning engine, the division needed an effective strategy for managing the inventory of a component of network servers at their factory and distribution centers. In particular, they wanted to reduce their logistics costs while maintaining high availability at the distribution centers by finding the right balance of inventory in their supply chain and by using different shipment modes (air and ocean) efficiently. Several factors complicate the process of managing this inventory, including highly non-stationary demand with large random fluctuations, rapid depreciation, high risk of obsolescence, short product life-cycles, alternative shipment modes with different associated cost and lead times, and long supply lead times.

This project was replete with the challenges that plague modern supply chain management. First, while the many complicating factors in the NSD supply chain are addressed in the literature, no previous work addresses all of them simultaneously. Indeed, it is the combination of all of them that made this problem an exciting research challenge. The authors used an approach that synthesized the learnings from several simpler inventory models. Additional constraints were that the authors had to recommend inventory policies that could be computed quickly, implemented within Red Pepper, understood by planners, and driven by the data that was available. A final difficulty was that the window of opportunity was short; NSD needed a solution quickly. Thus, heuristics offering speed and tractability were favored over optimality. As a result of the aggressive project timeline, many interesting research questions have remained open since the immediate goal of the project was met.

The selected approach, which offers an appropriate compromise between tractability and effectiveness, combines three elements. The first element is the computation of a parameterized set of candidate inventory policies at each location. These candidate policies, computed independently for each DC and warehouse, are myopic order-up-to inventory policies with a single replenishment mode and time and location varying costs. The second element of the approach presented here is supply chain simulation. This simulation, when given as input a combination of candidate policies (one for each location), provides estimates of the implied cost and service level, among other metrics, that this policy combination would produce when implemented. The final component of our approach is a search procedure that is used to select the best combination of locations' inventory policies. The search procedure is guided by the results of the simulation. When the approach is applied, the outcome is a supply chain inventory policy in the form of order-up-to targets for each time period and location. The shipment mode decisions are determined by the inventory targets and the realization of demand, rather than being explicit decision variables in this analysis.

The contents of this chapter are as follows. Section 2 describes the status of the NSD supply chain at the project's inception and the problem that the authors were asked to address. A summary of related literature is given in Section 3. In Section 4 we describe the HP Labs' model of the NSD supply chain. This model is the foundation of an approach used to find effective supply chain inventory policies, described in Section 5. Section 6 describes the results obtained at NSD through the implementation of this project. We conclude in Section 7.

2. Problem Description

2.1. Overview

NSD manufactures PC-based Windows NT® network servers. An important component of servers, called a MOD0 box, is pre-assembled in Singapore and then shipped to four distribution centers (DCs) throughout the world. The final assembly and configuration of servers is done according to re-sellers' specifications at the DCs just before the product is shipped to them. Shipments between the factory in Singapore and the distribution centers can be made by either air or ocean, with different transportation times and freight costs associated with each shipment mode. A central planning organization decides upon production and shipment quantities on a weekly basis.

NSD planners were faced with the perennial challenge of finding an inventory strategy to balance high service level objectives with supply chain costs. They wanted to guarantee a high level of off-the-shelf availability of MOD0 boxes while minimizing freight and inventory-related costs. The goal of this project was to develop a decision support tool to help NSD planners determine effective inventory replenishment strategies for MOD0 boxes with respect to these objectives.

2.2. The product

At the time the project commenced, NSD had ten active types of MOD0 boxes, which are subassemblies for approximately 30 product lines. MOD0's are a high-level subassembly of servers, and contain most of the non-configurable parts of a computer of a given product line, including the chassis, power supply, base motherboard (including minimal memory), control panel, and terminator. MOD0 boxes do not contain processor chips, additional memory, or disk drives; these are added to servers later at the DCs according to customer specifications. The cost of the MOD0 boxes constitutes about 50% of the cost of the finished product. MOD0 boxes stand out among all components of network

servers due to their bulk and their relatively small dollar value per unit
of volume. In addition, although prone to depreciation like all electronics products, MOD0 boxes do not depreciate as fast as hard drives
or processors. These properties make them good candidates for ocean
shipments.

2.3.   The supply chain

As mentioned above, MOD0 boxes are assembled at the Singapore
Factory (Warehouse)[1]. Components are procured from external and internal suppliers and stored in the Singapore component inventory. Procurement leadtimes for components vary but can be as long as 8 weeks.
MOD0 boxes are assembled in the Singapore factory and then shipped
to one of the four distribution centers: Roseville, California; Grenoble,
France; Guadalajara, Mexico and Singapore. Assembly takes only a few
hours, and assembly capacity is usually available when needed. The
Singapore DC is located in the same complex as the factory and therefore there is no shipment leadtime nor any choice of shipment modes
required. Shipments to the remaining three DCs can either be made by
air or sea. Air shipments to all the DCs take about one week. Ocean
shipments take, including loading, unloading, and subsequent land transportation to the DC, about four weeks to Grenoble and Roseville and
five weeks to Guadalajara. (Transshipments between DCs are made only
under extraordinary circumstances, amounting to less than 2% of total
shipments.)

Each DC serves demand in a pre-assigned region only (North America,
Asia Pacific, Europe, Latin America). For example, European customers
cannot receive shipments directly from the Roseville DC.

2.4.   Costs and performance measures

The four types of costs associated with the supply chain operation
are inventory holding costs, depreciation, obsolescence costs, and transportation costs.

Because inventory in the supply chain must be financed, holding costs
are applied to inventory at any location or in transit between nodes.
These finance charges are proportional to the cost of inventory held and
expressed as a percentage of the standard material cost per year.

If a server cannot be sold before the end of the product life cycle,
revenue loss is substantial. The server can then only sold in "fire sales,"
cannibalized for its parts or, in the worst case, written off. Components, on the other hand, are often used in subsequent products, and

so are less vulnerable to obsolescence. A given type of MOD0 box is typically used for multiple products within its product line, so its own life cycle is longer than that of any particular product, currently about 20–24 months. When an assembled MOD0 box does become obsolete, only a small fraction of its standard material cost can be recovered. Therefore, obsolescence cost equal to the difference of the original standard material cost and the recovered cost is incurred for every assembled MOD0 box left at the end of its life cycle. Obsolescence cost for lower-level component inventory is much lower than obsolescence cost for assembled boxes since many of the parts can be used in successor products.

Prices for the components needed to build a MOD0 box decline rapidly over the course of its lifecycle. Therefore, every MOD0 box held as inventory from one week to the next could have been built one week later at a lower price. The price difference is expressed in a depreciation cost applied to each unit of inventory in each period and location.

Transportation costs include freight cost, insurance, and costs of loading and unloading. Ocean shipment costs are charged per container shipped, whereas air shipment cost is charged per pallet. Ocean freight costs are about one-fifth of air freight costs. This cost advantage must be weighed against the higher inventory, depreciation and obsolescence cost associated with ocean shipments, as well as diminished agility as compared to air shipments.

The performance of the supply chain is currently measured by the off-the-shelf availability for finished servers (the percentage of orders filled within a week or less) and the total cost. Availability of components is currently not used of a performance measure.

2.5. Supply chain operations

At the project's inception, planners in the NSD World Wide Planning division in Santa Clara created weekly production and shipment plans using rules that had been programmed into spreadsheets. Production orders and shipments were based on on-hand inventory targets for all the DCs and the Singapore factory. When projected inventory on hand fell below the on-hand target, an order was triggered. Planners generally chose air as the shipment mode; only when a MOD0 box was in surplus due to repeated high forecasts was an ocean shipment made. As a result, the majority of shipments (over 85%) were made by air.

At that time, inventory targets were determined by using independent single-location, single-replenishment-mode inventory models at each location. A myopic order-up-to policy was computed based on the required service level. These calculations did not take into account non-stationarity of the demand, interactions between the DCs and the factory, nor potential savings from utilizing different shipment modes appropriately. The end of the product life cycle was handled on an manual exception basis. Inventory targets were reduced during this phase, creating lower availability (or longer response times) in order to avoid excessive obsolescence cost caused by leftover inventory.

At the same time, NSD was in the process of implementing Red Pepper as its planning engine. This system would automate the production and shipment decisions that planners were using spreadsheets to make. This automation would include the following rule for determining the allocation of air and sea shipments. First, let $airLT$ and $seaLT$ respectively denote the air and sea shipment leadtimes for a given DC. For a given period t and this DC, Red Pepper compares the projected inventory (based on on-hand and in-transit inventory and forecasts) in period $t + airLT$ with the DC's target inventory for that period. Red Pepper triggers an air shipment to the DC if the projected inventory falls short of the target. Next, it considers making a sea shipment to the DC. To avoid making a sea shipment now that will adversely impact the ability to make air shipments that would arrive sooner, Red Pepper first "reserves" inventory at the factory that is needed to make air shipments to all DCs that would arrive before the sea shipments in this period. If factory inventory remains after reserving for air shipments, then a sea order is triggered to bring projected inventory in period $t + seaLT$ up to the target for that period. Whenever there is a shortage of inventory at the factory, it is allocated to the DCs in proportion to the size of their requested shipments. This shipment mode selection rule is illustrated in Figure 8.1.

This selection rule, coupled with the relative magnitude of factory and DC targets, determines the fraction of shipments that are made by air. Thus, shipment decisions were not directly within the scope of the decision tool developed at HP Labs. Instead, these decisions were to be a by-product of the recommended inventory strategy.

The total inventory of MOD0 boxes in the NSD supply chain, including in-transit between the factory and DCs, was approximately 8 weeks of supply when the project began. The off-the-shelf availability was approximately 85% for MOD0 boxes and 75% for servers. NSD hoped to ultimately satisfy 95% of demand for the finished product within one week response time. Such an improvement would require, among other things outside the scope of this project, a dramatic increase of MOD0 box availability.

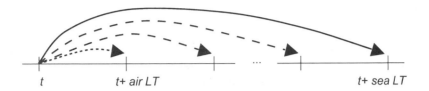

---

1. Make air shipments in period t. For each DC, determine desired air shipment using projected inventory and on-hand target in $t+airLT$. Make DC air shipments by allocating warehouse inventory in proportion to DC desired air shipments.

---

2. Plan for air shipments in $t+1$, ..., $t+seaLT - airLT$-1. For each DC, reserve factory inventory for air shipments that will arrive to meet DC targets in periods $t+airLT$+1, ... , $t+seaLT$-1. Account for shipments that will arrive at the factory in time to make these planned air shipments.

---

3. Make sea shipments in period t. For each DC, Determine desired sea shipments to meet on-hand target in $t+seaLT$. Allocate unreserved factory inventory in proportion to DC desired sea shipments to make shipments.

Figure 8.1. Selection of shipment mode

2.6. Anticipated project benefits

There were three ways in which we hoped to reduce NSD's supply chain costs while improving availability. The first was in redistribution of inventory. We hoped that we might improve availability per unit of inventory by rebalancing inventory between the factory and DCs. The second way we hoped to improve operations was to reduce overall supply chain inventory. A reduction of total MOD0 box inventory by 20% would save an equal percentage in inventory and depreciation cost annually. The third goal was increased used of ocean shipments. By shifting the proportion of shipments from 85% air and 15% ocean to the reverse proportions, NSD anticipated savings of about 30% of freight costs. Of course, this savings would be partly offset by higher inventory, depreciation and obsolescence cost; this tradeoff would have to be considered by the model.

Naturally, the potential cost savings was inversely related with service level goals. Thus, we wanted to develop an approach that would quantify the trade off between service level goals and cost, to enable NSD managers to strike the balance they desired.

The customers (re-sellers) have the right to return a certain percentage of the products they previously bought to the DC. Returns play a particularly important role close to the end of the life cycle of the product since they contribute to potential excess inventory and can incur high obsolescence cost for HP. On the other hand, even if inventory at the resellers is not returned to the DC, price protection cost, which is based on the reseller inventory, constitutes a big contribution to NSD's supply chain cost. An increase in service in the form of shorter and more reliable response times can impact both problems positively. If the DCs are more responsive resellers do not need to hold large inventory and returns as well as price protection expenses may drop.

2.7.   The data

NSD World Wide Planning receives demand forecast on a monthly basis from NSD Marketing. These monthly forecasts are broken down into weekly "buckets" based on an empirical distribution of historical orders over the month. Using the bill of material, the component demand is determined. Historical data of actual demand back to 1996 is available. Forecasts have a relatively high error; coefficients of variation of 30–60% are not uncommon.

Cost data for freight and capital cost are available and relatively reliable. Depreciation cost and obsolescence cost can be estimated from historical data and specific business knowledge available from the planners and the marketing department.

3.   Related Literature

Three features of the NSD supply chain problem complicate finding an optimal inventory policy. The first is the existence of two replenishment modes. The second is the fact that the supply chain in question is a two-echelon distribution system. The third is the non-stationarity of demand. Each of these features has been extensively studied independently. In what follows, we briefly summarize the previous contributions in each of those areas.

A number of authors have considered single location inventory models with two supply modes. One setting in which the form of the optimal inventory control policy has been characterized is under periodic review, with the critical assumption that the two modes' leadtimes differ by exactly one period. Many authors, notably Fukuda [12] , Bulinskaya [4], [5], Daniel [8], Neuts [21], Veinott [26], Wright [28], and Whittmore and Saunders [27] have presented results for this case. In those papers, it is shown that the optimal policy is a generalization of the traditional order-up-to type policy, in which there are two order-up-to levels in each period, for the short and long leadtime mode inventory positions, respectively.

Without the assumption that the leadtimes differ by one period, the form of the optimal policy is not known. Today's shipment decision is certain to depend on the amount of inventory on hand as well as the quantities due to arrive in each period between the short and long leadtimes from now. Moreover, it can be seen from a dynamic programming formulation of the problem than an optimal strategy will not only depend on the total on-hand and pipeline inventory (inventory position) but also on the period in which each fraction of the pipeline inventory is going to arrive. Work in this more general setting, both in periodic and continuous review, has been done by authors such as Allen and D'Esopo [2], Moinzadeh and Nahmias [19], Moinzadeh and Schmidt [20], Aggarwal and Moinzadeh [1], Pyke and Cohen [22], and Chaing and Gutierrez [6]. These papers propose heuristic policies, the latter two in a multi-echelon setting. The work presented in this chapter also takes the approach of using a form of heuristic policy that is more practical to implement than an optimal policy would be, and searching for the best among such heuristic policies.

There are numerous papers on multi-echelon inventory systems with the same supply chain layout as NSD's. Optimal policies have not been characterized in this setting, but instead, practical policies have been proposed and evaluated. One exception is in serial systems, where Clark and Scarf [7] gave the form of the optimal policy. Others such as Federgruen and Zipkin [10], [11], Jackson [15], and Graves [14] propose heuristics for the periodic review case. Authors who consider continuous review (r, Q)-type policies in the multi-echelon setting include Sherbrooke [23], Deuermeyer and Schwarz [9], Graves [13], Moinzadeh and Lee [18], Lee and Moinzadeh [17], Svoronos and Zipkin [24], and Axsater [3]. This list is not exhaustive, but fairly representative of the literature in this area. Like previous approaches, our tactic is to consider a class of inventory policies that we expect to be effective and practical to implement, and to search for the best policy within this class.

Finally, there have been many different models of nonstationary demand in single location inventory models. The papers most relevant to NSD's business and most related to the approach taken here are those of Karlin [16] and Veinott [25]. Karlin showed that order-up-to policies are optimal in simple one node inventory systems even when demand is nonstationary, but these order-up-to levels are not easy to calculate in general. Veinott showed that if the demands were stochastically nondecreasing, then the optimal order-up-to levels were in fact the myopic ones (the newsvendor solution) and so are easy to compute.

4.   The Model

In this section we describe the model used to represent NSD's supply chain. In order to make the model tractable, we made some simplifying assumptions. We felt that these assumptions were general enough reflect the main features of the original supply chain closely, so that the recommendations made would remain valid.

Due to the relatively high capacity for assembly at the Singapore factory and the linear cost structure, different products do not interact in our model. For that reason we can treat each of the products in a separate one-product model.

4.1.   Product flow

Since transshipments between DCs are rare and NSD does not wish to plan for them, we assume that transshipments are not available as replenishment mechanisms. The supply chain resulting from this assumption is a distribution system. Customer demand is filled by the DC assigned for the region, DCs receive their shipments from the Singapore factory, which in turn uses the components from the Singapore inventory. Figure 8.2 shows the supply chain layout.

Each inventory location in the supply chain is characterized by its replenishment leadtimes, i.e., the time between the instant a replenishment order is placed until the ordered amount arrives in the inventory

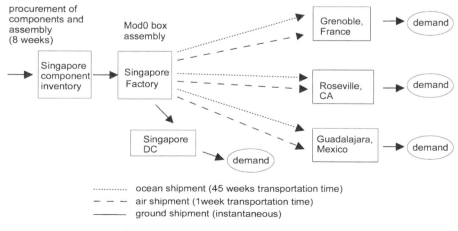

Figure 8.2.   The simplified supply chain

given the supplying entity has stock available. For the DCs in France, the US and in Mexico, these leadtimes are the air and sea transportation times, which are inputs to the model. For the Singapore DC we assume the leadtime to be zero due to its proximity to the factory. The leadtime for the Singapore component inventory is taken to be the component procurement time plus the assembly time. This absorption of the assembly time into the leadtime requires that assembly capacity is not restrictive.

4.2.   Demand

Demand occurs only at the DCs. We assume that the weekly demands can be described as sequences of independent non-negative random variables. Weekly demands are not assumed to be identically distributed; indeed they have been historically been highly non-stationary, due to life cycle issues. Non-negativity of demand is a technical assumption needed to make the mathematical analysis of the model more tractable. This assumption is not very restrictive for most of the product's life where actual demands vastly exceed returns. In fact, if demand is determined as sell-through+to^2, non-negativity is virtually guaranteed.

Earlier investigations at NSD showed that Weibull distributions fit historical demand data reasonably well, and so we used them to model demand. The distribution parameters are determined by using the forecast as the distribution's mean and the empirical coefficient of variation (obtained from historical data) as the distribution's coefficient of variation. Our analysis does not depend on this particular choice of distribution type. In fact, the software tool that implements our approach allows the user to choose distributions characterizing demand.

4.3.   Cost and performance measures

Although our analysis allows time- and location-dependent rates for holding and depreciation, NSD chose to apply constant, universal rates to all inventories and products in transit. In particular, NSD lacked the data required to support a nonlinear depreciation rate model. Both costs are expressed as a constant percentage of the standard material cost.

A one-time obsolescence cost is applied to the products in inventory at the end of the planning horizon. Currently the obsolescence cost is taken to be a fixed percentage of standard material cost for all products, and it is only applied to DCs, since inventory there represents assembled MOD0s whereas factory inventory represents more versatile components. These choices are not required for the analysis. In practice, the actual value that can be recovered at the end of a product's life will vary from product to product, but it is difficult to estimate these numbers. We use the historical average for similar products.

Transportation cost is assumed to be proportionate to the amount shipped. The rate can vary across product, location and shipment mode. The proportionality assumption ignores the effects of batching. This assumption is justified by the fact that NSD combines several different products in one container, making batch sizes small compared to total volume shipped. This practice enables them to ship full containers only.

The performance of the system is measured by a type II service level for each of the products across all DCs. A type II service level is simply a fill rate, i.e., the mean of the fraction of demand satisfied off-the-shelf.

Our goal is to minimize the total cost incurred over the life cycle while guaranteeing a minimum service level for each product.

5.   Approach

In what follows we describe the approach we used to find cost-effective inventory management policies. This approach, based on the model assumptions outlined above, is comprised of three major components. The first component is a refinement of the class of inventory strategies that we consider, and a parameterization of this class of strategies to afford efficient search among them. This is described in Section 5.1. The second element of our approach is a simulation engine that emulates and evaluates how policies in our candidate class perform with respect to the metrics we have established. The simulation is described in Section 5.2 The third component, the subject of Section 5.3, is a search procedure that relies on the parameterization of strategies and the simulation results to select the best inventory strategy.

5.1.   Candidate inventory replenishment strategies

5.1.1   Implementable strategies.   In every time period the inventory manager makes decisions about how much to order for the factory and how much to ship to each of the DCs. All order and shipment quantities are non-negative. Naturally, the decision must be based only on the information available at the time it is made. In any period, the state of the supply chain is completely described by the on-hand and in-transit quantities for each DC and the factory. Since demand is assumed

to be independent between periods, there is an optimal feedback policy, i.e., a policy that only depends on the current state rather than the complete history of the system.

Unfortunately, the optimal strategy is unlikely to have a simple structure, like that of a order-up-to or a two-bin policy. Because there are two shipment modes with different lead times, the optimal replenishment decisions at the DCs will depend on the inventory on hand as well as shipment quantities due to arrive in several different periods. The limitations of the Red Pepper system at that time prohibited implementing such complex rules.

Even if Red Pepper were flexible enough to allow more complex replenishment rules, computing an optimal policy may be difficult. To integrate our policy computation into a decision support tool that would interface with the Red Pepper planning engine, lengthy computation times would be unacceptable.

For these reasons, and because of the project urgency, we decided to restrict our attention to the class of order-up-to policies. These policies are described by an on-hand inventory target for each location and each period. This choice is in accordance with current practice at NSD, and seemed most fitting to ensure a timely implementation of the project. Since the class of order-up-to policies does not, in general, contain the optimal policy, we plan to consider more general policies at a later stage of the project.

5.1.2 Further reduction of the class of strategies.
A general order-up-to strategy requires inventory targets for the factory and each DC for each period. For the current supply chain layout with one factory and four DCs, and a typical planning horizon of 52 weeks, we would need 260 parameters to describe a order-up-to strategy completely. To reduce complexity of the optimization problem, we propose a heuristic approach that reduces the number of decision variables while still including a broad and sensible class of strategies.

In what follows we describe a class of strategies parameterized by two numbers, instead of one decision variable for each location and time period. As we will illustrate, these two numbers correspond respectively to factory and DC service levels in independent single location inventory models. To motivate this parameterization, recall that in a single location inventory model with one replenishment mode, the optimal inventory targets are determined by the service level desired. These targets reflect the changing magnitude of demand as well as the changing variability, expressed as its standard deviation, over time. The parameterization of policies allows us to construct a sensible inventory strategy

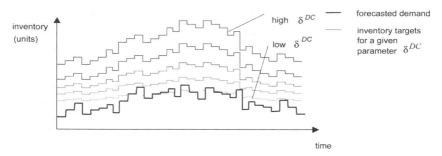

Figure 8.3.   Inventory targets at one DC for different values of δ^{DC}

for the entire supply chain from targets derived from simple, single location models.

To illustrate this approach, consider the following class of strategies,
described by two parameters δ^{DC} and δ^{WH}, each between zero and one:

- For each DC, use δ^{DC} as a type I service level[3] for a single location
  inventory with only the longer leadtime mode available (ocean) to
  determine a sequence of inventory targets for each period of the
  horizon of the problem $(S_1^{DCi}, S_2^{DCi}, \ldots, S_N^{DCi})$, $i = 1, \ldots, M$. For
  details on how the inventory targets are calculated, see Section
  5.1.4. Figure 8.3 illustrates how different values of δ^{DC} produce
  different sets of inventory targets at a particular DC.

- For the warehouse, use δ^{WH} as a service level for a single node
  inventory model with a fixed leadtime (the warehouse replenishment leadtime) to determine a sequence of inventory targets for
  each period of the horizon of the problem $(S_1^{WH}, S_2^{WH}, \ldots, S_N^{WH})$.
  The warehouse demand used for this calculation is the sum of the
  DC demands offset by the DCs ocean leadtime.

For each pair of parameters $(\delta^{DC}, \delta^{WH})$, the above calculations yield a
sequence of inventory targets for each of the DC and the warehouse and
each period. We call δ^{DC} and δ^{WH} the shape parameters of the strategy.
This strategy, used as described in the previous section by Red Pepper
and applied to a particular stream of realizations of demand, determines
shipments and replenishment orders.

The class of strategies described above is certainly less general then
the class of all possible combinations of on-hand inventory targets in the
DCs and the warehouse and, therefore, the best among these strategies
may not include the best of all possible order-up-to strategies. On the
other hand, the strategies within this class have the following desirable

properties:

- Targets reflect changes in magnitude and variability of demand over time.[4]

- Service is balanced over all DCs.

- Service is balanced over time, except for at the end of life, during which we impose a controlled decline in service to avoid expensive excess inventory; see Section 5.1.3.

Moreover, changing the parameters δ^{DC} and δ^{WH} changes the total amount of supply chain inventory, the balance between warehouse and DC inventory, and in turn the ratio of air to ocean shipments. While hard to prove, we feel that the generality of the strategies considered ensures that the best strategy among them will perform closely to the overall best strategy in terms of cost. In Section 5.3 we demonstrate how we select a strategy among those considered.

5.1.3 Changing service level requirements. Up to this point, we assumed the shape parameters of a particular strategy to be constant over the whole horizon. This assumption corresponds to a constant level of service for all periods. Under some circumstances it may be desirable to distribute service unequally over the horizon. For example, in the phase of new product introduction availability of the product might be crucial and one may want to achieve a higher level of service (lower probability of stock-outs) during this phase. On the other hand, at the end of life one might be willing to sacrifice service in order to clear the supply chain of the product and minimize excess inventory. We accommodate these requirements by introducing a service level shape that can be specified by the user. This service level shape is a multiplicative parameter for each period. If the service level shape for period t is α_t and for period t' it is $\alpha_{t'}$ then the required service level in period t' is $\alpha_{t'}/\alpha_t$ times the required service level in period t. For a given sequence of service level shape parameters $(\alpha_t)_{t=1,\ldots,N}$ we modify the class of candidate policy $(\delta^{DC}, \delta^{WH})$ by using the nonstationary sequence of type I service levels $\alpha_1 \delta^{DC}, \ldots, \alpha_N \delta^{DC}$ at the DCs and the service level δ^{WH} in every period at the warehouse.

We apply the service level shape to the DC targets only for two reasons. The first reason is that NSD incurs obsolescence cost only for excess inventory at the DCs. The second is that the level of service at the DCs is a major management objective while the service at the factory is merely a means to an end. Figure 8.4 depicts a typical service level shape for an end-of-life situation.

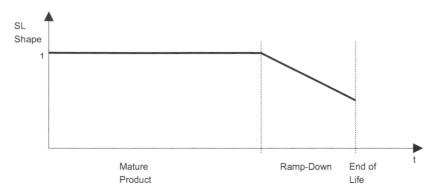

Figure 8.4.   Typical service level shape

5.1.4   Calculating candidate inventory policies.   The inventory targets for a given shape parameter δ^{DC} (respectively, δ^{WH}) are derived from a single-location, single-shipment-mode inventory model, with leadtime equal to the longer shipment leadtime. For the DC case, let y_t be the inventory position in period t after demand but before an order is placed, and let D_k denote the demand in period k. The optimal myopic policy in period t is computed by finding the inventory position target y_t^* to satisfy

$$P(y_t^* - D_{t+1} - \cdots - D_{t+seaLT} > 0) = \alpha_t \delta^{DC}. \tag{8.1}$$

If the inventory position after ordering in period t is equal to y_t^* then the expected on-hand inventory after demand in period $t + seaLT$ is equal to

$$S_{t+seaLT} = y_t^* - E(D_{t+1}) - \cdots - E(D_{t+seaLT}). \tag{8.2}$$

The values $S_{t+seaLT}$, $t = 0, \ldots, N - seaLT$, are the on-hand inventory targets for these periods.

To determine warehouse inventory targets, we use essentially the same approach. The major difference lies in the structure of the demand. Warehouse demand is determined by the shipment requests from the DCs. Using DC demands as approximations for DC shipment requests (a reasonable assumption if inventory targets are reasonably stable from week to week) and assuming DCs request only ocean shipments, we estimate the warehouse demand in period t to be

$$D_t^{WH} = \sum_i D_{t+seaLT_i}^{DCi}. \tag{8.3}$$

Inventory position and on-hand targets are computed as in the DC case with respect to this demand stream.

5.2. Simulating inventory policies

The second major component of the approach is a simulation engine. The simulation takes, as input, an inventory strategy characterized by an order-up-to level at each location and time period in the horizon. (For example, this strategy could be one calculated as in Section 5.1.4 for a given pair of parameters $(\delta^{DC}, \delta^{WH})$. It also requires demand distribution information (forecast, coefficient of variation, and distribution type) for each period and location, as well as structural supply chain data and costs. From this input, the program simulates how Red Pepper would make shipment decisions given random demand drawn from the specified distributions. In the process, it measures the costs associated with a policy and its overall service level (the percentage of demand satisfied immediately from stock) for a given demand outcome. This is repeated for many randomly chosen demand scenarios, and the resulting costs and service levels for all runs are averaged. These averages are an approximation for the expected cost and service level of the given set of inventory targets. Confidence intervals for these approximations are also computed.

5.3. Searching for optimal policy parameters

Section 5.2 describes how we evaluate the expected cost and order fulfillment performance of an arbitrary set of inventory targets in the NSD supply chain. In particular, the simulation can be used to evaluate strategies in the class parameterized by δ^{DC} and δ^{WH}. In this section we discuss how we use the simulation to guide a search for the $(\delta^{DC}, \delta^{WH})$ expected to perform best according to NSD's cost and while meeting the overall service level criterion. To that end, define a point $(\delta^{DC}, \delta^{WH})$ in the grid as infeasible if its associated policy has an expected service level below the desired one.

The pair $(\delta^{DC}, \delta^{WH})$ lies in the square $[0,1) \times [0,1)$. We begin by discretizing this region into a grid of arbitrarily small grid size. We use three observations to enable efficient search of points in this grid. The first observation is that if a given point $(\delta^{DC}, \delta^{WH})$ is infeasible, then any point smaller in both coordinates is also infeasible. This follows from the fact that decreasing either parameter corresponds to decreasing inventory targets uniformly at all locations, see Figure 8.5. The next two observations were made based on empirical evidence rather than proven analytically. For a fixed δ^{WH}, the cost is empirically observed to be strictly increasing in δ^{DC}. Therefore, the least feasible δ^{DC} is best for a fixed δ^{WH}. For a fixed δ^{DC}, the cost is quasi-convex in δ^{WH}. Thus, as you increase δ^{WH} the cost never decreases after it increases.

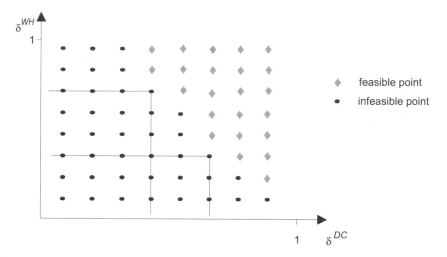

Figure 8.5.   Search grid for shape parameters

We use an efficient search algorithm that starts in the lower right hand corner of the grid and traces the boundary of the feasible region. The three observations about the cost function are used to fathom sections of the grid before they have been evaluated. In practice, the algorithm evaluates only about 6% of the grid before terminating with its recommended policy. It should be noted that even if all of these observations had been proven analytically, our approach would still be a heuristic method for finding the optimal $(\delta^{DC}, \delta^{WH})$ that is feasible. The reason is that estimates of expected cost and service level, rather than exact values, guide the search. By adjusting the search to account for confidence intervals, we could improve the likelihood that it finds an optimal $(\delta^{DC}, \delta^{WH})$ under the condition whenever the observations are true. Even without this adjustment, because we use many simulation trials and produce very small confidence intervals, we are confident that this procedure finds optimal solution in the vast majority of cases.

6.   Sample Results

This section contains a representative sample of results obtained for one of the MOD0 boxes based on actual NSD data. The horizon of the model is 47 weeks. Two of the DCs have sea shipments available (Roseville, CA and Grenoble, France). The others use only one shipment mode. The results of the optimization and simulation with our tool are compared to the previous inventory targets used at NSD.

Figure 8.6 compares the cost for different minimum service levels, the NSD strategy in effect at the time we first implemented the tool, (NSD

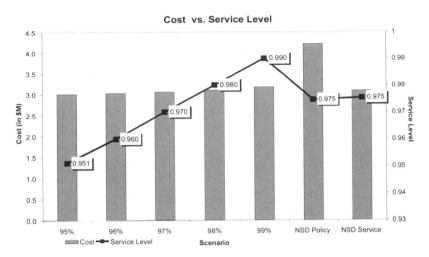

Figure 8.6. Cost for different scenarios

Strategy) and the best strategy found by our tool that achieves the same service level as the NSD strategy (NSD Service). Using our strategy, the simulation predicts a 27% cost improvement while maintaining the current service level. This savings is partly due to a reduction of average total supply chain inventory.

A peculiar feature of Figure 8.6 is that cost appears to be rather insensitive to service level. This deceptive phenomenon is due to the fact that a significant portion of supply chain costs over the horizon was largely independent of the inventory policy. In particular, costs such as holding costs associated with initial pipeline inventory and air shipment costs associated with early-horizon demand are unavoidable and thus unaffected by the policy. The effect of service level goals on cost is muted by the dominance of these fixed costs.

Figures 8.7 and 8.8 depict total cost and service level for candidate policies feasible with respect to the fill rate 95%. More specifically, in each Figure, each point on the grid gives the cost vs. service level for a feasible strategy corresponding to one point $(\delta^{DC}, \delta^{WH})$ in the search grid. The lower envelope of the curve corresponds to the least cost strategy for a given service level. It is an "efficient frontier" in the sense that for all points along it, increasing service requires increasing cost. For comparison we also plotted the cost vs. the service level of NSD's current policy at the time, represented by the point in the top center of the graph.

In Figure 8.7 we highlighted sets of policies that keep δ^{WH} constant and, therefore all use the same targets at the Factory. Consider, for example, the set of points marked by a square, corresponding to

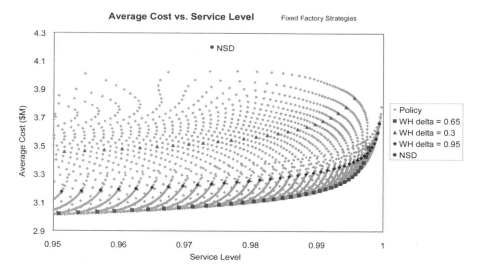

Figure 8.7.   The effect of changing DC strategies

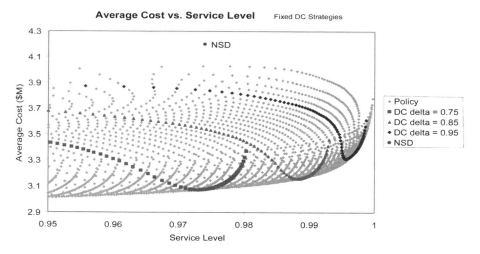

Figure 8.8.   The effect of changing factory strategies

$\delta^{WH} = 0.65$. From left to right δ^{DC} increases, i.e., DC targets increase. As a result, as you move from left to right in the set of points marked by a square, service improves at the expense of monotonically increasing cost. The use of different shipment modes is primarily dependent on the responsiveness of the factory, which is kept constant. Therefore, increased targets at the DC's increase the total supply chain inventory and with it the cost. An interesting observation is that $\delta^{WH} = 0.65$ appears to be close to optimal for a wide range of desired service levels for this particular product. Since δ^{WH} is the main driver for the percentage of shipments made by sea, this percentage is almost constant with respect to the required service level and is approximately equal to 30%.

By fixing δ^{DC} and varying δ^{WH} as in Figure 8.8, we observe a different behavior. In this Figure, the group of points marked by a square corresponds to $\delta^{DC} = 0.75$, and δ^{WH} increases as we move left to right. In general, increasing δ^{WH} will increase responsiveness of the Factory and therefore decrease shipment cost. On the other hand, it will also increase total supply chain inventory and with it inventory and possibly obsolescence cost. For a small δ^{WH}, i.e., low Factory targets, DCs are starved of inventory, and a high percentage of the products must be shipped by air as emergency shipments. High air shipment cost dominates the total cost. Increasing δ^{WH} will reduce the fraction of air shipments made and in doing so increase the in-transit inventory. Also, higher Factory targets mean higher inventory at the factory. This explains why in the group of points marked by a square (and similarly for other symbols), total cost first decreases and then increases as δ^{WH} increases, corresponding to moving from left to right on the graph.

The implementation of our results at NSD began in mid-1998. It is difficult to measure actual savings resulting from better inventory strategies, since the business situation changes rapidly. However, there are reliable estimates based on historical data and simulations. NSD attributes to our tool a reduction in the use of air shipments from 85% to 30% since implementation. They calculate the savings associated with this change at $300,000 per month to date, which corresponds to $3.5 million in annual freight cost savings. Furthermore, planners used the tool's recommendations to achieve a 30% reduction in total supply chain inventory. This cost reduction translates to an additional $3 million savings annually. These savings were obtained while achieving an estimated 95% service level for MOD0 boxes.

7. Conclusion

Prior to the commencement of this project, NSD used simple single-location, single-supply-mode inventory models to aid in the determination of inventory targets. These models also assume demand to be stationary. This chapter describes an approach that models the true two-tier, two-mode nature of the NSD supply chain and the interaction between these tiers. In particular, it takes into account the effects of shortages at the warehouse and the competition between DCs for warehouse supply. The model also accommodated product life-cycles by not requiring demand or service level goals to be stationary over time.

Our results demonstrate that this approach produced more effective inventory strategies than their previous method. This can explained by the fact that both the amount of supply chain inventory and its distribution between the factory and the DCs are critically important in determining overall costs and level of service. NSD's previous approach to finding inventory strategies was not general enough to address these factors.

Acknowledgments

The authors would like to acknowledge the support that was critical for the success of the project. During his tenure at HP Labs, Krishna Venkatraman was an invaluable contributor to all aspects of this work. Shailendra Jain, the manager of the Decision Technology Department at HP Labs, was instrumental in getting the project established and seeing it along. Our partners Roger Smith, Zawadi Pettes and Steve Kohler at NSD helped tremendously in identifying and refining the model and provided continuous feedback. We also thank Jeff McKibben, Amy Shao, Keiichiro Ichiryu, Darren Johnson, and Thomas Zscherpel in HP's Supply Chain Information Systems for helping us to build a graphical user interface for our tool. Finally, we are grateful to the referee for his or her helpful suggestions for improving the exposition of this work.

Notes

1. Throughout the chapter we will use Factory and Warehouse synonymously. Since capacity constraints are not a concern in the assembly of MOD0 boxes, the Factory functions in our model as an ordinary central storage location that in the inventory literature is usually called Warehouse.

2. For the purposes of this analysis, 'sell-through+to' can be thought of as end-customer demand.

3. A type I service level is the probability of not stocking out in a given period.

4. Restricting the search to stationary strategies, e.g., would simplify the analysis but seems like too strong a restriction for highly non-stationary demand of products like computers.

References

[1] AGGARWAL, P. AND K. MOINZEDEH. (1994) Order Expedition in Multi-Echelon Production/Distribution Systems. *IIE Transactions*, Vol. 26, 86–96.

[2] ALLEN, S. AND D. ESOPO. (1968) An Ordering Policy for Stock When Delivery Can Be Expedited. *Operations Research*, Vol. 16, 830–833.

[3] AXSATER, S. (1993) Exact and Approximate Evaluation of Two-Level Inventory Systems. *Operations Research*, Vol. 41, 777–785.

[4] BULINSKAYA, E. (1964) Some Results Concerning Optimal Inventory Policies. *Theory of Probability and Applications*, Vol. 9, 389–403.

[5] BULINSKAYA, E. (1964) Steady-State Solutions in Problems of Optimal Inventory Control. *Theory of Probability and Applications*, Vol. 9, 502–507.

[6] CHIANG, C. AND G. GUTIERREZ. (1996) Periodic Review Inventory System with Two Supply Modes. *European Journal of Operational Research* 94, 527–547.

[7] CLARK, A.J. AND H. SCARF. (1960) Optimal Policies for a Multi-Echelon Inventory Problem. *Management Science*, Vol. 6, 475–490.

[8] DANIEL, K. (1962) A Delivery-Lag Inventory Model with Emergency Order. In: *Multistage Inventory Models and Techniques*, Scarf, Gilford, Shelly (Eds.), Stanford University Press, Stanford, California.

[9] DEUERMEYER, B. AND L. SCHWARZ. (1981) A Model for the Analysis of System Service Level in Warehouse Retailer Distribution Systems. In: *Multi-Level Production/Inventory Control Systems: Theory and Practice*, Schwartz, L. (Ed.), TIMS Studies in Management Science, 16, Amsterdam, North-Holland, 51–67.

[10] FEDERGRUEN, A. AND P. ZIPKIN. (1984) Allocation Policies and Cost Approximations for Multilocation Inventory Systems. *Naval Research Logistic Quarterly*, Vol. 31, 97–129.

[11] FEDERGRUEN, A. AND P. ZIPKIN. (1984) Approximations of Dynamic Multilocation Production and Inventory Problems. *Management Science*, Vol. 30, 69–84.

[12] FUKUDA, Y. (1960) Optimal Policies for the Inventory Problem with Negotiable Leadtime. *Management Science*, Vol. 10, 690–708.

[13] GRAVES, S. (1985) A Multi-Echelon Inventory Model for a Repairable Item with One-For-One Replenshment. *Management Science*, Vol. 31, 1247–1256.

[14] GRAVES, S. (1996) Multiechelon Inventory Model with Fixed Replenishment Intervals. *Management Science*, Vol. 42, 1–18.

[15] JACKSON, P. (1988) Stock Allocation in a Two-Echelon Distribution System, or 'What To Do Until Your Ship Comes In'. *Management Science*, Vol. 34, 880–895.

[16] KARLIN, S. (1960) Dynamic Inventory Policy with Varying Stochastic Demands. *Management Science*, Vol. 6, 231–258.

[17] LEE, H. AND K. MOINZADEH. (1987) Two Parameter Approximations for Multi-Echelon Repairable Inventory Models with Batch Ordering Policy. *IIE Transactions*, Vol. 19, 140–149.

[18] MOINZADEH, K. AND H. LEE. (1986) Batch Size and Stocking Levels in Multi-Echelon Repairable Systems. *Management Science*, Vol. 32, 1567–1581.

[19] MOINZADEH, K. AND S. NAHMIAS. (1988) A Continuous Review Mode for an Inventory System with Two Supply Modes. *Management Science*, Vol. 34, 761–773.

[20] MOINZADEH, K. AND C. SCHMIDT. (1991) An $(S-1, S)$ Inventory System with Emergency Orders. *Operations Research*, Vol. 39, 308–321.

[21] NEUTS, M. (1964) An Inventory Model with an Optional Time Lag. *SIAM*, Vol. 12, 179–185.

[22] PYKE, D. AND M. COHEN. (1994) Multiproduct Integrated Production-Distribution Systems. *European Journal of Operations Research*, Vol. 74, 18–49.

[23] SHERBROOKE, C. (1968) METRIC: Multi-Echelon Technique for Recoverable Item Control. *Operations Research*, Vol. 16, 122–141.

[24] SVORONOS, A. AND P. ZIPKIN. (1988) Estimating the Performance of Multi-Level Inventory Systems. *Operations Research*, Vol. 36, 57–72.

[25] VEINOTT, A. (1965) Optimal Policy for a Multi-Product, Dynamic, Nonstationary Inventory Problem. *Management Science*, Vol. 12, 206–222.

[26] VEINOTT, A. (1966) The Status of Mathematical Inventory. *Management Science*, Vol. 12, 745–777.

[27] WHITTMORE, A. AND S. SAUNDERS. (1977) Optimal Inventory Under Stochastic Demand with Two Supply Options. *SIAM Journal of Applied Mathematics*, Vol. 32, 293–305.

[28] WRIGHT, G. (1968) Optimal Policies for a Multi-Product Inventory System with Negotiable Leadtimes. *Naval Research Logistics Quarterly*, Vol. 15, 375–401.

Chapter 9

INVENTORY ALLOCATION AT A SEMICONDUCTOR COMPANY

Alexander O. Brown
Manugistics, Inc.
901 Mariner's Island Blvd., San Mateo, CA 94404
abrown@manu.com

Markus Ettl
IBM Research Division
T.J. Watson Research Center, Yorktown Heights, NY 10598
msettl@us.ibm.com

Grace Y. Lin
IBM Research Division
T.J. Watson Research Center, Yorktown Heights, NY 10598
gracelin@us.ibm.com

Raja Petrakian
Xilinx, Inc.
2100 Logic Drive, San Jose, CA 95124
raja.petrakian@xilinx.com

David D. Yao*
IEOR Department, 302 Mudd Building
Columbia University, New York, NY 10027
yao@ieor.columbia.edu

*Research undertaken while an academic visitor at IBM Research Division, T.J. Watson Research Center.

1.   Introduction

Properly managing inventory is a challenging but critical problem for high-technology manufacturers. Due to rapidly declining manufacturing costs and high obsolescence risk, holding inventory is very expensive and risky. Customers often require near immediate availability of a large variety of products with highly uncertain demand and long lead times. Thus, manufacturers must hold high levels of inventory despite the expense and risk. Xilinx, a firm that produces programmable logic integrated circuits (ICs) for sale to original equipment manufacturers, faces this situation. In the programmable logic industry, manufacturing costs decline at rates of anywhere from 20% to 75% annually, and obsolescence costs are on the order of 5–10% of the gross inventory. As of 1999, Xilinx had upwards of 10,000 different finished products, only about 5% of which would be classified as having stable demand; total manufacturing lead times were 2–3 months; and a significant fraction of customer orders were requests to be filled same-day. A high level of serviceability is critical since although an IC may only be one component of hundreds in a customer's product, delays in delivery may shutdown a production line. Thus, service is a critical characteristic in supplier selection and retention. To meet this high level of service with long lead times and uncertain demands, leading programmable logic manufacturers in 1998 were holding very large inventories, 80 to 150 dollar days-of-inventory, despite the risk and expense. (The inventory measured in dollar days is the net inventory divided by the average cost of goods sold per day.)

To help lower the required inventory and maintain or increase serviceability, IC manufacturers like Xilinx use product and process design strategies like postponement (e.g., Brown *et al.* [4], Lee [12]). Under a postponement strategy, inventory is held at an intermediate point in a generic, non-differentiated form and is only differentiated when demand is better known. To take full advantage of the postponement strategy, one must determine the inventory levels in the intermediate (generic) and finished goods stocking points that allow for the best service at the lowest cost. This paper describes an inventory model developed for Xilinx to do just that.

After a brief review of the relevant literature in the rest of this section, we describe in section 2 the production-inventory planning process at Xilinx, and discuss the inventory management objectives. In section 3, we formulate the inventory allocation model, using an analytical approximation to compute backorders in the multi-echelon environment. The objective of the model is to determine the safety stock levels at each stocking point that maximize a serviceability measure with constraints on inventory holding cost by echelon. In section 4, we present an efficient algorithm for solving the optimization problem. In section 5, we extend the model and the algorithm to allow an additional set of constraints on the safety factor at each stocking location. In section 6, we conduct numerical experiments to test the performance of the analytical approximation and optimization procedure and to provide insights into how inventory should be allocated between echelons. Finally, we demonstrate the overall performance gains using a Xilinx data set.

1.1.   Related literature

There is an extensive literature on multi-echelon inventory models. See Axsaeter [3] and Federgruen [7] for reviews of this literature. A number of papers have been written that develop models and document their use in practice. One of the first was Sherbrooke [14] who developed the classic continuous review, multi-echelon inventory model to manage repairable item inventory for the U.S. Air Force. The model consists of two echelons, the upper one consisting of one single depot supplying repaired parts, and the lower one consisting of n identical stocking locations. The objective is to identify a stocking policy that minimizes backorders at the lower echelons subject to a constraint on inventory investment. Sherbrooke[15] improved the accuracy of the model.

Cohen *et al.* [5] developed a system called Optimizer that allowed IBM to make strategic changes to the configuration and control of a spare parts inventory network. Lee and Billington [11] describe the use of a multi-echelon inventory model to evaluate the performance of a printer supply chain at Hewlett Packard. Graves *et al.* [9] developed a model for requirements planning in a multi-echelon inventory system. The analysis is based on a deterministic model for a single echelon that serves as a building block for modeling multi-echelon systems. The model was used to determine inventory placement in a film manufacturing line at Kodak. Graves and Willems [10] model the supply chain as a spanning tree, develop an optimization algorithm, and describe the application at Kodak to determine strategic placement of safety stock in the Kodak supply chain. Ettl *et al.* [6] described a model to characterize the inventory-service tradeoff in multi-echelon supply networks. The model features performance evaluation and optimization with service level constraints. It captures the interdependence of inventory levels at different stocking locations, as well as their effect on overall system performance. Feigin [8] discusses an example of the application of this model to a personal computer manufacturing company. Other optimization models for supply and distribution networks include Arntzen *et al.* [2] and Andersson *et al.* [1].

2.   Production-Inventory Planning at Xilinx

2.1.   Process description

Xilinx produces programmable logic devices (PLDs), ICs that are used in a variety of products such as computer peripherals and networking or telecommunication equipment. Unlike traditional logic devices, known as Application Specific Integrated Circuits (ASICs), the specific logic and functionality of a PLD is not configured during physical fabrication. Instead, the specific function is programmed after fabrication by the customer through software. PLDs are not completely generic: they are differentiated by a variety of physical characteristics such as speed, voltage, gate count, and packaging types. However, since they are not custom made, there is an expectation of high availability, as opposed to ASICs that are often built-to-order with lead times on the order of months.

As illustrated in Figure 9.1, the manufacturing process consists of two major stages: fabrication and assembly & test. In fabrication, many identical die are created on a raw silicon wafer through a complex process of adding and removing patterned layers of various materials. Each die on the wafer will correspond to a given device. In assembly and test, the wafers are sliced into the individual die, and each die is assembled with a particular package type and tested for functionality and speed. Xilinx, like many semiconductor manufacturers, contracts out most of the manufacturing. Fabrication is performed at vendors in Taiwan and Japan. The wafers are then shipped to assembly vendors in Korea, Taiwan, and the Philippines, where they are held in die bank inventory until needed.

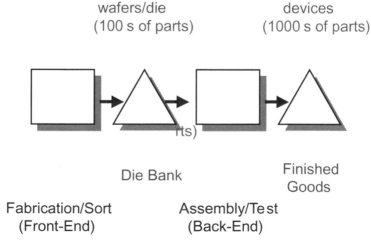

Figure 9.1. The Xilinx manufacturing process

Although separate die banks are maintained at each assembly vendor, transfer of die between vendor locations is done as needed. Thus, the die bank is modeled as a single inventory stocking location. Assembled parts are usually shipped to Xilinx facilities in San Jose or Ireland for testing, where they are held in finished goods inventory. Due to tax reasons, San Jose finished goods are not typically used to service areas typically serviced by the Ireland facility and vice-versa. Thus, the finished goods inventories at San Jose and Ireland are modeled as distinct inventory stocking locations.

The inventory held in die bank is fairly generic, consisting of about 200 different die types based on features such as voltage, gate count, and design architecture. Differentiation of the device into finished goods occurs during assembly and test. With different package types, grades, pin counts, speeds, and customer unique specifications, the total number of finished good types is very large. Currently, there are approximately 10,000 different finished goods parts that are active in Xilinx planning systems.

Since die bank inventory is more generic, it serves as an inventory postponement point. The die bank inventory is managed using a wafer starts planning package. Given a desired inventory target, this package determines the total amount to start at the fabrication contractor using a monthly base-stock policy. That is, the total on-hand inventory and work-in-process is netted out of total requirements (forecast plus inventory target) for all finished goods that use each type of die. Finished goods inventories are managed using a planning system based on assembly starts, with separate systems being run in Ireland and San Jose. Finished goods are classified as build-to-order or build-to-stock. For build-to-order parts, assembly requests are made according to customer orders - die are pulled from the die bank and started into assembly. For build-to-stock parts, assembly starts are determined using a weekly base-stock inventory control policy, driven by a forecast. The classification is adjusted periodically using an automated process based on volume, demand variability, cycle time, holding cost, and backorder cost. In 1998, about 80% of parts were build-to-order, but this represented only 10% of the total volume. Although considered in the actual model that we implemented, for the model presented in this paper, we do not consider the finished goods parts that are built-to-order. In the problem formulation in section 3, we discuss how such parts affect the results and how we include such parts in the actual implementation of the model at Xilinx.

The inventory optimization is integrated with Xilinx's current MRP planning system as follows. Demand, inventory, cost, price, and other

part specific data are collected from various databases and processed using database programming scripts. A graphical user interface (GUI) is
used to collect various user-specified information such as the inventory
budget. The data is transformed into relational data files necessary to
run the inventory optimization code. The output of the optimization
is a recommended target level for each inventory buffer. This output is
translated into projected service and inventory levels. Using this output
report, the supply chain planner can adjust various parameters in the
GUI and re-run the model. The final inventory targets are loaded into the
MRP-like planning system described earlier. The planning system uses
these targets along with forecasts, order backlog, current on-hand inventory levels, and WIP inventory levels to determine production starts to
be issued to the contract manufacturers.

2.2.   The inventory allocation problem

As with most inventory decision problems, choosing the appropriate
objective function is difficult and highly dependent on the business environment. A commonly used method is to find the minimum inventory
levels so that each finished good part achieves an expected level of service, e.g., 95% fill rate. At Xilinx, however, managers felt uncomfortable
with that approach since they wanted parts to have different service levels, but were very unsure about what those service levels should be. More
importantly, they wished to be able to specify an input that would tie
directly to key measures in the company's overall financial statements:
the corporate or business-unit inventory. Thus, it was determined that
the problem to be solved is an inventory allocation problem: for a given
inventory budget, what allocation provides the highest level of overall
service?

The key measure of service at Xilinx was total costed "delinquencies"
(costed backorders) across all parts, with each part having a different
unit delinquency cost. The individual unit delinquency costs were determined (and updated quarterly) using a percentage of unit revenue
based on input from the sales department using the following factors:
stage in product life-cycle, impact of product availability on future sales
of other products, competitive nature of product, and proportion of the
product's demand due to key customers. Since the inventory model
allocates an inventory budget across all the parts, the key is the relative sizes of the delinquency costs, not so much the absolute cost for
each part. Although there was no minimum service level requirement for
each product, minimum inventory levels for build-to-stock parts were set.

Xilinx's managers had a good feel about the minimum inventory levels, and they effectively act as an approximate surrogate for a minimum service level.

Separate inventory budgets were to be specified for different echelons (die bank versus finished goods) and for different locations within an echelon (San Jose finished goods and Ireland finished goods). Die bank inventory is often used as a buffer against unforeseen events such as capacity shortages at subcontractors or quality problems. So managers liked to be able to control the total die bank inventory to account for these additional risks. Management required that the inventory budgets be expressed in dollar days-of-inventory ($DOI) since they have a good intuitive feel for its value and since it is the measure used to present inventory results in key financial statements. However, parts have different holding cost rates depending on the product type, the stage in the product life-cycle, etc. Thus, a more appropriate measure is inventory *holding cost* budget, representing the total amount of money the company is willing to spend holding inventory. Using a weighted-average holding cost and current inventory targets, the inventory system automatically converts the inventory budget in $DOI into an approximate holding cost budget. This holding cost budget is used as a constraint in the optimization.

3. Problem Formulation

The inventory system consists of two echelons. The upstream echelon is die bank (DB) inventory and the downstream is finished goods (FG) inventory. The FG inventory is distributed at a set of locations, indexed by $i \in \mathcal{M} := \{1, \ldots, M\}$. In the Xilinx case, the locations were San Jose finished goods and Ireland finished goods. Each location i supplies a set of end products, $j \in \mathcal{S}_i := \{1, \ldots, N_i\}$. A separate inventory is kept for each product to satisfy its own demand stream. Let D_{ij} denote the demand (per time unit) for product j at location i, which is assumed to follow a normal distribution. For simplicity, we shall refer to such product at a given location as type ij product.

Let 0 index the die bank location. There are N_0 types of die, each with its own inventory, indexed by $d \in \mathcal{S}_0 := \{1, \ldots, N_0\}$. The relationship between the two stages is a one-to-many mapping: each type of die is used to make one or more end products, but each end product uses only a single type of die. Hence, for each $d \in \mathcal{S}_0$, let $\mathcal{S}_d$ denote the set of end products that use type d die. Then, the demand for type d die is: $\sum_{(i,j) \in \mathcal{S}_d} D_{ij}$. (One unit of end product uses one unit of die.)

Let L_d be the production leadtime for each die d. Let L_{ij} be the leadtime to transform die d into type ij product. The replenishment lead time for FG inventory (end products) is this nominal leadtime, L_{ij}, plus a delay time which takes into account the possible stockout of die bank inventory. This *actual* leadtime, denoted $\tilde{L}_{ij}$, has expected value

$$E(\tilde{L}_{ij}) = E(L_{ij}) + \tau_d p_d, \qquad (9.1)$$

where p_d is the stockout probability of type d die inventory used to make type ij product, and τ_d is the expected additional delay when this stockout occurs. Both quantities depend on the inventory levels of type d die. The derivation of the term $\tau_d p_d$ in (9.1) will be deferred to the next section.

The following quantities will play a key role in our analysis:

$$\mu_d := E(L_d) \sum_{(i,j) \in \mathcal{S}_d} E(D_{ij}), \qquad \sigma_d^2 := E(L_d) \sum_{(i,j) \in \mathcal{S}_d} \mathsf{Var}(D_{ij}), \quad (9.2)$$

for each die bank location d and

$$\tilde{\mu}_{ij} := E(D_{ij})E(\tilde{L}_{ij}), \qquad \tilde{\sigma}_{ij}^2 := \mathsf{Var}(D_{ij})E(\tilde{L}_{ij}), \qquad (9.3)$$

for each product ij. Note that these quantities are closely related to the mean and the variance of the number of jobs in an $M^X/G/\infty$ queue, which we use as our basic model for both the die and FG inventories, such that the number of jobs in the queue models the number of outstanding orders in the corresponding inventory system. Refer to the detailed derivation in [6]. (Also note that these quantities are not necessarily equal to the mean and the variance of demand over leadtime, unless the leadtime is a deterministic constant; refer to [6].)

For each ij product, let h_{ij} be the unit inventory holding cost (per time unit), and let s_{ij} be the unit backorder cost. For each type d die, let h_d be the unit inventory holding cost. Our decision variables are x_{ij}, the base-stock level for each type ij product's FG inventory, and x_d, the base-stock level for each type d die inventory. For all $d \in \mathcal{S}_0$, and all $i \in \mathcal{M}$, $j \in \mathcal{S}_i$, we can express the base-stock levels as

$$x_d = \mu_d + \sigma_d k_d, \qquad x_{ij} = \tilde{\mu}_{ij} + \tilde{\sigma}_{ij} k_{ij}, \qquad (9.4)$$

where k_d and k_{ij} are known as safety factors. This way, the decision variables are transformed from the base-stock levels to the safety factors.

Let B_{ij} and I_{ij} denote, respectively, the backorder and the on-hand inventory of type ij product. Similarly, let B_d and I_d denote the backorder and the on-hand inventory of type d die. Following the analysis in

[6], we can derive

$$\mathsf{E}(B_d) = \sigma_d G(k_d), \qquad \mathsf{E}(I_d) = \sigma_d H(k_d); \tag{9.5}$$

and

$$\mathsf{E}(B_{ij}) = \tilde{\sigma}_{ij} G(k_{ij}), \qquad \mathsf{E}(I_{ij}) = \tilde{\sigma}_{ij} H(k_{ij}); \tag{9.6}$$

where the functions $G(\cdot)$ and $H(\cdot)$ are defined as follows:

$$G(x) := \mathsf{E}[Z - x]^+ = \int_{z \geq x} (z - x)\phi(z)dz = \phi(x) - x\bar{\Phi}(x), \tag{9.7}$$

and

$$H(x) := \mathsf{E}[x - Z]^+ = x + G(x) = \phi(x) + x\Phi(x), \tag{9.8}$$

where Z denotes the standard normal variate, $\phi(x)$ and $\Phi(x)$ denote, respectively, the density function and distribution function of Z, and $\bar{\Phi}(x) := 1 - \Phi(x)$.

We can now present our optimization problem as follows:

$$\min \sum_{i \in \mathcal{M}} \sum_{j \in \mathcal{S}_i} s_{ij}\mathsf{E}(B_{ij})$$

$$\text{s.t.} \sum_{j \in \mathcal{S}_i} h_{ij}\mathsf{E}(I_{ij}) < C_i, \quad i \in \mathcal{M};$$

$$\sum_{d \in \mathcal{S}_0} h_d\mathsf{E}(I_d) \leq C_0;$$

where C_i denotes the inventory holding cost budget for location i, and C_0 is the inventory holding cost budget for die inventory.

Substituting the derived backorder and inventory expressions given by (9.5) and (9.6) into the above formulation, we have

$$\min \sum_{i \in \mathcal{M}} \sum_{j \in \mathcal{S}_i} s_{ij}\tilde{\sigma}_{ij}G(k_{ij}) \tag{9.9}$$

$$\text{s.t.} \sum_{j \in \mathcal{S}_i} h_{ij}\tilde{\sigma}_{ij}H(k_{ij}) \leq C_i, \quad i \in \mathcal{M}; \tag{9.10}$$

$$\sum_{d \in \mathcal{S}_0} h_d\sigma_d H(k_d) \leq C_0. \tag{9.11}$$

As mentioned earlier, for the model presented in this paper we do not consider the finished goods parts that are built-to-order, although we do consider them in the actual model that was implemented. By definition, the build-to-order parts have a fixed base-stock level of zero. Their presence, however, does affect availability of inventory at the die bank. Thus, in the implementation of the model at Xilinx, the die bank demand given by (9.2) is adjusted to include the demand for build-to-order parts. Also, delinquency costs for build-to-order parts are added, where an order for such a part is assumed to be delinquent if fulfillment takes longer than the nominal lead time.

4.   The Optimization Algorithm

Note that in the above optimization problem, the constraints (9.10) and (9.11) are necessarily binding at optimality since $\tilde{\sigma}_{ij}$ is decreasing in k_d (to be justified below) and since, from (9.7) and (9.8), the function $G(\cdot)$ is decreasing whereas $H(\cdot)$ is increasing.

The optimization problem is difficult to solve since $\tilde{\sigma}_{ij}$ is a complex function of the decision variable k_d, as we shall explain below. Therefore, we first consider the problem in which the values of k_d are given, and focus on finding the optimal values of k_{ij}. Since the k_d's are given, we can ignore the constraint in (9.11). The problem is then separable over the locations. That is, for each location $i \in \mathcal{M}$, we can solve a separate optimization problem as follows:

$$\min \ \sum_{j \in \mathcal{S}_i} s_{ij} \tilde{\sigma}_{ij} G(k_{ij}) \qquad (9.12)$$

$$\text{s.t.} \ \sum_{j \in \mathcal{S}_i} h_{ij} \tilde{\sigma}_{ij} H(k_{ij}) = C_i. \qquad (9.13)$$

From (9.7) and (9.8), it is obvious that $G(\cdot)$ and $H(\cdot)$ are both convex functions (note in particular that $(x)^+$ is convex in x, and expectation is a linear operator). Thus, a standard Lagrangian multiplier approach can be applied, leading to the following equations which are necessary and sufficient for optimality:

$$s_{ij} \tilde{\sigma}_{ij} G'(k_{ij}) + \lambda_i h_{ij} \tilde{\sigma}_{ij} H'(k_{ij}) = 0, \quad j \in \mathcal{S}_i; \qquad (9.14)$$

$$\sum_{j \in \mathcal{S}_i} h_{ij} \tilde{\sigma}_{ij} H(k_{ij}) = C_i; \qquad (9.15)$$

where λ_i is the Lagrangian multiplier.

From (9.7) and (9.8), we have

$$G'(x) = -\bar{\Phi}(x), \qquad H'(x) = \Phi(x).$$

Combining with (9.14) yields

$$k_{ij} = \Phi^{-1}\left(\frac{s_{ij}}{s_{ij} + \lambda_i h_{ij}}\right) \qquad j \in \mathcal{S}_i. \tag{9.16}$$

Expression (9.15) then becomes

$$\sum_{j \in \mathcal{S}_i} h_{ij}\tilde{\sigma}_{ij} H\left(\Phi^{-1}\left(\frac{s_{ij}}{s_{ij} + \lambda_i h_{ij}}\right)\right) = C_i, \tag{9.17}$$

from which $\lambda_i > 0$ can be obtained. Since both $H(\cdot)$ and $\Phi^{-1}(\cdot)$ are increasing functions, the left hand side is decreasing in λ_i. Thus, we can simply solve using bisection. Newton's method can also be used, in particular to take advantage of the explicit form of the derivative, since

$$\frac{d}{dy}H(\Phi^{-1}(y)) = \Phi(\Phi^{-1}(y)) \cdot \frac{1}{\phi(\Phi^{-1}(y))} = \frac{y}{\phi(\Phi^{-1}(y))}.$$

Next, we discuss how to determine the optimal values of k_d. Note that $\tilde{\sigma}_{ij}$ depends on k_d if and only if $(i,j) \in S_d$, i.e., if and only if d is the die type that is used for the type ij product. Following the analysis in [6], we have the following approximation for τ_d in (9.1):

$$\tau_d \approx \frac{\mathsf{E}(L_d)\mathsf{E}(B_d)}{p_d x_d},$$

where, recall, $x_d = \mu_d + \sigma_d k_d$ is the base-stock level of type d die. Substituting the above approximation into (9.1), and writing $\ell_d := \mathsf{E}(L_d)$ and $\ell_{ij} := \mathsf{E}(L_{ij})$, we have

$$\mathsf{E}(\tilde{L}_{ij}) \approx \mathsf{E}(L_{ij}) + \frac{\mathsf{E}(L_d)\mathsf{E}(B_d)}{x_d} = \ell_{ij}\left[1 + \frac{\ell_d \sigma_d G(k_d)}{\ell_{ij} x_d}\right].$$

Hence, taking into account (9.3),

$$\tilde{\sigma}_{ij}^2 = \mathsf{Var}(D_{ij})\mathsf{E}(\tilde{L}_{ij}) \approx \sigma_{ij}^2\left[1 + \frac{\ell_d \sigma_d G(k_d)}{\ell_{ij} x_d}\right], \qquad (i,j) \in S_d. \tag{9.18}$$

From the above expression, it is evident that $\tilde{\sigma}_{ij}$ is convex in k_d, since both $G(k_d)$ and $1/x_d$ are decreasing and convex in k_d (and both are positive). Therefore, the objective function and the constraints in (9.9)–(9.11) are all convex (and separable) in the decision variables k_d, allowing us once again to use the Lagrangian approach to derive the optimality conditions. (Recall that $H(\cdot)$ is a convex function.)

We have the following optimality equations, from which the optimal k_d's and the Lagrangian multiplier λ_0 can be obtained:

$$\sum_{(i,j)\in\mathcal{S}_d} [s_{ij}G(k_{ij}) + \lambda_i h_{ij}H(k_{ij})]\frac{d\tilde{\sigma}_{ij}}{dk_d} + \lambda_0 h_d\sigma_d\Phi(k_d) = 0, \quad (9.19)$$

for each $d \in \mathcal{S}_0$, and

$$\sum_{d\in\mathcal{S}_0} h_d\sigma_d H(k_d) = C_0. \tag{9.20}$$

Taking derivatives on both sides of (9.18) with respect to k_d, we find:

$$\frac{d\tilde{\sigma}_{ij}}{dk_d} = -\frac{\sigma_{ij}^2\sigma_d\ell_d}{2\tilde{\sigma}_{ij}\ell_{ij}x_d}[\bar{\Phi}(k_d) + \frac{\sigma_d}{x_d}G(k_d)]. \tag{9.21}$$

Substituting into (9.19), we have

$$\frac{\lambda_0 h_d\Phi(k_d)}{[\bar{\Phi}(k_d) + \frac{\sigma_d}{x_d}G(k_d)]}$$

$$= \frac{\ell_d}{2x_d}\sum_{(i,j)\in\mathcal{S}_d}\frac{\sigma_{ij}^2}{\ell_{ij}\tilde{\sigma}_{ij}}[s_{ij}G(k_{ij}) + \lambda_i h_{ij}H(k_{ij})] := \delta_d. \tag{9.22}$$

As the above equation is quite intractable, we use an approximate iterative approach to solve it. We evaluate the right hand side δ_d using the value of k_d from the previous iteration (note, x_d, and $\tilde{\sigma}_{ij}$, are functions of k_d). Thus, we solve the equation

$$\frac{\lambda_0 h_d\Phi(k_d)}{[\bar{\Phi}(k_d) + \frac{\sigma_d}{x_d}G(k_d)]} = \delta_d,$$

with δ_d being a constant, independent of k_d. In addition, direct verification will show that $G(k)$ is a higher-order infinitesimal than $\bar{\Phi}(k)$, i.e., $G(k)/\bar{\Phi}(k) \approx 0$, when k is large. Hence, we ignore the $G(k_d)$ term in the denominator of the left hand side of (9.22). This way, the equation in (9.22) is simplified to the following:

$$\lambda_0 h_d\Phi(k_d) = \delta_d\bar{\Phi}(k_d);$$

and hence,

$$k_d = \Phi^{-1}\left(\frac{\delta_d}{\delta_d + \lambda_0 h_d}\right), \qquad d \in \mathcal{S}_0. \tag{9.23}$$

Substituting the above into (9.20), we have

$$\sum_{d \in \mathcal{S}_0} h_d \sigma_d H \left(\Phi^{-1} \left(\frac{\delta_d}{\delta_d + \lambda_0 h_d} \right) \right) = C_0, \qquad (9.24)$$

from which λ_0 can be solved in exactly the same manner as solving the equations in (9.17).

Of course, we can also choose to keep the $G(k_d)$ term in the denominator of the left hand side of (9.22), if we treat x_d, and $\tilde{\sigma}_{ij}$ as given. Our numerical experience indicates, however, that this may cause difficulties in convergence, which never occur when the $G(k_d)$ term is ignored. Furthermore, ignoring the $G(k_d)$ term appears to have minimal impact on the optimality, as we have observed from our numerical studies.

We can now summarize the algorithm that solves the optimization problem in (9.9)–(9.11) as follows:

Algorithm 1:

1. (Initialization). Set $k_d = 0$ for all $d \in \mathcal{S}_0$, i.e., set all die safety stocks to zero.

2. For each location $i \in \mathcal{M}$, perform a line search on equation (9.17) to determine λ_i. Set k_{ij} as in (9.16), for all $j \in \mathcal{S}_i$. Set δ_d as defined in (9.22) using the current values of k_{ij} and k_d.

3. Using the value of δ_d just determined, perform another line search on equation (9.24) to determine λ_0. Set the new value of k_d to that given by (9.23), for all $d \in \mathcal{S}_0$.

4. Iterate between Steps 2 and 3 until convergence.

5. Safety-Stock Constraints

To minimize overstocking risk for various products, managers at Xilinx wanted to be able to place upper limits on the level of inventory allowed. For example, for a mature product nearer the end of its product life, setting a high inventory target would not be wise. Although we can, in part, deal with this problem by setting a higher unit holding cost for the product, the managers felt that this additional level of control was important. Thus, we add the following constraints on the safety factors to the optimization problem in (9.9)–(9.11):

$$k_{ij} \leq \bar{k}_{ij}, \quad j \in \mathcal{S}_i, i \in \mathcal{M}; \qquad k_d \leq \bar{k}_d, \quad d \in \mathcal{S}_0; \qquad (9.25)$$

where $\bar{k}_{ij}$ and $\bar{k}_d$ are positive upper limits.

One way to deal with these additional constraints is to first ignore them, obtain the optimal solution as in the last section, and then check if any of the decision variables has exceeded the upper limit. If so, set all such variables at their upper limits, remove them from the problem, and adjust the C_i and C_0 values correspondingly — subtract the inventory costs associated with those removed variables, and resolve the problem involving the remaining variables.

However, in general this will lead to a suboptimal solution (unless in the very special case of a single upper limit constraint). Therefore, we propose the following marginal allocation algorithm. Starting from some initial solution, we choose one of the safety factors and increase its current value by an incremental amount ϵ. We choose the safety factor that provides the greatest decrease in the objective function with the least increase in the inventory holding cost. (Recall $G(\cdot)$ is a decreasing function, while $H(\cdot)$ is increasing; and both are convex.) Specifically, note the following:

$$G(k + \epsilon) - G(k) \approx \epsilon G'(k) = -\epsilon \bar{\Phi}(k),$$
$$H(k + \epsilon) - H(k) \approx \epsilon H'(k) = \epsilon \Phi(k).$$

Hence, the safety factor to increment is the one that achieves the minimum below:

$$\min \left\{ -\frac{s_{ij} \bar{\Phi}(k_{ij})}{h_{ij} \Phi(k_{ij})}; \ j \in \mathcal{S}_i, \ i \in \mathcal{M} \right\}. \tag{9.26}$$

The k_d's have to be treated differently in this marginal allocation since incrementing k_d contributes to the objective function in (9.9) via the factor $\tilde{\sigma}_{ij}$, for $(i,j) \in \mathcal{S}_d$. Therefore, to evaluate the corresponding change in the objective function, we need to first evaluate the change in the $\tilde{\sigma}_{ij}$ values. Then, we solve the minimization problem in (9.12) and (9.13) given the new $\tilde{\sigma}_{ij}$ values, along with the upper limit constraint (on k_{ij}) in (9.25). This minimization problem itself is solvable by the marginal allocation scheme outlined above. In particular, use the criterion in (9.26) to choose the variable k_{ij} at each step for awarding the increment ϵ. We need to solve this minimization problem for each location i where type d die is used.

The following is a formal description of the algorithm:

Algorithm 2:

1. (Initialization) Set $k_d = 0$ for all $d \in \mathcal{S}_0$. Set $\mathcal{S}_0' = \mathcal{S}_0$.

2. For each $d \in \mathcal{S}_0'$, perform the following steps:

   2.1. Evaluate $\tilde{\sigma}_{ij}$ for all $(i, j) \in \mathcal{S}_d$;

   2.2. For each i, such that $(i, j) \in \mathcal{S}_d$, perform the following steps:

      2.2.1. Set $k_{ij} = 0$ for all $j \in \mathcal{S}_i$. Set $\mathcal{S}_i' = \mathcal{S}_i$.

      2.2.2. Find the index j' that achieves the minimum below:

$$\min \left\{ -\frac{s_{ij} \bar{\Phi}(k_{ij})}{h_{ij} \Phi(k_{ij})}; \ j \in \mathcal{S}_i' \right\}.$$

      If $k_{ij'} + \epsilon > \bar{k}_{ij'}$, set $\mathcal{S}_i' \leftarrow \mathcal{S}_i' - \{j'\}$; else set $k_{ij'} \leftarrow k_{ij'} + \epsilon$.

      2.2.3. Repeat 2.2.2 until $\mathcal{S}_i'$ becomes empty or the constraint in (9.13) becomes binding, whichever happens first. Evaluate the objective function in (9.12) corresponding to the obtained solution.

   2.3. Sum up all the objective values from 2.2.3, i.e., over all i such that $(i, j) \in \mathcal{S}_d$.

3. For each $d \in \mathcal{S}_0'$, set $k_d + \epsilon$ and repeat 2. Denote the sums obtained in 2.3 as $f_d(k_d)$ and $f_d(k_d + \epsilon)$, respectively, for the two iterations of 2.

4. Identify d', such that it achieves the maximum below:

$$\max \left\{ \frac{f_d(k_d) - f_d(k_d + \epsilon)}{h_d \sigma_d \Phi(k_d)}; \ d \in \mathcal{S}_0' \right\}.$$

   If $k_{d'} + \epsilon > \bar{k}_{d'}$, set $\mathcal{S}_0' \leftarrow \mathcal{S}_0' - \{d'\}$; else set $k_{d'} \leftarrow k_{d'} + \epsilon$.

5. Repeat 3 and 4 until $\mathcal{S}_0'$ becomes empty or the constraint in (9.11) becomes binding, whichever happens first.

Clearly, the above algorithm is quite demanding computationally. The marginal allocation problem for each FG inventory location will have to be repeated solved, at every step of deciding the marginal allocation of the k_d values. We therefore propose the following heuristic approach, which works much faster.

Algorithm 3:
   Treat the k_d variables just like the k_{ij}'s. Specifically, at every step, we consider all k_d's, in parallel with all k_{ij}'s, and pick one of them to award

the increment ϵ. While the choice of the k_{ij}'s still follow the criterion in (9.26), the choice of k_d is the one that achieves the following minimum:

$$\min_{d \in \mathcal{S}_0} \left\{ \frac{1}{h_d \sigma_d \Phi(k_d)} \sum_{(i,j) \in \mathcal{S}_d} s_{ij} G(k_{ij}) \frac{d\tilde{\sigma}_{ij}}{dk_d} \right\}$$

$$= \min_{d \in \mathcal{S}_0} \left\{ -\frac{\ell_d}{2h_d x_d} \left[\frac{\bar{\Phi}(k_d)}{\Phi(k_d)} + \frac{\sigma_d G(k_d)}{x_d \Phi(k_d)} \right] \sum_{(i,j) \in \mathcal{S}_d} \frac{s_{ij} \sigma_{ij}^2 G(k_{ij})}{\ell_{ij} \tilde{\sigma}_{ij}} \right\}$$

$$= \min_{d \in \mathcal{S}_0} \left\{ -\frac{\ell_d}{2h_d x_d^2 \Phi(k_d)} [\mu_d \bar{\Phi}(k_d) + k_d \phi(k_d)] \sum_{(i,j) \in \mathcal{S}_d} \frac{s_{ij} \sigma_{ij}^2 G(k_{ij})}{\ell_{ij} \tilde{\sigma}_{ij}} . \right\}$$

The contest between the best k_d variable and the best k_{ij} variable is determined by the smaller of the two.

6.   Numerical Results

In this section, we describe numerical experiments that were used for the following four purposes:

(A) To validate the accuracy of the analytical approximation by comparing its results to those obtained from simulations;

(B) To validate the effectiveness of the optimization procedure by comparing its results to the solutions found by a local neighborhood search procedure;

(C) To illustrate how a fixed inventory holding cost budget should be split between the die bank and finished goods inventories in order to minimize delinquency costs;

(D) To quantify the improvements that can be achieved with using this model by comparing the optimized inventory policy to the 1996 inventory policy at Xilinx.

For the experiments (A) to (C), we use a system with three die and nine end products. Each type of die is used to make three end products as illustrated in Figure 9.2. All FG inventory for end products is kept at the same location. The inventory holding cost budget is C_0 for the die bank and C_1 for finished goods.

We represent the customer orders per time unit as i.i.d. normal random variables with mean $\mathsf{E}(D_{ij}) = 100$ and coefficient of variation

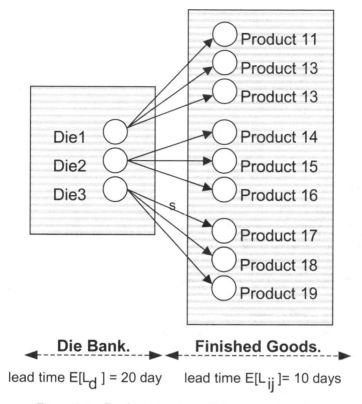

Die Bank.   **Finished Goods.**

lead time $E[L_d] = 20$ day   lead time $E[L_{ij}] = 10$ days

Figure 9.2. Product structure of the example system

$cv(D_{ij}) = 0.5$. The manufacturing lead times are deterministic with $E(L_d) = 20$ time units for type d die, and $E(L_{ij}) = 10$ time units for type ij end product. For experiments (A) and (B), we assume identical unit holding cost among all die types ($h_d = 1.0$) and among all end products ($h_{ij} = 1.75$). The unit backorder costs s_{ij} range from 0.875 to 4.25 as shown in Table 9.1, assuming identical backorder costs for end products that are manufactured from the same die type.

Table 9.1. Inventory holding costs and backorder costs for the example system

	Holding Costs						Backorder Costs		
h_1	h_2	h_3		h_{11-13}	h_{14-16}	h_{17-19}	s_{11-13}	s_{14-16}	s_{17-19}
1.0	1.0	1.0		1.75	1.75	1.75	0.875	1.75	4.25

(A) Validation of analytical approximation.   Using the optimization procedure described in section 4, we first ran Algorithm 1 to find the optimal base-stock levels x_d^* and x_{ij}^*. We determined the values of the expected number of backorders and the expected on-hand inventory for each product in the system under the analytical approximation (eqns. (9.5) and (9.6)). Second, using these same base-stock levels, we determined the values of the backorders and on-hand inventory using simulation. The simulation point estimates and confidence intervals were obtained by the batch means method, dividing every run into 20 batches of 10,000 arrivals per batch. Tables 9.2 and 9.3 report the results for two cases with different values of C_0 and C_1.

The numerical agreement between simulations and the analytical approximation is good, and in line with the agreement we found in our other experiments. If the simulation results are treated as exact, the relative error of the on-hand inventory approximation is less than 2 percent. The analytical backorder queue lengths are within the confidence interval in most cases.

Tables 9.2 and 9.3 also illustrate the impact of holding and backorder costs on safety stock. As one would expect, the optimal reorder point of

Table 9.2.   Backorder queue length and on-hand inventory comparisons; $C_0 = 1500$; $C_1 = 3000$

d	k_d^*	x_d^*	$E(B_d)$ analy.	$E(B_d)$ sim.	$E(I_d)$ analy.	$E(I_d)$ sim.
1	1.16	6467	22.9	23.6 ± 2.7	464.6	464.9 ± 10.8
2	1.27	6507	18.4	18.1 ± 1.4	500.3	$498.7 \pm\ \ 8.8$
3	1.37	6546	14.8	15.5 ± 1.5	535.2	540.6 ± 10.0

ij	k_{ij}^*	x_{ij}^*	$E(B_{ij})$ analy.	$E(B_{ij})$ sim.	$E(I_{ij})$ analy.	$E(I_{ij})$ sim.
11	0.75	1127	20.5	21.8 ± 1.4	136.5	136.9 ± 2.9
12	0.75	1127	20.5	22.4 ± 1.5	136.5	138.6 ± 3.0
13	0.75	1127	20.5	22.5 ± 1.2	136.5	135.7 ± 2.6
14	1.13	1186	10.0	10.7 ± 0.9	186.0	187.1 ± 3.0
15	1.13	1186	10.0	10.9 ± 0.6	186.0	185.4 ± 3.0
16	1.13	1186	10.0	10.7 ± 0.7	186.0	186.3 ± 3.0
17	1.58	1254	3.8	4.1 ± 0.4	248.9	251.7 ± 2.5
18	1.58	1254	3.8	4.6 ± 0.5	248.9	249.8 ± 3.8
19	1.58	1254	3.8	4.7 ± 0.5	248.9	247.8 ± 3.1

Table 9.3. Backorder queue length and on-hand inventory comparisons; $C_0 = 2000$; $C_1 = 5000$

d	k_d^*	x_d^*	$\mathsf{E}(B_d)$ analy.	sim.	$\mathsf{E}(I_d)$ analy.	sim.
1	1.69	6666	7.1	7.3 ± 1.1	648.0	647.7 ± 12.0
2	1.74	6686	6.3	5.9 ± 0.8	666.4	665.5 ± 9.4
3	1.79	6705	5.5	6.1 ± 0.9	685.6	690.1 ± 10.2

ij	k_{ij}^*	x_{ij}^*	$\mathsf{E}(B_{ij})$ analy.	sim.	$\mathsf{E}(I_{ij})$ analy.	sim.
11	1.73	1274	2.7	2.8 ± 0.3	270.2	270.3 ± 3.4
12	1.73	1274	2.7	3.1 ± 0.3	270.2	271.7 ± 3.6
13	1.73	1274	2.7	3.0 ± 0.3	270.2	268.6 ± 3.1
14	2.02	1320	1.2	1.3 ± 0.2	314.8	315.7 ± 3.4
15	2.02	1320	1.2	1.4 ± 0.2	314.8	314.1 ± 3.1
16	2.02	1320	1.2	1.4 ± 0.2	314.8	315.1 ± 3.2
17	2.37	1373	0.5	0.5 ± 0.1	367.3	370.3 ± 2.7
18	2.37	1373	0.5	0.6 ± 0.1	367.3	367.9 ± 3.9
19	2.37	1373	0.5	0.6 ± 0.1	367.3	365.8 ± 3.3

a finished goods product increases with its backorder cost, or, in other words, the higher the backorder cost of a product, the more safety stock should be allocated to that product. Similarly, for the die bank, it is advantageous to allocate the largest safety stock to those die types that supply finished goods with the highest backorder cost (provided all other parameters are the same).

(B) Validation of optimization procedure. As it is computatio-nally intractable to exhaustively search for the best base-stock policy, we performed a limited enumeration as follows. First, we run a neighbor-hood search to generate 1,000,000 solutions around the base-stock policy found by Algorithm 1, using the analytical approximation to find the cor-responding die bank and FG holding costs. Second, if that solution is feasible (i.e., the holding costs at die bank and finished goods do not exceed the constraints C_0 and C_1), we evaluate the objective value with simulation. Finally, we pick the best solution and compare it against the solution of our algorithm.

Table 9.4 shows the objective value and optimal base-stock policy for several values of C_0 and C_1. The results of the optimization algorithm

Table 9.4. Comparisons between optimal base-stock policies from optimization and enumeration

	C_0	C_1	z^*	Die Bank			Finished Goods		
				x_1^*	x_2^*	x_3^*	x_{11-13}^*	x_{14-16}^*	x_{17-19}^*
Opt.	1300	4000	74.0 ± 4.2	6400	6431	6462	1208	1260	1319
Enum.	1300	4000	70.5 ± 4.2	6390	6433	6464	1218	1250	1320
Opt.	1400	4200	59.0 ± 3.4	6441	6469	6498	1222	1272	1330
Enum.	1400	4200	56.0 ± 3.6	6480	6453	6465	1211	1279	1332
Opt.	1500	4400	47.1 ± 2.8	6481	6507	6533	1235	1284	1341
Enum.	1500	4400	44.4 ± 3.0	6496	6529	6485	1236	1280	1342
Opt.	1600	4600	37.5 ± 2.2	6519	6544	6568	1248	1296	1352
Enum.	1600	4600	35.4 ± 2.4	6502	6544	6534	1244	1299	1353
Opt.	1700	4700	32.9 ± 1.9	6557	6580	6603	1254	1302	1357
Enum.	1700	4700	30.7 ± 2.0	6524	6615	6590	1257	1303	1352
Opt.	1800	4800	28.2 ± 1.7	6594	6615	6638	1261	1308	1362
Enum.	1800	4800	27.4 ± 2.1	6607	6625	6602	1258	1310	1362
Opt.	1900	4900	25.4 ± 1.4	6630	6651	6672	1267	1314	1367
Enum.	1900	4900	24.2 ± 1.5	6589	6654	6687	1263	1318	1366
Opt.	2000	5000	22.3 ± 1.3	6666	6686	6705	1274	1320	1373
Enum.	2000	5000	20.8 ± 1.3	6711	6637	6703	1272	1329	1365

and the enumeration are reported in the rows labeled "Opt." and "Enum.", respectively. The objective values z^* were obtained from simulation for each system under study. The base-stock policies found by the enumeration result in only slightly lower delinquency costs (the maximum relative deviation between the optimization and enumeration is 7.2%). Notice that the confidence intervals for the objective values overlap in all cases.

The number of iterations required to reach convergence in Algorithm 1 depends on the size of the problem (number of die, number of end products) and on the termination criterion. The termination criterion used in our experiments required that the L_∞-norm of the difference of two consecutive solution vectors be less than 10^{-8}. For the examples shown in Table 9.4, the number of iterations to convergence varied between 6 and 8.

We next analyze the performance of Algorithms 2 and 3. We keep the same parameter values as in experiment (A), except that we now introduce safety-stock constraints. For simplicity, we assume that the

Table 9.5. Comparisons between optimal base-stock policies from Algorithms 2 and 3 (with additional safety-stock constraints)

	C_0	C_1	$\bar{k}_{ij}$	z^*	Die Bank			Finished Goods		
					k_1^*	k_2^*	k_3^*	k_{11-13}^*	k_{14-16}^*	k_{17-19}^*
Alg.1	1500	3000	–	171.9	1.16	1.27	1.37	0.75	1.13	1.58
Alg.2	1500	3000	1.20	202.9	0.96	1.20	1.62	1.09	1.20	1.20
Alg.3	1500	3000	1.20	203.2	0.93	1.20	1.65	1.09	1.20	1.20
Alg.2	1500	3000	1.30	187.5	1.14	1.12	1.53	0.91	1.28	1.30
Alg.3	1500	3000	1.30	188.0	1.17	1.09	1.53	0.91	1.28	1.30
Alg.2	1500	3000	1.40	178.2	1.19	1.18	1.43	0.85	1.23	1.40
Alg.3	1500	3000	1.40	178.7	1.23	1.15	1.41	0.85	1.23	1.40
Alg.2	1500	3000	1.50	174.2	1.25	1.23	1.32	0.80	1.17	1.50
Alg.3	1500	3000	1.50	174.2	1.30	1.22	1.29	0.80	1.17	1.50
Alg.1	2000	5000	–	22.3	1.69	1.74	1.79	1.73	2.02	2.37
Alg.2	2000	5000	2.00	30.6	1.40	1.72	2.08	2.00	2.00	2.00
Alg.3	2000	5000	2.00	30.6	1.39	1.72	2.09	2.00	2.00	2.00
Alg.2	2000	5000	2.10	25.4	1.55	1.64	2.02	1.92	2.10	2.10
Alg.3	2000	5000	2.10	25.4	1.55	1.64	2.02	1.92	2.10	2.10
Alg.2	2000	5000	2.20	23.5	1.67	1.64	1.90	1.81	2.10	2.20
Alg.3	2000	5000	2.20	23.5	1.67	1.64	1.90	1.81	2.10	2.20
Alg.2	2000	5000	2.30	22.6	1.74	1.70	1.78	1.76	2.05	2.30
Alg.3	2000	5000	2.30	22.6	1.74	1.70	1.78	1.76	2.05	2.30

same upper limit, $\bar{k}_{ij}$, is imposed on all nine end products, whereas the three die types are unconstrained (i.e., $\bar{k}_d = \infty$). Table 9.5 shows the optimal delinquency costs, z^*, as well as the optimal base-stock policies, in terms of the safety factors k_d^* and k_{ij}^*, for different values of $\bar{k}_{ij}$. The top half of the table summarizes the results for budget values of $C_0 = 1{,}500$ and $C_1 = 3{,}000$, and the bottom half for budget values of $C_0 = 2{,}000$ and $C_1 = 5{,}000$. For comparison, the table also shows the optimal base-stock policy without safety stock constraints as generated by Algorithm 1. As in the previous examples, we evaluated the delinquency cost of each approach through simulation.

When safety factors are constrained, it is still optimal to allocate the largest amounts of safety stock to products with the highest backorder costs, although the differences become smaller when $\bar{k}_{ij}$ becomes smaller. On the other hand, it is not always optimal to give preference to die types that supply products with high backorder cost. For example, when $\bar{k}_{ij} = 1.30$, both algorithms suggest that k_1 be larger than k_2, even though the backorder costs of products manufactured from die type 1 are lower than the backorder costs of products manufactured from die type 2.

Notice that the difference between the objective values found by Algorithms 2 and 3 are very minor, as are the optimal base-stock policies. However, we observed that the heuristic approach of Algorithm 3 offers substantial advantages in terms of computation times. With a step size of $\epsilon = 0.005$, Algorithm 3 consistently produced solutions in less than one second on a 300 MHz workstation, compared to computation times in excess of 400 sec for Algorithm 2. The difference becomes even more significant in the Xilinx manufacturing environment where computational performance is important.

(C) Investigation of split of inventory budget between echelons.
One key question we wish to investigate is how the total inventory holding cost budget $C_0 + C_1$ should be split between the die bank and end products in order to minimize the delinquency cost. Figure 9.3 shows the optimal delinquency cost z^* as a function of $C_0/(C_0 + C_1)$, i.e., the relative amount of inventory holding cost budget allocated to the die bank, for four scenarios. The inventory holding cost for all die types is held constant at $h_d = 1.0$ across scenarios, but the values of h_{ij} are changed. For the four scenarios we used h_{ij} values of 1.0, 1.75, 5.25, and 10.50 and total inventory holding cost budget of 3,000, 4,000, 8,000, and 12,000 dollars.

From Figure 9.3, we learn that most of the inventory budget should be allocated to end products. For instance, when $h_{ij} = h_d = 1.0$, the optimal operating point is 0.2, suggesting that 20 percent of safety stock should be kept at the die bank, and 80 percent should be kept in finished goods. Table 9.6 shows the delinquency cost z^* and the optimal split C_0^* and C_1^* of the total inventory cost budget for each of the scenarios under study. In additional to reporting the budgets, the table also shows the total safety stock in units allocated to the die bank and finished goods. We observe that the while percent of inventory cost budget allocation to die bank actually decreases as the cost of holding FG stock increases, the number of *units* allocated to the die bank $\mathsf{E}(I_0^*)$ versus the number of units allocated to finished goods $\mathsf{E}(I_1^*)$ increases. For example, when $h_{ij} = 1.0$, about 20 percent of safety stock inventory should be kept in the die bank, whereas when $h_{ij} = 10.50$ the amount in die bank should be increased to roughly 55 percent. When h_{ij} is large, allocating a larger amount of inventory to the die bank is optimal since multiple units of die can be stocked at the same cost as, say, one unit of finished goods.

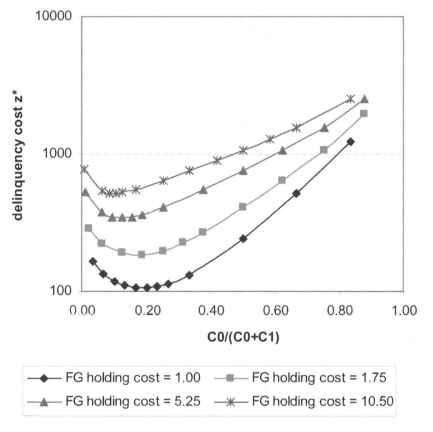

Figure 9.3. Delinquency cost vs. relative budget allocated to the die bank

Table 9.6. Optimal split of safety stock inventory between die bank and finished goods

h_{ij}	$C_0 + C_1$	C_0^*	C_1^*	z^*	Die (in units) $E(I_0^*)$	Finished Goods (in units) $E(I_1^*)$
1.00	3,000	600	2,400	106	605	2423
1.75	4,000	750	3,250	185	755	1880
5.25	8,000	1,000	7,000	343	1005	1349
10.50	12,000	1,250	10,750	519	1255	1035

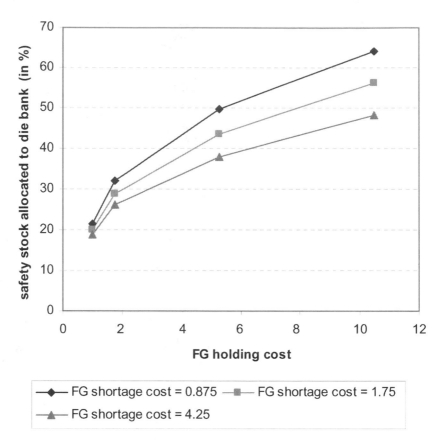

Figure 9.4.   Relative amount of safety stock (in units) allocated to the die bank

Figure 9.4 shows the percent of safety stock (in units) allocated to die bank as a function of FG holding cost h_{ij}, for three different values of the FG backorder cost s_{ij}. The amount kept in the die bank increases monotonically with h_{ij}. When s_{ij} is high, the relative amount of safety stock that should be kept at the die bank becomes smaller. For example, in the specific case of $h_{ij} = 10.50$, it is optimal to keep 64 percent die stock when $s_{ij} = 0.875$, 56 percent when $s_{ij} = 1.75$, and 48 percent when $s_{ij} = 4.25$.

(D) Quantifying improvements at Xilinx.   In order to project the improvements that can be achieved with optimization, we compared the optimized policy with that of the original policy implemented at Xilinx in 1996 when we began the development of the model. The 1996 policy is well-defined: all finished goods part inventories (for parts that were build-to-stock) had the same target level in days-of-inventory, and each of the die bank inventories had the same target level of in days-of-inventory. (Although the 1996 policy is no longer used and the new policy is considerably different, Xilinx has requested that we not reveal the exact targets in the old policy.) We used a full data set from 1998 in order to compare the 1996 policy and the optimized policy. The data set consisted of 104 die bank parts, 314 finished goods parts held at the Xilinx facility in Ireland, and 1194 parts held at the facility in San Jose. Comparisons between the two policies were made in two different ways. First, the total inventory holding cost budget was held at the same level for both policies, and the projected improvement in delinquency cost was found. Thus, the inventory holding cost that resulted from the 1996 policy was calculated and used as the holding cost budget constraint for the inventory optimization. The ideal split of this inventory holding cost budget between die bank and finished goods was found by running the inventory optimization under a number of different splits. Under the resulting inventory targets, the total delinquency cost was reduced by 54%. The huge improvement was primarily due to reallocation of the inventory among the various finished goods parts. Inventory targets for stable finished goods parts with low delinquency costs and high holding costs were set lower, while inventory targets for less stable parts with higher delinquency costs were set higher.

Second, the total delinquency cost was held constant for both policies and the projected improvement in inventory holding cost budget was found. Thus, the total delinquency cost for the data set was found using the 1996 policy. Then, the inventory optimization was run with a number of different settings of the inventory holding budget until the budget was found that yielded approximately the same delinquency cost. The optimized policy was found to yield a policy with a 19.9% reduction in overall inventory holding cost (which includes work-in-process) and a 51.2% reduction in inventory holding cost budget when only safety stock is considered.

7. Summary and Conclusions

In this paper, we explored an inventory allocation model for use in the semiconductor industry. Using an analytical approximation to determine backorders in the multi-echelon setting, we developed an efficient procedure to perform the allocation. Using numerical experiments, we demonstrated that the analytical approximation and solution procedure performed well, provided insights into how inventory should be allocated between the two echelons, and demonstrated the results for a Xilinx data set.

Implementation of the system at Xilinx was completed in 1999, but many of the gains were already achieved through the insights derived from running and testing this model. For example, Xilinx found that the 1996 inventory target for die bank was far too high relative to the overall budget. The ability to replenish the pipeline to finished goods (i.e., the fill rate of parts held at the die bank) is an increasing but concave function of the die bank inventory target. A large fraction of the parts had inventory targets set at a level past the point of diminishing returns, and the target inventory level could be dropped significantly with a very marginal impact on the die fill rate. Another insight was that the inventory mix among products within an echelon was very suboptimal. The old adage "it's not that we have too little inventory, we just don't have the right inventory" was certainly the case at Xilinx under the old inventory policies. Just using the insights derived from this model on how inventory target levels should depending on the cost and demand uncertainty characteristics allowed Xilinx to significantly improve performance.

Acknowledgments

The authors thank Luis Diaz of Xilinx for his work in the installation and implementation of the inventory allocation model at Xilinx. We appreciate the support and guidance of Chris Wire of Xilinx and Alan Spangler of IBM during the course of development and implementation. We also thank the anonymous referees for very helpful comments. The research of Alex Brown was undertaken while at Owen Graduate School of Management, Vanderbilt University, Nashville, TN. Part of David Yao's work was undertaken while on leave at the Department of Systems Engineering and Engineering Management, the Chinese University of Hong Kong, and supported in part by HK/RGC Grant CUHK4376/99E.

References

[1] ANDERSSON, J., S. AXSAETER, AND J. MARKLUND. (1998) Decentralized Multi-Echelon Inventory Control. *Production and Operations Management*, Vol. 7, 370–386.

[2] ARNTZEN, B.C., G.G. BROWN, T.P. HARRISON, AND L.L. TRAFTON. (1995) Global Supply Chain Management at Digital Equipment Corporation. *Interfaces*, Vol. 25, 69–93.

[3] AXSAETER, S. (1993) Continuous review policies. In: *Logistics of Production and Inventory*. S. Graves, A. Rinooy Kan, and P. Zipkin (Eds.), North Holland.

[4] BROWN, A., H. LEE, AND R. PETRAKIAN. (2000) Xilinx improves
     its Semiconductor Supply Chain using Product and Process Post-
     ponement, *Interfaces*, Vol. 30, No. 4, 65–80.

[5] COHEN, M., P.V. KAMESAM, P. KLEINDORFER, H. LEE, AND
     A. TEKERIAN. (1990) Optimizer: IBM's Multi-Echelon Inventory
     System for Managing Service Logistics, *Interfaces*, Vol. 20, No. 1,
     65–82.

[6] ETTL, M., G. FEIGIN, G. LIN, AND D. YAO. (2000) A Supply
     Network Model with Base-Stock Control and Service Requirements.
     Operations Research, Vol. 48, 216–232.

[7] FEDERGRUEN, A. (1993) Centralized Planning Models for Multi-
     Echelon Inventory Systems Under Uncertainty. In: *Logistics of Pro-
     duction and Inventory*. S. Graves, A. Rinooy Kan, and P. Zipkin
     (Eds.), North Holland.

[8] FEIGIN, G. (1998) Inventory Planning in Large Assembly Supply
     Chains. In: *Quantitative Models for Supply Chain Management*. S.
     Tayur, R. Ganeshan, and M. Magazine (Eds.), Kluver Academic
     Publishers.

[9] GRAVES, S., D. KLETTER, AND W. HETZEL. (1998) A Dynamic
     Model for Requirements Planning with Application to Supply Chain
     Optimization. *Operations Research*, Vol. 46, S35–S49.

[10] GRAVES, S. AND S. WILLEMS. (2000) Optimizing Strategic Safety
     Stock Placement in Supply Chains. *Manufacturing and Service Op-
     erations Management*, Vol. 2, 68–83.

[11] LEE, H. AND C. BILLINGTON. (1993) Material Management in
     Decentralized Supply Chains. *Operations Research*, Vol. 41, 835–
     847.

[12] LEE, H. (1996) Effective Inventory and Service Management
     Through Product and Process Redesign, *Operations Research*, Vol.
     44, No. 1, 151–159.

[13] NAHMIAS, S. (1993) *Production and Operations Analysis*. 2nd ed.
     Irwin.

[14] SHERBROOKE, C. (1968) METRIC: A Multi-Echelon Technique for
     Recoverable Item Control. *Operations Research*, Vol. 16, 122–141.

[15] SHERBROOKE, C. (1986) VARI-METRIC: Improved Approxima-
     tions for Multi-Indenture, Multi-Echelon Availability Models. *Op-
     erations Research*, Vol. 34, 311–319.

Chapter 10

LEADTIME, INVENTORY, AND SERVICE LEVEL IN ASSEMBLE-TO-ORDER SYSTEMS

Yashan Wang
MIT Sloan School of Management, Room E53-353
Cambridge, MA 02142
yawang@mit.edu

1.   Introduction

Supply chain design and management has recently gained unprecedented recognition as a vital factor in the success of a business. Among the most fundamental characteristics that determine the performance of a supply chain are capacity, inventory, delivery leadtime, and customer service level. The first two quantities are under direct management control and are associated with the cost in a supply chain; the last two are the effects of managerial decisions and are associated with customer expectation or requirement. While the qualitative relationship among them is often obvious, understanding their quantitative relationship is much more difficult yet increasingly important in today's business environment. This chapter intends to introduce a set of tools and results to help better understand the quantitative relationship among these fundamental measures and make supply chain decisions more effectively.

Our investigation focuses on a special type of supply chain, namely the assemble-to-order (ATO) system. ATO systems arise in many industries and businesses, with the well-known Dell-direct mode in personal computer industry being a typical example. The main feature of these systems is that multiple components are produced and kept in inventories to be assembled into final products according to customer orders. No (or very little) inventory is held to provide maximum flexibility for customer order configurations. On the other hand, since component procurement leadtimes are often long and uncertain, some inventory is maintained at the component level to hedge against uncertainties in demand and in component procurement and be more responsive to demand. There have been a lot of research efforts recently (Cheng et al. [8], Gallien and Wein [19], Glasserman and Wang [23, 24], Song [36], Song et al. [37], Song and Yao [38], Wang [42]) on the evaluation and design of ATO systems.

Unlike many inventory models where unlimited production capacity is assumed for each component, and component procurement leadtimes are taken to be constant, or stochastic but independent of the rest of the model, we consider the more realistic case where component production capacities are limited. We make an important assumption in our models that all components operate under a *base-stock* or *one-for-one* replenishment policy. Base-stock policies have been shown to be optimal in various other types of systems including: the multistage uncapacitated system of Clark and Scarf [10], its generalization to the uncapacitated assembly system of Rosling [32], the capacitated single-item system of Federgruen and Zipkin [16], [17], and its generalization to system with periodic demand of Aviv and Federgruen [3] and Kapucinski and Tayur [27]. Although we do not suggest that a base-stock policy remains optimal in our multi-component capacitated systems — it is probably not, and the true optimal policy is not yet known and probably has a very complex structure — its simplicity and popularity in practice make it an attractive candidate for our study, at least as a benchmark. Other related papers on multi-item production-inventory systems include Bessler and Veinott [4], Eppen and Schrage [13], Federgruen and Zipkin [15], Glasserman and Tayur [22], and Graves [25].

In much of the literature in stochastic inventory theory, customer service level is modeled indirectly through a penalty cost, a cost that is incurred when an order is not filled immediately. A higher penalty cost would yield, after cost minimization, inventory policies under which finished goods are more readily available. In practice, however, purchasing contracts rarely stipulate explicitly imposing a penalty on suppliers for not delivering orders on time. The cost for back order is often intangible (loss of good will, for example) and difficult to estimate. Although the qualitative relationship between penalty cost and customer service level is intuitively obvious, specifying a penalty cost does not reveal exactly how timely customer orders are fulfilled. Therefore, working with an explicitly defined service-level requirement is a more natural way to model reality than the penalty cost approach.

For systems with no capacity limit under periodic review policy, several authors (Bookbinder and Tan [5], Cohen et al. [11], Deuermeyer and Schwarz [12], Nahmias [30], Schneider and Ringuest [33], and Tijms and Groenevelt [41]) have studied customer service levels as measured in terms of the proportion of periods in which all demand is met from stock or the proportion of demand satisfied immediately from inventory; Agrawal and Cohen [1] analyze the cost-service performance in an assembly system with component commonality and constant procurement leadtime; Boyaci and Gallego [6] minimize inventory holding cost directly in a serial system subject to a lower bound on the time average fill rate; Zhang [43] studies the computational issues and bounds on the order fill rate in an ATO system with deterministic leadtimes and normally distributed demands.

In uncapacitated continuous-review models, Boyaci and Gallego [6] minimize inventory holding cost directly in a serial system subject to a lower bound on the time average fill rate; Gallien and Wein [19] examine the problem of optimally procuring components in an assembly system with Poisson demand and component procurement leadtimes having special (Gumbel) distribution; Song [36] calculates the exact off-the-shelf order fill rate in a related setting, under the assumptions of Poisson demand and deterministic production times; Song et al. [37] extend the model of [36] to allow for random (exponential) production times; Song and Yao [38] study performance bounds and use them as surrogates in several optimization problems that seek the best trade-off between inventory and service level.

Glasserman [20] initiates the recent study of capacitated production-inventory systems at high service levels. He develops bounds and approximations for setting base-stock levels in a system with zero leadtime. Glasserman and Liu [21] study the rare event simulation for performance evaluation at high service levels. Glasserman and Tayur [22] develop a simple method for multistage systems by approximating the echelon inventory by a sum of exponential random variables. Glasserman and Wang [23] characterize the quantitative trade-offs between delivery leadtime and inventory. Glasserman and Wang [24] identify measures of a facility's propensity to constrain service in a multi-stage ATO network and demonstrate that a facility with a minimal measure is a bottleneck for the service level.

We measure customer service level by the fill rate — proportion of orders filled within a target delivery leadtime — and characterize the fill rate first on the component level and then on the finished goods level. We show that at fixed high fill rates there is an approximately linear relationship between inventory and delivery leadtime. The effect of capacity (measured in terms of average unit production time) and its interplay with leadtime and inventory is also illustrated. Intuitively, the higher the capacity, the higher the fill rate; if the product capacity is decreased while everything else of the demand and production processes remains changed, then more inventory or a longer target leadtime or both are needed to achieve the same fill rate. However, capacity itself does only solely represent the impact of the product process on the service level. Variability, or more generally, the stochastic behavior of the production process plays an important role in determining the service level. This chapter goes beyond qualitative arguments and gives specific quantitative recipes. See in particular Subsection 3.2.

Building on the understanding of the quantitative relationships, we further explore how to efficiently manage inventory in ATO systems. Instead of measuring service level indirectly through a backorder penalty cost, as is the approach often used in the literature but difficult to apply in practice, we incorporate a fill rate requirement directly in our model. Based on an approximate formulation that is asymptotically exact as the fill rate becomes high, we develop a closed-form solution to the problem of minimizing inventory holding cost while satisfying a fill rate requirement. Extensive numerical studies show that the closed-form solution is very effective. In addition, we give insights on which system characteristics influence the decision on where to place inventory optimally. (See Remarks (b) and (c) in Subsection 4.2.)

The rest of the chapter is organized as follows. The next section describes the models we consider in more detail and gives results on the asymptotic fill rate in three progressively more complex systems. Section 3 identifies the asymptotically linear trade-offs between leadtime and inventory. Section 4 shows a fast and efficient approximate solution in closed form to the cost optimization problem under a service level constraint. Section 5 gives the proofs to some keys results in Sections 2 and 4 and introduces some technical tools used in our analysis. Section 6 contains some concluding remarks.

2.   The Model and the Asymptotic Fill Rate

2.1.   Basic setting and notation

Figure 10.1 illustrates the basic setting we consider. Multiple components are produced one at a time on dedicated facilities (the circles) and kept in inventories (the triangles). Each component operates under a continuous-review base-stock policy under which demand for a unit of a component triggers a replenishment order for that component. Production capacity is limited and is modeled as a single server queue. A *product* is a collection of a possibly random number of components of each type. Production intervals, interarrival times, and order sizes are random variables, all independent of each other and over time, but

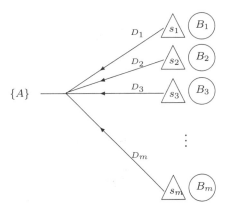

Figure 10.1. Multiple components assembled to order into a final product. The demand process has a generic interarrival time A. An order requires a generic component portfolio: D_1 units of component $1, \ldots, D_m$ units of component m. Component i has a generic unit production time B_i and a base-stock level s_i

we allow dependence among the order sizes of different components required by an order. We allow the batch sizes and the production times to have general distributions. To simplify the formulation of our problem, we assume throughout that the intcrarrival times have a continuous distribution. (Interarrival times with lattice distributions can be easily accommodated.) Demands are filled on a first-come-first-served basis; unsatisfied demands are backlogged.

We use the following notation,

$$A = \text{order interarrival time;}$$
$$B = \text{unit production interval;}$$
$$D = \text{batch order size;}$$
$$I = \text{inventory on hand;}$$
$$R = \text{response time;}$$
$$h = \text{holding cost rate;}$$
$$s = \text{base-stock level;}$$
$$x = \text{delivery leadtime.}$$

Upper case letters are used to denote random variables, and lower case ones are deterministic parameters. A subscript i refers to component i, ranging from 1 to m except in Subsection 5.1. Input variables A, B, and D represent the generic versions of their respective sequences. System performance variables R and I both have the steady state distribution. An order index is placed in parentheses when needed. Production intervals, interarrival times, and batch sizes are all independent of each other, but we allow dependence among the batches of different components required by an order. In other words, the vectors $\{D_1(1), \ldots, D_m(1)\}, \{D_1(2), \ldots, D_m(2)\}, \ldots$ are i.i.d. across orders, but their elements may be dependent. (This is particularly important if different groups of components constitute different products, see the next subsection.) When there are multiple products, a superscript j refers to product j, ranging from 1 to d. The response time R is the time elapsed from when an order arrives until it is filled, and the fill rate can be expressed as

$$P(R \le x).$$

To emphasize the dependence of R on inventory, we often write it as $R(s)$. In this chapter, it is more convenient to work with the complement of the fill rate, $P(R(s) > x)$, called the *unfill rate*, which is the proportion of orders not filled within the leadtime.

2.2.   Asymptotic fill rate

Before stating our main results of this section, we need to introduce the following concept. For any random variable Y, the symbol ψ_Y denotes the function

$$\psi_Y(\theta) = \log E[e^{\theta Y}], \tag{10.1}$$

called the *cumulant generating function* (c.g.f.) of Y. The function ψ_Y is convex, and it is differentiable in the interior of its domain (the set of θ at which it is finite). The c.g.f.s of all our input random variables will be finite for some $\theta > 0$, and this implies $\psi_Y'(0) = E[Y]$ and $\psi_Y''(0) = Var[Y]$. See Chapter 3 of Kendall [28] for relevant background.

We begin with a result for the case of a single component. In this setting, we may omit the subscript. We require that $E[D]E[B] < E[A]$ so that the system is stable and a steady-state response time exists. Defining $X = \sum_{j=1}^{D} B(j) - A$, we have

$$\psi_X(\theta) = \psi_D(\psi_B(\theta)) + \psi_A(-\theta). \tag{10.2}$$

In the following result, the notation $\lim_{s+x \to \infty}$ refers to the limit as either $s \to \infty$ (through integer values), or $x \to \infty$, or both.

Theorem 10.1 *If there is a $\gamma > 0$ at which $\psi_X(\gamma) = 0$, then with $\beta = \psi_B(\gamma)$,*

$$\lim_{s+x \to \infty} e^{\gamma x + \beta s} P(R(s) > x) = C \tag{10.3}$$

for some constant $C > 0$.

The result in (10.3) suggests that when the fill rate is high, its complement, the unfill rate, has the approximation

$$P(R(s) > x) \approx Ce^{-\gamma x - \beta s}. \tag{10.4}$$

As the unfill rate becomes small, it decays exponentially in the target delivery leadtime and the base-stock level. The smaller the values of γ and β, the lower the fill rate at the same x and s; i.e., the smaller the values of γ and β, the larger x or s has to be to achieve the same fill rate. In this sense, the parameters γ and β represent the sensitivity of the fill rate to leadtime and inventory, respectively. Obviously, both of these parameters are determined by the stochastic characteristics of the demand (inter-demand time A and order size D) and production (unit production time B) processes. Graphically, γ represents the positive intersection of the curve $\psi_X(\theta) = \psi_D(\psi_B(\theta)) + \psi_A(-\theta)$ with with the θ axis, and β is the the the function $\psi_B(\theta)$ evaluated at $\theta = \gamma$. Loosely speaking, the more variable the processes are while the system utilization level is fixed, or the higher the utilization at fixed process variability, the smaller the values of γ and β are. More discussion on the intuitive explanations of these two parameters will be given in the next section, particularly through their two-moment approximations in Subsection 3.2.

If γ in Theorem 10.1 exists, it is unique because ψ_X is convex and $\psi_X(0) = 0$. Sufficient conditions for the existence of γ are that $\psi_X(\theta)$ be finite for some $\theta > 0$ and that $\psi_X(\theta)$ not jump to infinity as θ increases — i.e., if $\bar{\theta} = \sup\{\theta \geq 0 : \psi_X(\theta) < \infty\}$, then $\psi_X(\theta) \to \infty$ as $\theta \uparrow \bar{\theta}$. These conditions are met for virtually all commonly used distributions. (An exception is the lognormal distribution, which has no exponential moments.) To avoid repeating technical conditions, we assume throughout that all input random variables (A, B and D) satisfy these conditions. The proof of Theorem 10.1 is deferred to Section 5.

In general, only a partial characterization of the constant C in (10.3) is available (see Equation (10.38) in Subsection 5.1). However, in the important special case of Poisson arrivals, we obtain an explicit formula:

Proposition 10.2 *In the setting of Theorem 10.1, if arrivals are Poisson with rate λ, then*

$$C = \lambda^{-1}(\lambda + \gamma)(1 - \lambda E[D]E[B])(\psi_D'(\beta)\psi_B'(\gamma)(\lambda + \gamma) - 1)^{-1}.$$

The proof of Proposition 1 follows by adapting the calculation in Equation 8.48 of Siegmund [35] and can be found in Glasserman and Wang [23].

We can extend the setting in Theorem 10.1 to include an explicit assembly time U for an order after all the components it requires are available. We assume the assembly times to be i.i.d. bounded random variables. We also assume no congestion and no finished goods inventory at the assembly stage, so U acts as an extra delay in the response time. This changes only the constant C in (10.3), as we explain after the proof of Theorem 10.1 in Subsection 5.1.

Consider, next, a system with m components. Each component i has a base-stock level s_i, so we need to make an assumption about how these s_i scale to get a limiting result. Let $s = s_1 + \cdots + s_m$ and $k_i = s_i/s$, $i = 1, \ldots, m$; we hold these ratios constant as s increases. This assumes that the proportion of total inventory held in each component remains constant, though we could just as easily assume that, e.g., the proportion of work content or holding cost for each component remains constant; this would merely change the constants k_i in the subsequent analysis. For each component i define X_i from A, B_i, and D_i paralleling the definition of X just before (10.2) and et

$$\psi_i(\theta) = \psi_{D_i}(\psi_{B_i}(\theta)) + \psi_A(-\theta), \tag{10.5}$$

in analogy with (10.2). Clearly, Theorem 10.1 applies to each component separately. Suppose $\gamma_i > 0$ solves $\psi_i(\gamma_i) = 0$ and set $\beta_i = \psi_{B_i}(\gamma_i)$, $\alpha_i = k_i\beta_i$. Then Theorem 10.1 implies

$$\lim_{s+x \to \infty} e^{\gamma_i x + \alpha_i s} P(R_i(s) > x) = C_i$$

for some $C_i > 0$. The response time for the full order is the maximum of the response times for the individual components required. Its behavior is a bit more subtle, because of the interactions among the multiple components.

Let $\gamma = \min_i \gamma_i$ and $\mathcal{I}_x = \{i : \gamma_i = \gamma\}$; these are the set of *leadtime-critical* components in the sense that their individual fill rates increase most slowly as x increases to ∞. Let $\alpha = \min_i \alpha_i$ and $\mathcal{I}_s = \{i : \alpha_i = \alpha\}$; these are similarly the set of *inventory-critical* components because their fill rates increase most slowly as s increases to ∞. These sets of components determine the product fill rate when x or s becomes large. To exclude trivial cases, we assume that for any two components i and j in $\mathcal{I}_x$ or $\mathcal{I}_s$, we have $P(X_i(n) \neq X_j(n)) > 0$ for all $n \geq 1$. Indeed, the only case in which this fails is if the two components are always ordered in the same quantity and take the same time to produce, in which case they should be modeled as a single component.

Theorem 10.3 *Suppose the solutions $\gamma_1, \ldots, \gamma_m$ all exist. Then*

$$\lim_{x \to \infty} e^{\gamma x} P(R(s) > x) = \sum_{i \in \mathcal{I}_x} C_i e^{-\alpha_i s} \tag{10.6}$$

and if either $|\mathcal{I}_s| = 1$ *or* $\{D_i, i \in \mathcal{I}_s\}$ *are independent, then*

$$\lim_{s \to \infty} e^{\alpha s} P(R(s) > x) = \sum_{i \in \mathcal{I}_s} C_i e^{-\gamma_i x}. \tag{10.7}$$

The condition that the D_i be independent is far from necessary for (10.7), even if $|\mathcal{I}_s| > 1$. As long as the batch sizes in $\mathcal{I}_s$ are not too perfectly correlated, (10.7) holds true. (See Equation (35) of Glasserman and Wang [23] for a much weaker sufficient condition.)

If $\mathcal{I}_x = \mathcal{I}_s = \mathcal{I}$, Theorem 10.3 yields (without assuming any condition on $|\mathcal{I}_s|$ or D_i)

$$\lim_{s+x \to \infty} e^{\gamma x + \alpha s} P(R(s) > x) = \sum_{i \in \mathcal{I}} C_i \equiv C_{\mathcal{I}},$$

which provides a simpler counterpart to Theorem 10.1 and suggests the approximation

$$P(R(s) > x) \approx C_{\mathcal{I}} e^{-\gamma x - \alpha s}.$$

In the general case, $\mathcal{I}_x$ and $\mathcal{I}_s$ represent the sets of components that constrain the fill rate at long delivery intervals and high base-stock levels, respectively. When these are not the same, different trade-offs apply in different regions, depending on how the fill rate becomes high. If the fill rate becomes high due to long leadtime, then the fill rate is determined and most constrained by the components with the smallest γ; if the fill rate becomes high due to high inventory, then the fill rate is determined and most constrained by the component with the smallest β. The intuition of Theorem 10.3 is further explored in the next section. The proof of Theorem 10.3 is rather involved and technical so we omit it here. Interested readers should consult Glasserman and Wang [23].

The final variant we consider allows multiple sets of components to be combined into d distinct products. In this setting, we require that arrivals of orders for the various products follow independent (compound) Poisson processes. We need to vary our notation slightly here to distinguish products from components: we use superscripts for products and continue to use subscripts for components. Orders for product j arrive at rate λ^j, and each order of product j requires D_i^j units of component i.

Let $\mathcal{I}^j = \{i : P(D_i^j > 0) > 0\}$ be the set of components required by product j; let $\mathcal{P}_i = \{j : P(D_i^j > 0) > 0\}$ be the set of products requiring component i. For each component i, the effective demand process is the superposition of independent (compound) Poisson processes hence itself a compound Poisson process whose rate rate λ_i is equal to $\sum_{j \in \mathcal{P}_i} \lambda^j$ and whose batch size D_i is distributed as a mixture of $\{D_i^j\}$, i.e., with probability λ^j/λ_i, D_i is distributed as D_i^j, for $j \in \mathcal{P}_i$. With γ_i and α_i calculated just as before, Theorem 10.1 applies to R_i, the steady-state component-i response time. Let R^j be the steady-state response time for product j; then $R^j = \max_{i \in \mathcal{I}^j} R_i$ and the setting is reduced to that of Theorem 10.3. Define

$$\bar{\gamma}^j = \min_{i \in \mathcal{I}^j}\{\gamma_i\} \quad \text{and} \quad \mathcal{I}_x^j = \{i \in \mathcal{I}^j : \gamma_i = \bar{\gamma}^j\};$$

$\mathcal{I}_x^j$ is the set of leadtime-critical components for product j. Also define

$$\bar{\alpha}^j = \min_{i \in \mathcal{I}^j}\{\alpha_i\} \quad \text{and} \quad \mathcal{I}_s^j = \{i \in \mathcal{I}^j : \alpha_i = \bar{\alpha}^j\};$$

$\mathcal{I}_s^j$ is the set of inventory-critical components for product j. We now have

Theorem 10.4 *Suppose the solutions* $\gamma_1, \ldots, \gamma_m$ *all exist. Then for each product j*

$$\lim_{x \to \infty} e^{\bar{\gamma}^j x} P(R^j(s) > x) = \sum_{i \in \mathcal{I}_x^j} C_i e^{-\alpha_i s}, \qquad (10.8)$$

and if $|\mathcal{I}_s^j| = 1$ *or* $\{D_i^j, i \in \mathcal{I}_s^j\}$ *are independent, then*

$$\lim_{s \to \infty} e^{\bar{\alpha}^j s} P(R^j(s) > x) = \sum_{i \in \mathcal{I}_s^j} C_i e^{-\gamma_i x}. \qquad (10.9)$$

The same comments on the independence condition for the D's after Theorem 10.3 also apply here. As with Theorem 10.3 the cleanest version of this result applies when $\mathcal{I}_x^j = \mathcal{I}_s^j$; i.e., the leadtime-critical and inventory-critical components coincide. Because the result has been specialized to the case of Poisson arrivals, Proposition 10.2 applied to each component i yields an expression for each C_i.

3.   Leadtime-Inventory Trade-Offs

In this section, we study the quantitative trade-offs between leadtime and inventory levels in achieving a target fill rate. Based on the asymptotic results of the fill rate in Section 2, we address the question of how much a delivery leadtime can be reduced, per unit increase in inventory, at a fixed fill rate. We first illustrate the trade-offs in single-component systems and show some numerical results. Then, we discuss further approximations for the trade-off parameters. Finally, we address issues in applying the trade-off when there is interaction among multiple components.

3.1.   Single-component systems

A level curve of constant service is a set of (x, s) points for which $P(R(s) > x) = \delta$, for some $0 < \delta < 1$. It follows by taking logarithm on (10.4) that the level curve is given approximately by the set of solutions to $C \exp(-\gamma x - \beta s) = \delta$. In the positive (x, s) orthant, these solutions form the line

$$s = -\frac{\gamma}{\beta}x + \frac{1}{\beta}\log(C/\delta).$$

Thus, a leadtime reduction of Δx entails an increase in s of $(\gamma/\beta)\Delta x$, and a 1-unit increase in s buys a reduction of β/γ in x, if the fill rate is to remain unchanged. In other words, when the fill rate is high, varying the values of leadtime x and inventory level s according to the linear trade-off rule

$$\Delta s = -\frac{\gamma}{\beta}\Delta x \qquad (10.10)$$

entails little change in the fill rate. Indeed, the smaller the resulting change in the fill rate, the better the linear approximation to the trade-off. We will test how the linear approximation works through several examples. In each example, we first calculate γ and β according to Theorem 10.1. We study the systems from $x = 0$ and choose some $s > 0$ such that the actual fill rate is high (90% or higher). We then make a series of increases in x and decrease s according to the trade-off rule (10.10) until s drops to 0. This way we get a series of (x, s) pairs. At each pair the actual fill rate is estimated by Monte Carlo simulation. Plotting the fill rate against the leadtime x yields a curve; when this curve is nearly a horizontal line the linear approximation in (10.10) works well.

The procedure to estimate the fill rate is discussed after the proof of Theorem 10.1 in Subsection 5.1. We will test several commonly used distributions at different utilization levels, where by utilization we mean the ratio of the mean production time of a random batch to the mean interarrival time. By choosing appropriate s values at $x = 0$, we study the trade-offs when the fill rate is around 90%, 95%, and 99%. (At higher service levels, s has a larger value at $x = 0$ and reaches 0 at a larger x, so the resulting curve is longer.) The simulation results are graphed in Figures 10.2–10.7 below, with three curves corresponding to the three service levels in each figure. The captions specify the distributions of A, B and D; all cases have $\mathsf{E}[B] = 1$. We use c_A and c_B to denote the coefficient of variation for A and B, respectively, and use $P_D(n)$ to denote $P(D = n)$ and ρ to denote the system utilization.

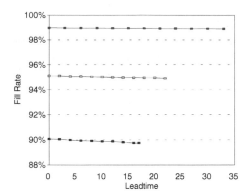

Figure 10.2. Compound Poisson demand process, $\mathsf{E}[A] = 2.4$; $P_D(1) = 0.7$ and $P_D(4) = 0.3$; Erlang production time with $c_B^2 = 0.2$; $\rho = 80\%$

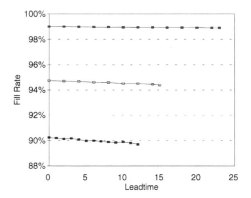

Figure 10.3. Compound Poisson demand process, $\mathsf{E}[A] = 2.7$; $P_D(1) = 0.7$ and $P_D(4) = 0.3$; Erlang production time with $c_B^2 = 0.2$; $\rho = 70\%$

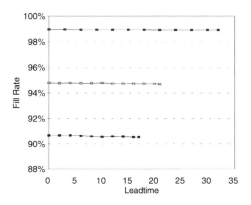

Figure 10.4. Compound Poisson demand process, $\mathsf{E}[A] = 2.4$; $P_D(1) = 0.7$ and $P_D(4) = 0.3$; normally distributed production time with $\mathsf{Var}[B] = 0.09$; $\rho = 80\%$

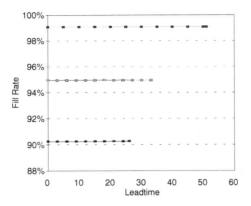

Figure 10.5. Compound Poisson demand process, $\mathsf{E}[A] = 2.5$; $P_D(1) = P_D(2) = P_D(3) = 0.3$, $P_D(4) = 0.1$; deterministic production time; $\rho = 90\%$

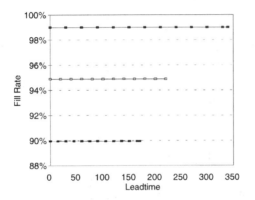

Figure 10.6. Compound Poisson demand process, $\mathsf{E}[A] = 2.24$; $P_D(1) = P_D(2) = P_D(3) = 0.3$, $P_D(4) = 0.1$; deterministic production time; $\rho = 98\%$

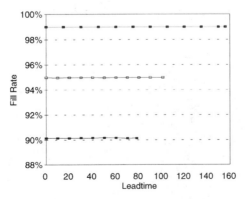

Figure 10.7. Hyperexponential interarrival time, $\mathsf{E}[A] = 2.5$, $c_A^2 = 4$; $P_D(1) = P_D(2) = P_D(3) = 0.3$, $P_D(4) = 0.1$; deterministic production time; $\rho = 88\%$

From the examples, we make the following observations:

1 In all the examples studied the simulated fill rate curves are very close to flat, straight lines, regardless of the difference in distributions and utilization levels. This means that varying x and s according to the linear s-x trade-off rule (10.10) indeed yields approximately the same fill rate. Hence the linear limiting trade-off between x and s captures the essence of the relation between x and s.

2 The approximation works very well when the service level is reasonably high, and the higher the fill rate, the better the results we get. When the fill rate is around 99%, the linear trade-off rule is virtually exact. This is consistent with the fact that at higher service level the system is closer to the limit regime of Theorem 10.1.

3 Along each fill rate curve in the figures, x and s vary over a very wide range (in fact, over the entire possible range from $x = 0$ to $s = 0$), yet the curves remain nearly flat and straight. Thus, the trade-off rule (10.10) is not just a local property. Changes in delivery leadtime and inventory level, Δx and Δs, do not have to be very small; they can be made to be very big as long as $x + \Delta x$ and $s + \Delta s$ remain nonnegative.

4 We observe a decline of the 90% fill rate curve as x increases in Figures 10.2 and 10.3. This is mainly due to a rounding effect: for some changes in x, the corresponding change in s given by (10.10) may not be an integer. But s has to be integral because of its physical meaning, so we round the nonintegral change to the nearest integer. In these examples, it happens that we always round down so the fill rate curves go down. We do not see the decline of the fill rate in Figures 10.4 and 10.5 where rounding is not necessary.

3.2.   Two-moment approximation

Calculating the trade-off parameters γ and β requires knowledge of the distributions of A, B and D, which may not always be available. If we only have partial knowledge of the distributions — specifically, the means and variances — we approximate $\gamma > 0$ through a two-moment approximation for the ψ_X in (10.2), i.e., we set $\psi_X(\theta) \approx \mathsf{E}[X]\theta + (1/2)\mathsf{Var}[X]\theta^2 = 0$ and solve to get

$$\gamma \approx -\frac{2\mathsf{E}[X]}{\mathsf{Var}[X]} \equiv \gamma_{2\text{-}m}, \tag{10.11}$$

where $\mathsf{E}[X] = \mathsf{E}[B]\mathsf{E}[D] - \mathsf{E}[A]$ and $\mathsf{Var}[X] = \mathsf{E}[D]\mathsf{Var}[B] + \mathsf{Var}[D](\mathsf{E}[B])^2 + \mathsf{Var}[A]$. Similarly by the two-moment approximation for ψ_B we get

$$\beta = \psi_B(\gamma) \approx \mathsf{E}[B]\gamma + \frac{1}{2}\mathsf{Var}[B]\gamma^2$$

$$\approx \mathsf{E}[B]\gamma_{2\text{-}m} + \frac{1}{2}\mathsf{Var}[B]\gamma_{2\text{-}m}^2 \equiv \beta_{2\text{-}m}. \qquad (10.12)$$

There is an interesting connection between our two-moment approximation in (10.11) and heavy-traffic approximation, although different assumptions are made and different tools are used in these two settings. Specifically, $1/\gamma_{2\text{-}m}$ would be the mean response time in a Brownian approximation. There does not appear to be an obvious counterpart or even an interpretation for $\beta_{2\text{-}m}$ in the heavy-traffic setting. From these approximations, it should be very obvious how the production capacity, as measured in terms of $\mathsf{E}[B]$, would affect γ hence affect the fill rate given that everything else is fixed. In addition, when both γ and β are small, any increase in the variability of either the demand process (A and D) or the production process (B) would effect in a smaller $\gamma_{2\text{-}m}$ and a smaller $\beta_{2\text{-}m}$, hence a lower fill rate.

We tested the linear trade-off of (10.10) on the same systems studied in the previous subsection, but replacing γ and β with their two-moment approximations (10.11) and (10.12). The resulting graphs are virtually indistinguishable from the previous ones and are therefore omitted. This indicates that the two-moment approximation may be adequate in practice.

3.3. Multiple-component systems

In Section 2 we state two limiting results (10.6) and (10.7) for the tail probability of the product response time, one on x and one on s. When the leadtime x is long, $P(R > x) \approx \sum_{i \in \mathcal{I}_x} C_i e^{-\gamma_i x - \beta_i s_i}$ and the product fill rate is constrained by the components with the smallest γ (leadtime-critical components). When the total inventory level s is high, $P(R > x) \approx \sum_{i \in \mathcal{I}_s} C_i e^{-\gamma_i x - \alpha s}$ and the product fill rate is constrained by the components with the smallest α (inventory-critical components). While it is sometimes possible to determine which regime applies, in many other cases, the dominating effect of the components in $\mathcal{I}_x$ or $\mathcal{I}_s$ may not be evident, since x and s are always finite in practice. We address this issue next.

When product fill rate is high, the fill rate of each component i must be as high or higher, and is approximately equal to $1 - C_i e^{-\gamma_i x - \beta_i s_i}$. Obviously, components with relatively small fill rates are the ones that constrain the product fill rate. So we propose the following criterion to determine a set $\mathcal{I}$ of constraining components. For each component i we calculate $\hat{p}_i \stackrel{\text{def}}{=} e^{-\gamma_i x - \beta_i s_i}$ as a surrogate for $P(R_i(s) > x)$ and take $\mathcal{I}$ to be the set of components with high $\hat{p}_i$. (The constant C_i is, of course, unknown in general.) When the leadtime x is changed, inventory levels of the components in $\mathcal{I}$ should be varied according to the component-level trade-off rule

$$\Delta s_i = -\frac{\gamma_i}{\beta_i} \Delta x \qquad (10.13)$$

to maintain the same product fill rate. Our experience from numerical studies is that when $\sum_{i \in \mathcal{I}} \hat{p}_i / \sum_{i=1}^{m} \hat{p}_i \geq 80\%$, we get satisfactory results, as will be illustrated through several examples.

We first study two-component systems with $\hat{p}_1$ and $\hat{p}_2$ close to each other so $\mathcal{I}$ contains both components. Much as in the single component case, we choose some $s_1 > 0, s_2 > 0$ at $x = 0$ such that the product fill rate is high (90% or higher). We then make a series of increases in x and decrease both s_1 and s_2 according to the component-level trade-off (10.13) until either s_i drops to 0. By estimating the product fill rate at each (x, s_1, s_2) triple we get a plot of fill rate against leadtime x. Again, a horizontal curve means the linear s-x trade-off relation and the proposed mechanism to identify constraining components work well. The results of the two-component systems are in Figures 10.8–10.13 below. In the captions, we use $P_D(m, n)$ to denote $P(D_1 = m, D_2 = n)$, and ρ_1, ρ_2 to denote the utilization level of components 1 and 2.

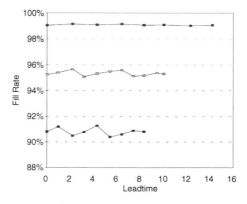

Figure 10.8. Poisson demand process, $E[A] = 2.4$; $P_D(1, 1) = 0.4$, $P_D(1, 2) = 0.3$, and $P_D(4, 3) = 0.3$; B_1, B_2 have Erlang distribution, $E[B_1] = 1$, $c_{B_1}^2 = 0.2$, $E[B_2] = 0.8$, $c_{B_2}^2 = 0.2$; $\rho_1 = 79\%$, $\rho_2 = 63\%$

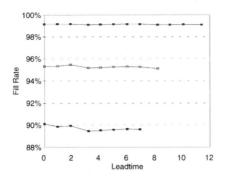

Figure 10.9. Poisson demand process, $\mathsf{E}[A] = 2.7$; $P_D(1,1) = 0.4$, $P_D(1,2) = 0.3$, and $P_D(4,3) - 0.3$; B_1, B_2 have Erlang distribution, $\mathsf{E}[B_1] = 1$, $c_{B_1}^2 - 0.2$, $\mathsf{E}[B_2] = 0.8$, $c_{B_2}^2 = 0.2$; $\rho_1 = 70\%$, $\rho_2 = 56\%$

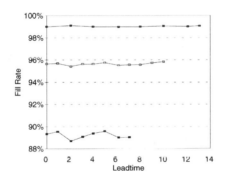

Figure 10.10. Poisson demand process, $\mathsf{E}[A] = 2.4$; $P_D(1,1) = 0.4$, $P_D(1,2) = 0.3$, and $P_D(4,3) = 0.3$; B_1, B_2 have normal distribution, $\mathsf{E}[B_1] = 1$, $\mathsf{Var}[B_1] = 0.09$, $\mathsf{E}[B_2] = 0.8$, $\mathsf{Var}[B_2] = 0.625$; $\rho_1 = 79\%$, $\rho_2 = 63\%$

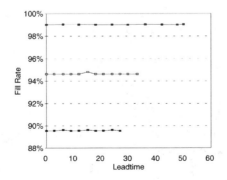

Figure 10.11. Poisson demand process, $\mathsf{E}[A] = 2.5$; $P_D(1,2) = P_D(1,3) = P_D(1,4) = P_D(2,2) = P_D(3,4) = P_D(4,1) = 0.1$, $P_D(2,3) = P_D(3,1) = 0.2$; deterministic production times $B_1 = 1$, $B_2 = 0.9$; $\rho_1 = 88\%$, $\rho_2 = 86\%$

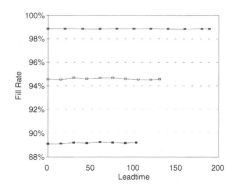

Figure 10.12. Poisson demand process, $\mathsf{E}[A] = 2.24$; $P_D(1,2) = P_D(1,3) = P_D(1,4) = P_D(2,2) = P_D(3,4) = P_D(4,1) = 0.1, P_D(2,3) = P_D(3,1) = 0.2$; deterministic production times $B_1 = 1$, $B_2 = 0.9$; $\rho_1 = 98\%$, $\rho_2 = 96\%$

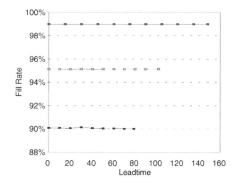

Figure 10.13. Hyperexponential interarrival time, $\mathsf{E}[A] = 2.5$, $c_A^2 = 4$; $P_D(1,2) = P_D(1,3) = P_D(1,4) = P_D(2,2) = P_D(3,4) = P_D(4,1) = 0.1$, $P_D(2,3) = P_D(3,1) = 0.2$; deterministic production times $B_1 = 1$, $B_2 = 0.9$; $\rho_1 = 88\%$, $\rho_2 = 86\%$

Finally, we consider a five-component system with a Poisson order process, having $\mathsf{E}[A] = 3$. The order quantities have the following probabilities: $P_D(2,3,4,2,1) = 0.2$, $P_D(3,2,2,4,3) = 0.3$, $P_D(4,4,1,1,3) = 0.2$ and $P_D(1,1,2,2,2) = 0.3$. Production times all have Erlang distributions with $c^2 = 0.2$, $\mathsf{E}[B_1] = \mathsf{E}[B_2] = \mathsf{E}[B_3] = 1$, $\mathsf{E}[B_4] = \mathsf{E}[B_5] = 0.9$. We choose different s_i values at $x = 0$ so that different sets of components are binding. The results are illustrated in Figures 10.14–10.17. In Figure 10.14, all five components have similar $\hat{p}$ values so the constraining set $\mathcal{I} = \{1,2,3,4,5\}$. In Figure 10.15 one component is constraining and $\mathcal{I} = \{1\}$. In Figures 10.16 and 10.17, the same three components are constraining, $\mathcal{I} = \{1,2,3\}$. The difference is that in Figure 10.16 the trade-offs are on the constraining components, whereas in Figure 10.17

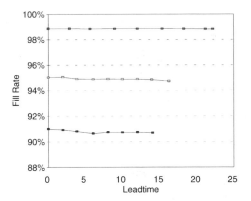

Figure 10.14. A five-item system. All five items are equally constraining and trade-offs are on five items

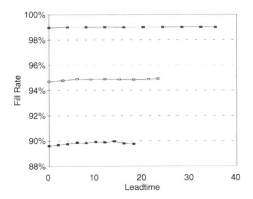

Figure 10.15. A five-item system. One item is constraining and the trade-off is on that constraining item

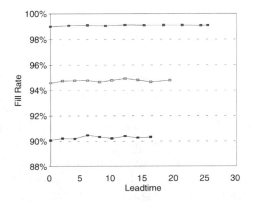

Figure 10.16. A five-item system. Trade-offs are on the three constraining items

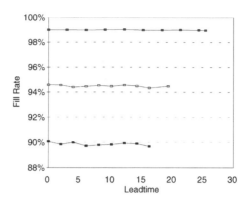

Figure 10.17. A five-item system. Same three items as in Figure 10.16 are constraining. But the trade-offs are on all five items for comparison with Figure 10.16

the trade-offs are on all five components — when x changes, s_4 and s_5 also change according to their trade-off equations, although components 4 and 5 are considered nonbinding. This is for comparison with Figure 10.16.

From Figures 10.2–10.17, we see that the proposed trade-off rule generally works well, regardless of the distribution and utilization level. In some examples, the lowermost (90%) curve is jagged. One of the main reasons is the rounding effect, which is more severe here than in the single-component case. For example on the 90% curve in Figure 10.8, the s_2 value should be 4.80 for the fifth point and 3.47 for sixth point according to (10.13). After rounding, the values become 5 and 3, respectively. A difference of $4.80 - 3.47 = 1.33$ increase to $5 - 3 = 2$, so we see a drop in the fill rate. Two-moment approximations for γ_i and β_i yield virtually the same graphs.

4.   Cost Optimization

In this section, we investigate how to set the component base-stock levels to minimize the total inventory holding cost while satisfying a high fill rate requirement. The difficulty with this problem, as should be clear from the previous two sections, is that generally there is no analytical expression for the fill rate and even determining the feasibility of an inventory policy usually involves (rare event) simulation. We derive closed-form solutions to the problem based on an approximate formulation, where the original fill rate requirement is replaced by a more stringent, but asymptotically tight surrogate. We also discuss further refinement to the approximate solution. To illustrate the effectiveness of our method, we compare inventory holding cost of the approximate solution with that of the optimal solution obtained from computationally expensive direct search. To keep our analysis tractable, we make the following simplifying assumption about the inventory cost that is commonly made in the assembly system literature: when an order is backlogged, it claims the available units of the components from inventory and waits for the rest from production, and no inventory holding cost is charged on those claimed units. In 88 numerical examples studied, the average gap from optimality is 2.8% and the worst is 14.8%.

Our objective is to set the component base-stock levels s_i to minimize the system-wide inventory holding cost while keeping a product unfill rate of at most δ for some small value δ. Formally, we formulate the inventory-planning problem as

$$\textbf{P1:} \qquad \min \ \sum_{i=1}^{m} h_i \mathsf{E}[I_i]$$

$$\text{s.t.} \quad P(R > x) \leq \delta, \qquad\qquad (10.14)$$

$$s_i \geq 0, s_i \text{ integers}.$$

All holding cost rates h_i are assumed to be positive.

While we focus on the case of a single product demand stream, it should be clear from the discussion just before Theorem 10.4 that the result can be readily extended to one with multiple product streams, if they are independent Poisson processes.

4.1. Approximate formulation

The optimization problem **P1** has a nonlinear objective function and nonlinear constraints. Moreover, there is generally no explicit expressions for either the objective function or the constraints in terms of the decision variables s_i. Solving **P1** directly is virtually impossible. We will construct an approximate formulation of **P1** based on the key results in Section 2.

Since a product order can be delivered only when all the components it requires are available, either from inventory or from production, and we have assumed the assembly time to be zero, it is obvious that $R = \max_{1 \leq i \leq m} R_i$. Therefore, the unfill rate for the product, $P(R > x)$, satisfies

$$P(R > x) = P\left(\left(\max_{1 \leq i \leq m} R_i\right) > x\right) = P\left(\bigcup_{i=1}^{m}(R_i > x)\right) \leq \sum_{i=1}^{m} P(R_i > x).$$

By Theorem 10.3 in Section 2, this upper bound becomes asymptotically tight when the product unfill rate decreases to zero.

The upper bound above and the component-level approximation (10.4) together suggest that when the service-level requirement is high (i.e., δ is small), the service-level requirement in (10.14) can be approximated by a more stringent, yet asymptotically tight one,

$$\sum_{i=1}^{m} C_i e^{-\gamma_i x - \beta_i s_i} \leq \delta.$$

This constraint is much easier to work with than the original requirement (10.14).

We next establish a relationship between the inventory I_i and the base-stock level s_i that will facilitate the analysis and calculation of the objective function. We often write I_i as $I_i(s_i)$ to emphasize the dependence of I_i on s_i. For each component i, consider an associated $G/G/1$ queue whose arrival process is the same as the batch demand process in the production-inventory system and whose service times are B_i. Let Q_i be the steady state queue length of the associated queue. Since the production system operates under a base-stock policy, Q_i is the number of outstanding orders (in production or waiting to be produced) for component i. Thus, there is a one-to-one correspondence between I_i and Q_i:

$$I_i(s_i) = (s_i - Q_i)^+ = s_i - Q_i + (Q_i - s_i)^+,$$

where y^+ denotes $\max(y, 0)$. Note that the distribution of Q_i does not depend on s_i. The long run average cost rate of the entire system is

$$\sum_{i=1}^{m} h_i \mathsf{E}[I_i(s_i)] = \sum_{i=1}^{m} h_i \mathsf{E}[s_i - Q_i + (Q_i - s_i)^+]$$

$$= \sum_{i=1}^{m} h_i s_i - \sum_{i=1}^{m} h_i \mathsf{E}[Q_i] + \sum_{i=1}^{m} h_i \mathsf{E}[(Q_i - s_i)^+]. \qquad (10.15)$$

Notice that $\mathsf{E}[Q_i]$ does not depend on our decision variables s_i so it can be removed from the objective function without changing the choice of optimal s_i. If the fill rate $P(R \leq x)$ is high not due to large x, but due to large s_i, then we can assume $\sum_{i=1}^{m} h_i \mathsf{E}[(Q_i - s_i)^+]$ to be negligible in the objective function compared with $\sum_{i=1}^{m} h_i s_i$. (More treatment of this issue will be given in Subsection 4.3.) We also ignore the integrality constraint on s_i since it is not the focus of our investigation in this chapter. As a result, we formulate the inventory control problem approximately as

$$\textbf{P2:} \qquad \min \quad \sum_{i=1}^{m} h_i s_i$$

$$\text{s.t.} \quad \sum_{i=1}^{m} C_i e^{-\gamma_i x - \beta_i s_i} \leq \delta, \qquad (10.16)$$

$$s_i \geq 0, \quad i = 1, 2, \ldots, m. \qquad (10.17)$$

4.2. Closed-form solutions

Before introducing our main result, we first sort and re-label the components such that

$$0 < \frac{h_1/\beta_1}{C_1 e^{-\gamma_1 x}} \leq \frac{h_2/\beta_2}{C_2 e^{-\gamma_2 x}} \leq \cdots \leq \frac{h_m/\beta_m}{C_m e^{-\gamma_m x}}. \qquad (10.18)$$

This objective can be easily achieved with an algorithm of complexity $m \log(m)$. For notational convenience, define

$$\frac{h_{m+1}/\beta_{m+1}}{C_{m+1} e^{-\gamma_{m+1} x}} \equiv \infty.$$

To exclude triviality, we assume that the condition

$$\sum_{i=1}^{m} C_i e^{-\gamma_i x} > \delta \qquad (10.19)$$

always holds. Otherwise, $s_i = 0$ for all i is obviously the optimal solution.

Theorem 10.5 *For $1 \leq j \leq m$, define*

$$\theta_j = \frac{\sum_{i=1}^{j} (h_i/\beta_i)}{\delta - \sum_{i=j+1}^{m} C_i e^{-\gamma_i x}}, \qquad (10.20)$$

with the convention $\sum_{i=m+1}^{m} \cdot = 0$. Then there exists a unique index $1 \leq k \leq m$ for which

$$\frac{h_k/\beta_k}{C_k e^{-\gamma_k x}} \leq \theta_k < \frac{h_{k+1}/\beta_{k+1}}{C_{k+1} e^{-\gamma_{k+1} x}}, \qquad (10.21)$$

and the optimal solution to problem **P2** *is given by*

$$\tilde{s}_i = \frac{1}{\beta_i} \log \frac{\theta_k C_i e^{-\gamma_i x}}{h_i/\beta_i}, \quad \text{for } i = 1, 2, \ldots, k, \quad \text{and} \qquad (10.22)$$

$$\tilde{s}_i = 0, \quad \text{for } i = k+1, \ldots, m.$$

Remarks

(a) Although our investigation is restricted to the class of base-stock policies, as far as we know, the results in (10.22) are the first explicit formulas for base-stock levels in a capacitated stochastic multi-component system. Moreover, all the parameters used in the formulas can be calculated quite easily from input information.

(b) Examination of (10.22) reveals that the optimal base stock levels to problem **P2** satisfy

$$\beta_1 \tilde{s}_1 \geq \beta_2 \tilde{s}_2 \geq \cdots \geq \beta_m \tilde{s}_m,$$

i.e., they are in the reversed order of the ratio $(h_i/\beta_i)/(C_i e^{-\gamma_i x})$. The numerator is the unit holding cost, adjusted by the fill rate sensitivity to inventory; the denominator represents the fill rate sensitivity to leadtime and the need for inventory on the component level — the larger the value of $C_i e^{-\gamma_i x}$, the greater the need on s_i to achieve a small component level unfill rate (cf. Equation (10.4)). There is no easy ranking on $\tilde{s}_i$'s themselves.

(c) If $C_i e^{-\gamma_i x} \leq \delta$ for some i, then it might be *possible* to achieve the desired fill rate $1 - \delta$ without holding inventory in *all* components. In this case, the priority of zero-inventory is first given to components with higher values of $(h_i/\beta_i)/(C_i e^{-\gamma_i x})$, i.e., those components with higher holding cost h_i, smaller β_i (less inventory sensitive), and smaller $C_i e^{-\gamma_i x}$ (less need to achieve the fill-rate requirement through inventory holding). (See the ranking of components in Equation (10.18).) The exact cutoff point of inventory vs. zero-inventory among components is determined by the calculation in (10.20) and (10.21). It can be shown that the optimal index k in (10.21) is non-increasing in δ — when δ becomes smaller and the fill-rate requirement is more stringent, then k decreases or remains the same, and the optimal set of components with positive inventory increases or remains the same. Once it is determined that positive inventory should be held for a component, the insight on how much inventory should be held is more complicated. Nevertheless, we can make the following observations. (1) Each positive $\tilde{s}_i$ is strictly decreasing in δ. The more stringent the fill rate requirement, the higher the optimal base-stock levels. (2) Suppose that only one of the unit holding costs h_j increases for some j, and everything else remains unchanged. Then, a little algebra shows that only $\tilde{s}_j$ decreases and all the other $\tilde{s}_i$'s for $i \neq j$ will increase.

(d) Under the assumption $C_i e^{-\gamma_i x} > \delta$ for all i, we have $\tilde{s}_i > 0$ for all i. In this case, as $h_j \to \infty$ for some component j,

$$\frac{\beta_j}{h_j} \sum_{i=1}^{m} (h_i/\beta_i) \to 1$$

and the optimal $\tilde{s}_j$ approaches its lower bound

$$(1/\beta_j) \log(C_j/\delta) - (\gamma_j/\beta_j)x. \qquad (10.23)$$

The lower bound is independent of the holding cost parameters — no matter how expensive a component's holding cost is, *some* inventory must be held for that component in order to meet the fill rate requirement on the final product. The lower bound also provides a stopping rule in the search for the optimal policy.

4.3.   Poisson demand process

We now revisit the optimization problem **P2** and consider the true holding cost $\sum_{i=1}^{m} h_i \mathsf{E}[I_i] = \sum_{i=1}^{m} h_i \mathsf{E}[(s_i - Q_i)^+]$ as the objective function instead of the surrogate $\sum_{i=1}^{m} h_i s_i$ used there.

To make the problem tractable, we assume that $C_i e^{-\gamma_i x} > \delta$ for all $i = 1, 2, \ldots, m$ so that all $s_i > 0$ and (10.16) is the only binding constraint in **P2**. The problem becomes

$$\min \quad \sum_{i=1}^{m} h_i \mathsf{E}[(s_i - Q_i)^+]$$

$$\text{s.t.} \quad \sum_{i=1}^{m} C_i e^{-\gamma_i x - \beta_i s_i} \leq \delta.$$

The (necessary and sufficient) Kuhn-Tucker conditions for optimality are

$$\left.\begin{array}{l} \theta \geq 0, \\[4pt] h_i P(Q_i \leq s_i) - \theta \beta_i C_i e^{-\gamma_i x - \beta_i s_i} = 0, \ \forall i, \quad \text{and} \\[4pt] \sum_{i=1}^{m} C_i e^{-\gamma_i x - \beta_i s_i} = \delta. \end{array}\right\} \qquad (10.24)$$

We can no longer obtain closed-form solutions for s_i from (10.24). But these nonlinear equations can be solved numerically once we have the cumulative distribution function $P(Q_i \leq \cdot)$. When the interarrival time A has a general distribution, the queue length distribution can be calculated by approximation (for example, using phase-type distribution), or by simulation. We focus on the case of Poisson demand, where the distribution function for each component can be calculated by using an algorithm developed in Tijms [40] for an $M^X/G/1$ queue (M^X means compound Poisson arrivals). We consider a generic component here, and a subscript does not represent a component any more. Denote $P(Q = j)$ as p_j. The algorithm to calculate $\{p_j\}$ is

$$p_0 = 1 - \lambda\mathsf{E}[B]\mathsf{E}[D],$$

$$p_j = \lambda p_0 \sum_{i=1}^{j} P(D = i)a_{j-i} + \lambda \sum_{i=1}^{j} \left(\sum_{l=0}^{i} p_l P(D \geq i+1-l) \right) a_{j-i},$$

$$j = 1, 2, \ldots,$$

where a_n is the expected amount of time that n individual jobs (not batches) are present in the $M^X/G/1$ queue during a service time that starts a busy period, $n = 0, 1, 2, \ldots$. We briefly explain below how to calculate $\{a_n\}$.

Without loss of generality, suppose the service starts at time 0 and there is no other job in the system at time 0. Let N_t be the number of Poisson batch arrivals in time interval $(0, t]$, then N_t has a Poisson distribution with mean $\frac{t}{\mathsf{E}[A]}$. The total number of arriving jobs in $(0, t]$ is

$$J_t = \sum_{i=1}^{N_t} D_i,$$

where D_i are i.i.d. batch sizes. From the cumulant generating function

$$\begin{aligned}
\psi_{J_t}(\theta) &= \log \mathsf{E}\left[e^{\theta J_t}\right] \\
&= \log \mathsf{E}\left[\mathsf{E}\left[e^{\theta \sum_{i=1}^{N_t} D_i} \big| N_t\right]\right] \\
&= \log \mathsf{E}\left[\mathsf{E}\left[e^{\theta D_1} e^{\theta D_2} \cdots e^{\theta D_{N_t}} \big| N_t\right]\right] \\
&= \log \mathsf{E}\left[\left(\mathsf{E}[e^{\theta D_1}]\right)^{N_t}\right] \\
&= \log \mathsf{E}\left[e^{N_t \cdot \psi_D(\theta)}\right] \\
&= \psi_{N_t}(\psi_D(\theta)),
\end{aligned}$$

we can calculate the distribution of J_t. Define $\chi_n(t) = 1$ if the service that starts at time 0 is still in progress at time t and n jobs are present in the queue; otherwise, define $\chi_n(t) = 0$. Then

$$a_n = \mathsf{E}\left[\int_0^\infty \chi_n(t)\, dt\right]. \tag{10.25}$$

Because of the independence of the service time B and the arrival process, we have

$$E[\chi_n(t)] = P(\chi_n(t) = 1) = P(B > t)P(J_t = n).$$

Exchanging the order of expectation and integration (by Tonelli's Theorem) in (10.25), we get

$$a_n = \int_0^\infty P(B > t)P(J_t = n)\, dt.$$

The actual calculation of $\{a_n\}$ may involve numerical integration when B has a general distribution. In the case that B has an Erlang distribution with $E[B] = r/\mu$ and coefficient of variation $\sqrt{1/r}$ (i.e., B is a sum of r i.i.d. exponentially distributed random variables each having mean $1/\mu$), we have the following closed-form recursion scheme

$$a_n = \phi_n^{(r)};$$
$$\phi_0^{(i)} = \frac{1}{\lambda + \mu} + \frac{\mu}{\lambda + \mu}\phi_0^{(i-1)}, \quad 1 \le i \le r;$$
$$\phi_n^{(i)} = \frac{1}{\lambda + \mu}\sum_{j=1}^n P(D = j)\phi_{n-j}^{(i)} + \frac{\mu}{\lambda + \mu}\phi_n^{(i-1)}, \quad 1 \le i \le r, n \ge 1;$$

where $\phi_n^{(0)} = 0$ by convention. The above recursion can be extended to the case where B has a generalized Erlang distribution. (A generalized Erlang distribution is a mixture of Erlang distributions with the same μ but different r's, which can be used to approximate arbitrarily closely the distribution of any positive random variable.) In case of a deterministic production time, we approximate it by an Erlang distribution with a large r.

Notice that we need to calculate the probability mass function $P(Q_i = j)$ only for $j \le s_i$ in (10.24). In our numerical studies we repeatedly use the algorithm above to compute the queue length distribution and the actual holding cost of a policy. In most examples, the solutions obtained by numerically solving the equations in (10.24) are the same as the closed-form ones in (10.22). In some examples, they improve over the the closed-form solutions by one to two units of base-stock level. Such a small difference is mainly due to the fact that the base-stock levels s_i are high enough in our examples so that the effect of $E[(Q_i - s_i)^+]$ is negligible.

4.4.   Numerical results

To see how well the approximate solutions work, we next perform some numerical studies. In each of the examples studied, the approximate base-stock levels $\tilde{s}_i$ are first calculated according to Equation (10.22). The actual fill rate at these base-stock levels is estimated by simulation to determine whether the service-level requirement (10.14) is satisfied. If it is, then the corresponding average holding cost $\sum_{i=1}^{m} h_i \mathsf{E}[I_i(\tilde{s}_i)]$ is computed and compared with that of the optimal policy $\{s_i^*\}$ obtained by directly searching over the feasible base-stock policies.

The search for optimal policy is done only among feasible policies that are not *dominated* by others. In a two-component system, for example, if (s_1, s_2) is a feasible policy, then $(s_1 + 1, s_2)$ and $(s_1, s_2 + 1)$ are also feasible but their costs are higher than that of (s_1, s_2). In this sense, $(s_1 + 1, s_2)$ and $(s_1, s_2 + 1)$ are dominated by (s_1, s_2). Dominated policies cannot be optimal so they are skipped in our search.

Using a two-component system as an illustrative example, we next propose a refinement to our closed-form solution (10.22). We write the closed-form solution $\tilde{s}_i$ as $\tilde{s}_i(\delta)$ to emphasize its dependence on δ. The constraint in (10.16) is an approximation to a service-level requirement that is more stringent than the original given one in (10.14). As a result, the base-stock policy $(\tilde{s}_1(\delta), \tilde{s}_2(\delta))$ almost always yields an unfill rate $P(R(\tilde{s}_1(\delta), \tilde{s}_2(\delta)) > x)$ strictly smaller than the required δ, at the expense of a higher holding cost. We wish to push this base-stock policy from the interior of the feasible region closer to its frontier. We propose to compute a new set of base-stock levels $(\tilde{s}_1', \tilde{s}_2')$ based on a "relaxed" requirement δ' larger than δ, with the expectation that the resulting unfill rate $P(R(\tilde{s}_1'(\delta'), \tilde{s}_2'(\delta')) > x)$ will be smaller than δ' but closer to δ. There is no obvious rule on how δ' should be chosen, and too large a δ' will cause the new policy $(\tilde{s}_1'(\delta'), \tilde{s}_2'(\delta'))$ to become infeasible to the original requirement δ. The ratio $\delta/P(R(\tilde{s}_1(\delta), \tilde{s}_2(\delta)) > x)$ gives a good indication of how much slack exists in the closed-form policy. Our numerical experience shows that inflating the target unfill rate by this ratio, i.e., setting

$$\delta' = \delta \cdot \left(\frac{\delta}{P\left(R(\tilde{s}_1(\delta), \tilde{s}_2(\delta)) > x\right)} \right) \qquad (10.26)$$

yields very good results. By doing so, we always get a lower cost and most of the time maintain feasibility. In the rare case that $P(R(\tilde{s}_1(\delta), \tilde{s}_2(\delta)) > x)$ exceeds δ and the initial closed-form solution is infeasible, δ' will become smaller than δ according to (10.26), which often makes the refined solution feasible.

In fact, Equation (10.26) can be extended to the following recursive scheme to guarantee feasibility. Suppose δ is the real unfill rate requirement, $\delta^{(n)}$ is the current target value used in the computation of $\tilde{s}_i$'s in Theorem 10.1, and $P(R(\tilde{s}_1(\delta^{(n)}), \ldots, \tilde{s}_m(\delta^{(n)})) > x)$ is the unfill rate attained by solution $\tilde{s}_i(\delta^{(n)})$. We start at $\delta^{(0)} = \delta$ set

$$\delta^{(n+1)} = \delta^{(n)} \cdot \left(\frac{\delta}{P(R(\tilde{s}_1(\delta^{(n)}), \ldots, \tilde{s}_m(\delta^{(n)})) > x)} \right). \qquad (10.27)$$

In all of our numerical experiments, the solution converges with one iteration of (10.27).

We measure the quality of the approximate solutions by the *optimality gap*, defined to be the percentage by which the cost associated with the refined solution $\tilde{s}_i'$ is above the optimal cost $\sum_{i=1}^{m} h_i \mathsf{E}[I_i(s_i^*)]$. We also list the optimality gaps of a feasible closed-form solution $\tilde{s}_i$ to show the improvement with the refinement scheme (10.27). When a closed-form solution violates the fill-rate requirement in simulation, the solution is labeled "×". In the numerical examples that will be presented next, we choose distributions that are commonly used in the literature (Poisson for the arrival process, Erlang and deterministic for the production times) and study the effectiveness of our method over a wide rages of values in h, δ, and the utilization level. Most of our experiments are focused on two-component systems. We report some results of a five-component system at the end of this section.

Example 1. The demand process is compound Poisson with mean interarrival time $\mathsf{E}[A] = 2.4$. The batch size distribution for the two components are $P_D(1, 1) = 0.4$, $P_D(1, 2) = 0.3$, and $P_D(4, 3) = 0.3$, with $P_D(i, j)$ denoting $P(D_1 = i, D_2 = j)$. The correlation between the batch sizes is 0.87. Production times B_1 and B_2 both have an Erlang distribution with $\mathsf{E}[B_1] = 1$, $\mathsf{E}[B_2] = 0.8$, and coefficients of variation $\sqrt{0.2}$. The utilization levels are $\rho_1 = 79\%$ and $\rho_2 = 63\%$ for the two components. Service-level requirement is on the off-the-shelf fill rate, $P(R > 0) \leq \delta$. Different combinations of δ and (h_1, h_2) values are used. The results are listed in Table 10.1. The base-stock level and average inventory holding cost of the closed-form solution and its refinement is compared with those of the optimal solution. Optimality gaps of the 21 parameter incidences studied are all under 6.4%, with an average of 2.0%.

Example 2. This example considers the same system as that in Example 1. The only difference is a non-zero target leadtime $x = 5$. The results are listed in Table 10.2. For the same δ value, base-stock levels $\tilde{s}_1$ and $\tilde{s}_2$ are smaller than the corresponding ones in Example 1.

Table 10.1. Results of Example 1. Comparison of the closed-form approximate solution and its refinement with the optimal solution

δ	h_1	h_2	Optimal Solution			Closed-Form			Refined Solution		
			s_1^*	s_2^*	cost	$\tilde{s}_1$	$\tilde{s}_2$	gap(%)	$\tilde{s}_1'$	$\tilde{s}_2'$	gap(%)
	1	1	19	10	20.9	19	11	4.7	19	11	4.7
	1	2	19	10	28.6	21	10	6.6	20	10	3.3
	2	1	18	12	33.3	18	13	5.9	18	13	5.9
0.1	1	5	22	9	45.8	25	9	5.8	25	9	5.8
	5	1	17	15	69.5	18	15	6.7	17	15	0.0
	1	10	29	8	80.7	29	8	0.0	29	8	0.0
	10	1	17	15	126.3	17	17	1.6	17	17	1.6
	lower bound								17	8	
	1	1	25	12	28.6	24	13	0.1	24	13	0.1
	1	2	27	11	38.3	26	12	2.6	25	12	0.0
	2	1	23	14	45.7	23	15	2.2	23	15	2.2
0.05	1	5	27	11	64.3	30	11	4.6	29	11	3.1
	5	1	22	17	95.0	22	17	0.0	22	17	0.0
	1	10	27	11	107.7	33	10	×	34	11	6.4
	10	1	22	17	175.4	22	19	1.1	22	19	1.1
	lower bound								22	10	
	1	1	35	18	44.5	36	18	2.2	35	18	0.0
	1	2	39	16	60.1	37	17	0.0	37	17	0.0
	2	1	35	18	73.3	35	20	2.7	35	20	2.7
0.01	1	5	39	16	101.0	41	16	2.0	41	16	2.0
	5	1	34	20	156.9	34	22	1.3	34	22	1.3
	1	10	45	15	165.2	45	15	0.0	45	15	0.0
	10	1	34	20	296.2	33	24	×	34	24	1.4
	lower bound								33	15	

Optimality gaps are generally larger than those in the previous example, with an average of 4.1%.

Example 3. For our next example, we study a two-component system with compound Poisson demand and $E[A] = 2.5$. The joint batch distributions are $P_D(1,2) = P_D(1,3) = P_D(1,4) = P_D(2,2) = P_D(3,4) = P_D(4,1) = 0.1$, and $P_D(2,3) = P_D(3,1) = 0.2$, with a correlation of -0.53. Production times are deterministic, $B_1 = 1$, and $B_2 = 0.9$. The utilization levels are $\rho_1 = 88\%$ and $\rho_2 = 86\%$. The results are listed in Table 10.3 under different combination of δ and (h_1, h_2). Optimality gaps of the 21 parameter incidences studied are all under 9.0%, with an average of 2.4%.

Example 4. Our last main example considers the same system as that in Example 3. The only difference is a non-zero target leadtime $x = 15$.

Table 10.2. Results of Example 2. Comparison of the closed-form approximate solution and its refinement with the optimal solution

δ	h_1	h_2	Optimal Solution			Closed-Form			Refined Solution		
			s_1^*	s_2^*	cost	$\tilde{s}_1$	$\tilde{s}_2$	gap(%)	$\tilde{s}_1'$	$\tilde{s}_2'$	gap(%)
	1	1	13	5	10.7	14	5	8.0	14	5	8.0
	1	2	14	4	13.2	16	4	13.6	15	4	6.7
	2	1	13	5	18.7	13	7	9.5	13	7	9.5
0.1	1	5	17	3	18.9	20	3	14.8	20	3	14.8
	5	1	13	5	42.1	13	9	8.7	13	9	8.7
	1	10	17	3	26.5	24	2	1.1	24	2	1.1
	10	1	12	11	78.1	12	11	0.0	12	11	0.0
	lower bound								12	2	
	1	1	18	8	18.1	19	7	0.0	19	7	0.0
	1	2	20	6	22.1	21	6	4.3	20	6	0.0
	2	1	18	8	30.4	18	9	3.1	18	9	3.1
0.05	1	5	22	5	31.4	25	5	9.2	24	5	6.1
	5	1	17	12	66.5	17	11	×	18	11	5.5
	1	10	22	5	46.8	29	4	×	29	5	14.6
	10	1	17	12	123.3	17	13	0.8	17	13	0.8
	lower bound								17	4	
	1	1	30	12	33.6	31	12	2.9	30	12	0.0
	1	2	30	12	43.2	33	11	2.3	32	11	0.0
	2	1	30	12	57.4	30	14	3.4	29	14	0.0
0.01	1	5	34	10	66.4	36	10	3.0	36	10	3.0
	5	1	29	14	126.2	29	16	1.6	29	16	1.6
	1	10	40	9	101.2	40	9	0.0	40	9	0.0
	10	1	29	14	240.7	29	18	1.6	29	18	1.6
	lower bound								28	9	

The results are listed in Table 10.4. For the same δ value, the base-stock levels $\tilde{s}_1$ and $\tilde{s}_2$ are smaller than the corresponding ones in the previous example. Optimality gaps are slightly larger than those in Example 3, with an average of 3.0%.

From these examples, we make the following observations:

1 The approximate solution works very well in two-component systems. Of the 84 incidences studied, the worst optimality gap is 14.8% and the average is 2.9%. (The average gap for the feasible closed-form solution itself is 5.5%.) This means that setting base-stock levels according to the simple closed-form formula (10.22) or its refinement (10.27) yields a cost very close to the optimal one that usually requires an astronomical computation effort. (The search for the optimal policy is computationally expensive because

Table 10.3. Results of Example 3. Comparison of the closed-form approximate solution and its refinement with the optimal solution

δ	h_1	h_2	Optimal Solution			Closed-Form			Refined Solution		
			s_1^*	s_2^*	cost	$\tilde{s}_1$	$\tilde{s}_2$	gap(%)	$\tilde{s}_1'$	$\tilde{s}_2'$	gap(%)
	1	1	30	31	42.3	33	32	9.0	32	31	4.5
	1	2	34	28	62.2	37	29	7.7	36	28	3.1
	2	1	29	32	61.8	30	37	10.9	28	35	1.6
0.1	1	5	38	27	118.	45	27	5.8	45	27	5.8
	5	1	27	38	117.	28	44	9.0	28	44	9.0
	1	10	44	26	205.	51	26	3.4	50	26	2.9
	10	1	27	38	205.	27	50	5.8	27	50	5.8
	lower bound								26	25	
	1	1	38	38	56.7	40	40	6.9	38	38	0.0
	1	2	42	35	83.3	45	37	8.2	43	35	1.2
	2	1	37	39	83.9	37	44	5.9	36	42	1.2
0.05	1	5	42	35	160.2	52	34	3.2	52	34	3.2
	5	1	34	48	159.9	35	51	4.9	34	50	1.2
	1	10	56	33	282.9	58	33	0.7	58	33	0.7
	10	1	34	48	281.4	34	57	3.2	34	57	3.2
	lower bound								33	32	
	1	1	55	56	91.3	57	56	2.2	57	55	1.1
	1	2	56	55	136.6	62	53	1.5	61	53	0.7
	2	1	55	56	136.2	54	60	1.5	53	60	0.0
0.01	1	5	64	51	260.8	69	51	1.9	69	51	1.9
	5	1	52	64	264.1	52	68	1.5	52	68	1.5
	1	10	70	50	463.6	75	50	1.1	70	50	1.1
	10	1	51	70	470.0	51	74	0.9	51	74	0.9
	lower bound								50	49	

for each candidate policy, the feasibility verification involves rare event simulation and the corresponding cost has to be evaluated numerically.)

2 Although the approximation is based on asymptotic results on the fill rate, it works well at reasonably high fill rate requirements. The higher the requirement (smaller δ), the better the results we get. The average optimality gaps are 4.4%, 3.1%, and 1.1% for fill rate requirements at 90%, 95%, and 99%, respectively. This is consistent with our analysis. At a smaller δ, both the product unfill rate $P(R > x)$ and component unfill rate $P(R_i > x)$ have to be smaller. As a result, $C_i e^{-\gamma_i x - \beta_i s_i}$ and $\sum_{i=1}^{m} P(R_i > x)$ are better approximations for the component unfill rate $P(R_i > x)$ and the product unfill rate $P(R > x)$, respectively, and the surrogate

Table 10.4. Results of Example 4. Comparison of the closed-form approximate solution and its refinement with the optimal solution

δ	h_1	h_2	Optimal Solution			Closed-Form			Refined Solution		
			s_1^*	s_2^*	cost	$\tilde{s}_1$	$\tilde{s}_2$	gap(%)	$\tilde{s}_1'$	$\tilde{s}_2'$	gap(%)
	1	1	15	15	15.0	18	16	21.3	16	14	0.1
	1	2	18	12	20.4	22	13	23.7	20	11	1.5
	2	1	14	16	21.6	15	20	22.4	13	18	0.8
0.1	1	5	20	11	34.7	30	10	16.6	29	10	13.9
	5	1	12	22	39.3	13	27	20.5	11	26	0.6
	1	10	24	10	55.0	36	9	9.2	36	9	9.2
	10	1	12	22	65.2	12	34	17.2	11	33	5.3
	lower bound								11	8	
	1	1	23	22	27.5	25	23	9.8	24	21	0.0
	1	2	27	19	39.4	30	20	11.5	28	19	2.4
	2	1	22	23	40.7	22	27	9.0	21	26	2.4
0.05	1	5	33	17	69.2	37	18	11.5	36	17	4.2
	5	1	20	28	76.1	20	34	7.5	19	34	1.9
	1	10	33	17	115.1	43	17	8.4	43	16	1.4
	10	1	19	34	130.6	19	41	5.2	19	40	4.5
	lower bound								18	16	
	1	1	40	39	59.6	42	40	4.9	41	38	0.0
	1	2	45	36	88.2	47	37	4.4	47	37	4.4
	2	1	38	41	87.8	39	44	5.6	39	43	4.5
0.01	1	5	50	34	163.3	54	34	2.4	53	34	1.8
	5	1	38	41	172.3	37	51	2.9	37	50	2.3
	1	10	58	33	284.9	60	33	0.7	60	33	0.7
	10	1	36	53	305.7	36	57	1.3	36	57	1.3
	lower bound								35	32	

service-level requirement (10.16) is a better approximation for the true requirement (10.14).

3 For the same requirement δ, the optimal base-stock levels and optimal costs are higher when the leadtime x is zero than when x is positive. Since the same amount of absolute difference will translate to a smaller (relative) optimality gap when the optimal costs themselves are higher, we do see better results in examples where the leadtime is zero than their counterparts with positive leadtime. Optimality gaps are larger at lower s_1, s_2 for another reason. The objective function of **P2**, $\sum_i h_i s_i$ is a surrogate of the true cost function $\sum_i h_i \mathsf{E}[I_i]$, ignoring the term $\sum_i h_i \mathsf{E}[(Q_i - s_i)^+]$. Obviously, the effect of this term is smaller and more negligible at higher s_i. In addition, all the s_i are treated as continuous variables

Table 10.5. Results of the five-component example*

	Optimal Solution						Refined Closed-Form Solution					
δ	s_1^*	s_2^*	s_3^*	s_4^*	s_5^*	cost	$\tilde{s}_1'$	$\tilde{s}_2'$	$\tilde{s}_3'$	$\tilde{s}_4'$	$\tilde{s}_5'$	gap(%)
.05	28	26	22	22	19	93.7	29	25	22	23	19	1.0
.01	45	38	32	33	27	151.4	42	36	30	31	26	0.7
.05	30	29	21	20	21	169.7	29	29	22	20	21	0.0
.01	42	40	30	28	28	253.1	41	40	30	29	28	0.4

*The first panel is for equal holding costs $h_1 = h_2 = h_3 = h_4 = h_5 = 1$, and the second for
the case $h_1 = 2$, $h_2 = 1$, $h_3 = 2$, $h_4 = 3$, $h_5 = 1$

in the approximate optimization problem **P2**. But they have to
be integers due to the physical interpretation. So we always round
the value from (10.22) to the nearest integer. In some cases, the
rounding alone can contribute a 3–5% increase in the optimality
gap.

4 The approximate solution in Equation (10.22) is also valuable if
the optimal policy has to be found. The solution itself provides an
excellent starting point for the search and the lower bound (10.23)
provides a useful stopping rule in the search.

Finally, we consider a five-component system, again with a Poisson order process, having $\mathsf{E}[A] = 3$. The order quantities have the
following probabilities: $P_D(2, 3, 4, 2, 1) = 0.2$, $P_D(3, 2, 2, 4, 3) = 0.3$,
$P_D(4, 4, 1, 1, 3) = 0.2$ and $P_D(1, 1, 2, 2, 2) = 0.3$. Production times all
have Erlang distributions with coefficient of variation $\sqrt{0.2}$, and $\mathsf{E}[B_1]$
$= \mathsf{E}[B_2] = \mathsf{E}[B_3] = 1$, $\mathsf{E}[B_4] = \mathsf{E}[B_5] = 0.9$. The leadtime is fixed at
zero. Numerical studies in this system are by far more difficult than in
the earlier two-component examples because of the size of the space that
needs to be searched over for the true optimal solution. The results are
again very good and are shown in Table 10.5.

5.   Proofs

5.1.   Proof of Theorem 10.1

The main purpose of this subsection is to prove Theorem 10.1, and
illustrate the technical tools in the single-component setting. With only
one component, we use subscripts to denote order indices. For example,
A_n is the time between the nth and $(n+1)$th order, D_n is the batch
size of the nth order, $B_{n,j}$ is the production time for the jth unit of
the nth order. To emphasize its dependence on the base-stock level, we

often write the nth response time R_n as $R_n(s)$. A useful first step is the characterization of the response time in terms of an associated queue driven by the same input variables. Specifically, consider a batch-arrival queue with interarrival times $\{A_n\}$, batch sizes $\{D_n\}$, and individual service times $\{B_{n,1}, \ldots, B_{n,D_n}\}$ in a batch. The queue is empty at time zero. The following sample-path result relates the response times $R_n(s)$ in the original system to the waiting times in the associated queue when both systems are driven by the same inputs:

Lemma 10.6 *Let W_n be the waiting time of batch n in the queue. Then*

$$R_n(s) = \left(W_{N_n} + \sum_{j=1}^{H_n} B_{N_n,j} - \sum_{j=1}^{n-N_n} A_{n-j} \right)^+, \qquad (10.28)$$

where

$$N_n = \sup \left\{ 1 \le k \le n : \sum_{j=k}^{n} D_j > s \right\}, \qquad (10.29)$$

and

$$H_n = \sum_{j=N_n}^{n} D_j - s. \qquad (10.30)$$

Proof. In the queue, let $\{t_n, n \ge 1\}$ be the batch arrival epochs, let $\{t'_n, n \ge 1\}$ be the batch departure epochs and let $\{\hat{t}(k), k \ge 1\}$ be the service completion epochs of the individual jobs. In the production-inventory system, the inventory level starts from s at time 0. The cumulative demand at time t_n is $\sum_{j=1}^{n} D_j$. For n with $\sum_{j=1}^{n} D_j \le s$, no waiting is necessary for demand n and $R_n(s) = 0$; equation (10.28) holds trivially because $N_n = 1$, $H_n \le 0$ and $W_1 = 0$. For n with $\sum_{j=1}^{n} D_j > s$,

$$R_n(s) = \left(\hat{t}\left(\sum_{j=1}^{n} D_j - s \right) - t_n \right)^+.$$

To see this, notice that at time $\hat{t}(\sum_{j=1}^{n} D_j - s)$ the system finishes producing $\sum_{j=1}^{n} D_j - s$ units. With the initial inventory of s units, there are cumulatively $\sum_{j=1}^{n} D_j$ units available to meet the first n demands at this time. By the definition of N_n,

$$t'_{N_n} - \sum_{j=1}^{D_{N_n}} B_{N_n,j} < \hat{t}\left(\sum_{j=1}^{n} D_j - s \right) \le t'_{N_n}.$$

A little careful bookkeeping shows that

$$\hat{t}\left(\sum_{j=1}^{n} D_j - s\right) = t'_{N_n} - \sum_{j=1}^{D_{N_n}} B_{N_n,j} + \sum_{j=1}^{H_n} B_{N_n,j},$$

and

$$\hat{t}\left(\sum_{j=1}^{n} D_j - s\right) - t_n = t'_{N_n} - t_{N_n} + t_{N_n} - t_n - \sum_{j=1}^{D_{N_n}} B_{N_n,j} + \sum_{j=1}^{H_n} B_{N_n,j}$$

$$= W_{N_n} + \sum_{j=1}^{H_n} B_{N_n,j} - \sum_{j=1}^{n-N_n} A_{n-j}.$$

The result (10.28) follows.   **QED.**

In words, equation (10.28) states that order n is filled when the system finishes producing all the units of order (N_n-1), plus the first H_n units of order N_n. By construction, W_{N_n} is determined by $\{A_j, (B_{j,1}, \ldots, B_{j,D_j}), D_j; j \leq N_n - 1\}$ and (N_n, H_n) are determined by $\{D_j, N_n \leq j \leq n\}$. Under our independence assumptions, $\{A_j, (B_{j,1}, \ldots, B_{j,D_j}), D_j; j \leq k - 1\}$ and $\{D_j, j \geq k\}$ are independent of each other for any k, and therefore

$$P(W_{N_n} \leq x | N_n = k) = P(W_k \leq x | N_n = k) = P(W_k \leq x), \qquad (10.31)$$

for any x. As $n \to \infty$, $N_n \to \infty$, a.s., and if the system is stable (i.e., $\mathsf{E}[D_1]\mathsf{E}[B_1] - \mathsf{E}[A_1] < 0$) the W_n converge weakly to a random variable W having the distribution of the steady-state batch waiting time in the queue. Using (10.31) in the first equation of the proof of Theorem 1.1.1 of Gut [26] we conclude that the W_{N_n} also converge in distribution to W. Indeed, a small extension of Gut's result shows that $(W_{N_n}, n - N_n, H_n)$ converge jointly in distribution to the steady state version $(W, N - 1, H)$ with

$$N = N(s) \stackrel{\text{def}}{=} \min\left\{k \geq 1 : \sum_{j=1}^{k} \bar{D}_j > s\right\}, \qquad (10.32)$$

and

$$H = H(s) \stackrel{\text{def}}{=} \sum_{j=1}^{N(s)} \bar{D}_j - s, \qquad (10.33)$$

and W independent of (N, H). (Here, the $\bar{D}_j$ are i.i.d. with the same distribution as the D_j.) In light of Lemma 10.6, the steady-state response time can therefore be represented as

$$R(s) = \left(W + \sum_{j=1}^{H(s)} \bar{B}_j - \sum_{j=1}^{N(s)-1} \bar{A}_j \right)^+, \tag{10.34}$$

where the $\bar{A}_j$ and $\bar{B}_j$ are i.i.d. with the same distributions as the original interarrival and unit production times.

Define $X_n = \sum_{j=1}^{D_n} B_{n,j} - A_n$ for all $n \geq 1$. It is obvious that this sequence of variables share the same c.g.f.

$$\psi_X(\theta) = \psi_D(\psi_B(\theta)) + \psi_A(-\theta).$$

The proof of Theorem 10.1 uses an exponential change of measure (also called exponential twisting), so we briefly review this concept; see Chapter XII of Asmussen [2] or Chapter VIII of Siegmund [35] for additional background. Suppose a random variable Z has distribution F and that

$$\psi_Z(\theta) = \log \int e^{\theta x} dF(x)$$

is finite in some non-degenerate interval containing 0. Then

$$dF_{(\theta)}(x) = e^{\theta x - \psi_Z(\theta)} dF(x)$$

defines a family of distributions indexed by θ. Let $P_{(\theta)}$ be the probability measure under which Z has distribution $F_{(\theta)}$; we say that $P_{(\theta)}$ is obtained by θ-twisting Z. For any integrable function g, the expectation of $g(Z)$ under the original measure can be evaluated under $P_{(\theta)}$ if we first multiply $g(Z)$ by the likelihood ratio $e^{-\theta Z + \psi_Z(\theta)}$, i.e.,

$$\mathsf{E}[g(Z)] = \mathsf{E}_{(\theta)}\left[g(Z) \cdot e^{-\theta Z + \psi_Z(\theta)} \right], \tag{10.35}$$

where $\mathsf{E}_{(\theta)}$ denotes expectation under $P_{(\theta)}$. From this, we get the following relation which will be used very often in our proofs:

$$\mathsf{E}_{(\theta)}[Z] = \mathsf{E}\left[Z e^{\theta Z - \psi_Z(\theta)} \right] = \psi_Z'(\theta).$$

A twist can be applied to a sequence of i.i.d. random variables $\{Z_n\}$. Let $\tilde{P}$ be the probability measure obtained by θ-twisting $Z_1, Z_2, \ldots$. Then for any fixed n and any function g of $\{Z_1, \ldots, Z_n\}$, (10.35) generalizes to

$$\mathsf{E}[g(Z_1, \ldots, Z_n)] = \tilde{\mathsf{E}}\left[g(Z_1, \ldots, Z_n) \prod_{j=1}^{n} e^{-\theta Z_j + \psi_Z(\theta)} \right],$$

where $\tilde{\mathsf{E}}$ denotes expectation under $\tilde{P}$. This identity extends to stopping times — i.e., for any stopping time T and any function g of $\{Z_1, \ldots, Z_T\}$, *Wald's likelihood ratio identity* (see Siegmund [35], p. 166 or Asmussen [2], p. 258) gives

$$\mathsf{E}[g(Z_1, \ldots, Z_T); T < \infty] = \tilde{\mathsf{E}}\left[g(Z_1, \ldots, Z_T) \prod_{j=1}^{T} e^{-\theta Z_j + \psi_Z(\theta)}; T < \infty\right].$$

We use this frequently. (A semicolon inside an expectation indicates that the expectation is evaluated over the event following the semicolon.)

Proof of Theorem 10.1. Because of the representation in (10.34), the distribution of $R(s)$ can be analyzed through W. To that end, let $S_0 = 0$ and $S_n = \sum_{j=1}^{n} X_j$. A classical result (see, e.g., Asmussen [2], p.80) states that W has the same distribution as $\max_{n \geq 0} S_n$. Thus,

$$P(R(s) > x) = P\left(W > \sum_{j=1}^{N(s)-1} \bar{A}_j - \sum_{j=1}^{H(s)} \bar{B}_j + x\right)$$

$$= P\left(\max_{n \geq 0} S_n > \sum_{j=1}^{N(s)-1} \bar{A}_j - \sum_{j=1}^{H(s)} \bar{B}_{N(s),j} + x\right).$$

If we set $L = \sum_{j=1}^{N(s)-1} \bar{A}_j - \sum_{j=1}^{H(s)} \bar{B}_{N(s),j} + x$ and $T = \inf\{n \geq 1 : S_n > L\}$, then

$$P(R(s) > x) = P(T < \infty). \tag{10.36}$$

Let $\tilde{P}$ be the measure obtained by γ-twisting $X_1, X_2, \ldots$, β-twisting $\bar{D}_1, \bar{D}_2, \ldots$, $(-\gamma)$-twisting $\bar{A}_1, \bar{A}_2, \ldots$ and γ-twisting $\bar{B}_{N,1}, \bar{B}_{N,2}, \ldots$. Notice that $N(s)$ is a stopping time for $\{\bar{D}_j\}$ and T is a stopping time for $\{X_n\}$. In this setting, Wald's identity gives

$$P(T < \infty) = \tilde{\mathsf{E}}\left[\prod_{j=1}^{T} e^{-\gamma X_j + \psi_X(\gamma)} \prod_{j=1}^{N} e^{-\beta \bar{D}_j + \psi_D(\beta)}\right.$$

$$\left. \times \prod_{j=1}^{N-1} e^{\gamma \bar{A}_j + \psi_A(-\gamma)} \prod_{j=1}^{H} e^{-\gamma \bar{B}_{N,j} + \psi_B(\gamma)}; T < \infty\right]$$

$$= \tilde{\mathsf{E}}\left[e^{-\gamma S_T - \beta \sum_{j=1}^{N} \bar{D}_j + N \psi_D(\beta) + \gamma \sum_{j=1}^{N-1} \bar{A}_j}\right.$$

$$\left. \times e^{(N-1)\psi_A(-\gamma) - \gamma \sum_{j=1}^{H} \bar{B}_{N,j} + H\beta}; T < \infty\right]$$

$$= e^{-\gamma x - \beta s + \psi_D(\beta)} \tilde{\mathsf{E}}\left[e^{-\gamma(S_T - L)}; T < \infty\right]. \tag{10.37}$$

Under $\tilde{P}$, $\tilde{\mathsf{E}}[X_1] = \mathsf{E}[e^{\gamma X_1} X_1] = \psi'_X(\gamma) > 0$, so $\tilde{P}(T < \infty) = 1$. The random level L is independent of $\{S_n\}$ and $L \to \infty$ as $s + x \to \infty$. A minor extension of a classical result in renewal theory (see Corollary 8.33 of Siegmund [35] or Theorem XII.5.2 of Asmussen [2]) shows that

$$C_1 = \lim_{s+x \to \infty} \tilde{\mathsf{E}}\big[e^{-\gamma(S_T - L)}\big]$$
$$= \tilde{\mathsf{E}}\big[e^{-\gamma Z_e}\big]$$

exists, where Z_e has the equilibrium distribution of the ascending ladder heights of the random walk under $\tilde{P}$. (See Chapter 12 of Feller [18], Chapter 1 of Prabhu [31], Chapter VII of Asmussen [2], or Chapter VIII of Siegmund [35] for background on ladder heights and related results from the theory of random walks.) So we have

$$\lim_{s+x \to \infty} e^{\gamma x + \beta s} P(R(s) > x) = e^{\psi_D(\beta)} C_1 = C. \qquad (10.38)$$

QED.

When we have an assembly time U_n, the steady-state response time becomes $\hat{R}(s) = R(s) + U$, where $R(s)$ is as in Theorem 10.1. By the simple relation $P(\hat{R}(s) > x) = P(R(s) > x - U)$ and conditioning on U, the argument in the proof of Theorem 10.1 shows that

$$\lim_{s+x \to \infty} e^{\gamma x + \beta s} P(\hat{R}(s) > x | U) = \lim_{s+x \to \infty} e^{\gamma x + \beta s} P(R(s) > x - U | U) = C e^{\gamma U}.$$

When $s \to \infty$ but x stays finite in the limit above, we require x to be such that $P(U \leq x) = 1$. (If this condition is not met, then the unfill rate $P(R(s) > x)$ will never decrease to 0 no matter how large s is.) Invoking the dominated convergence theorem, we get

$$\lim_{s+x \to \infty} e^{\gamma x + \beta s} P(\hat{R}(s) > x) = C \mathsf{E}\big[e^{\gamma U}\big],$$

which means that the assembly time changes only the constant, and does not alter the asymptotic trade-off between x and s. Obviously, this is also true when assembly times are added in multiple-component systems.

The change of measure introduced above is also useful in estimating the service level for given x and s through relation (10.36). When $x + s$ is large, L is also large. But for a stable production system the random walk $\{S_n\}$ has negative drift, so $\{T < \infty\}$ is a rare event and it becomes increasingly rare as the service level increases. Straightforward simulation is not efficient, if possible at all. Working with the new measure $\tilde{P}$, we appeal to (10.37) and estimate $P(T < \infty)$ by averaging i.i.d.

replications of

$$e^{-\gamma(S_T - L) + \psi_D(\beta) - \gamma x - \beta s}$$

generated under $\tilde{P}$.

5.2.   Proof of Theorem 10.5

The purpose of this subsection is to prove Theorem 10.5 and show the algorithm for identifying the optimal index. To simplify our notation, we define $d_i = C_i e^{-\gamma_i x}$, $y_i = \beta_i s_i$, and $\bar{h}_i = h_i / \beta_i$. Problem **P2** can be stated in terms of decision variables y_i and parameters d_i, $\bar{h}_i$ as

$$\textbf{P2':} \qquad \min \ \sum_{i=1}^{m} \bar{h}_i y_i$$

$$\text{s.t.} \ \sum_{i=1}^{m} d_i e^{-y_i} \leq \delta,$$

$$y_i \geq 0, \quad i = 1, 2, \ldots, m.$$

The components are ordered the same way as in (10.18)

$$0 < \frac{\bar{h}_1}{d_1} \leq \frac{\bar{h}_2}{d_2} \leq \cdots \leq \frac{\bar{h}_m}{d_m} < \frac{\bar{h}_{m+1}}{d_{m+1}} \equiv \infty. \tag{10.39}$$

The parameters satisfy

$$\sum_{i=1}^{m} d_i > \delta, \tag{10.40}$$

the same assumption as (10.19) to exclude the trivial solution $y_1 = y_2 = \cdots = y_m = 0$.

Because the objective function and the constraints in **P2'** are either linear or strictly convex, the Kuhn-Tucker conditions are both necessary and sufficient for optimality. These conditions are

$$\bar{h}_i - \theta d_i e^{-y_i} - \alpha_i = 0, \quad i = 1, 2, \ldots, m;$$

$$\theta \left(\sum_{i=1}^{m} d_i e^{-y_i} - \delta \right) = 0;$$

$$\alpha_i y_i = 0, \quad i = 1, 2, \ldots, m;$$

$$\theta \geq 0; \quad \alpha_i \geq 0, \quad i = 1, 2, \ldots, m;$$

$$y_i \geq 0, \quad i = 1, 2, \ldots, m;$$

$$\sum_{i=1}^{m} d_i e^{-y_i} - \delta \leq 0.$$

With the new notation, the definition of θ_j (cf. Equation (10.20)) becomes

$$\theta_j \equiv \frac{\sum_{i=1}^{j} \bar{h}_i}{\delta - \sum_{i=j+1}^{m} d_i}.$$

If there exists an index k for which

$$\frac{\bar{h}_k}{d_k} \le \theta_k < \frac{\bar{h}_{k+1}}{d_{k+1}}, \tag{10.41}$$

then setting

$$\left. \begin{array}{l} \theta = \theta_k, \quad \text{and} \\ y_i = \log \frac{\theta d_i}{\bar{h}_i}, \alpha_i = 0 \ \text{ for } i = 1, 2, \ldots, k, \\ y_i = 0, \alpha_i = \bar{h}_i - \theta d_i \ \text{ for } i = k+1, \ldots, m, \end{array} \right\} \tag{10.42}$$

satisfies the Kuhn-Tucker conditions. Notice that Equation (10.41) is equivalent to Equation (10.21), and Equation (10.42) gives the same solution as Equation (10.22) of Theorem 10.1. We next illustrate through a backward induction algorithm that such an index k exists and can be found very efficiently.

Algorithm

1 Starting at $j = m$, $0 < \theta_m < \frac{\bar{h}_{m+1}}{d_{m+1}} \equiv \infty$ is obviously true.

2 Backward induction. Suppose $0 < \theta_j < \frac{\bar{h}_{j+1}}{d_{j+1}}$.

- If $\frac{\bar{h}_j}{d_j} \le \theta_j$, then (10.41) is satisfied and the algorithm terminates with the optimal index $k = j$.

- Otherwise, $0 < \theta_j < \frac{\bar{h}_j}{d_j}$ implies $0 < \theta_{j-1} < \frac{\bar{h}_j}{d_j}$. To see this, observe the following fact: for constants $A, B, C, D > 0$, if

$$\frac{A}{B} < \frac{C}{D} \quad \text{and } A > C,$$

  then

$$0 < \frac{A - C}{B - D} < \frac{C}{D}.$$

  Applying this fact to

$$0 < \theta_j \equiv \frac{\sum_{i=1}^{j} \bar{h}_i}{\delta - \sum_{i=j+1}^{m} d_i} < \frac{\bar{h}_j}{d_j}$$

yields

$$0 < \theta_{j-1} \equiv \frac{\sum_{i=1}^{j-1} \bar{h}_i}{\delta - \sum_{i=j}^{m} d_i} < \frac{\bar{h}_j}{d_j}.$$

3 If the algorithm has not found an optimal index at $j = 2$, i.e., if $0 < \theta_2 < \frac{\bar{h}_3}{d_3}$ but $\theta_2 < \frac{\bar{h}_2}{d_2}$, then

$$0 < \frac{\bar{h}_1}{d_1} < \frac{\bar{h}_1}{\delta - \sum_{i=2}^{m} d_i} \equiv \theta_1 < \frac{\bar{h}_2}{d_2},$$

where the second inequality is due to (10.40), and the first and the last inequalities are based on the backward induction. The algorithm terminates at optimal index $k = 1$.

Finally, we show the uniqueness of the optimal solution. Suppose k is the first optimal index found by the above algorithm with

$$\theta_k \equiv \frac{\sum_{i=1}^{k} \bar{h}_i}{\delta - \sum_{i=k+1}^{m} d_i} \geq \frac{\bar{h}_k}{d_k}, \qquad (10.43)$$

we will show that any index smaller than k cannot be an optimal one.

1 If $\delta - \sum_{i=k+1}^{m} d_i < d_k$, then for $1 \leq j \leq k - 1$,

$$\theta_j \equiv \frac{\sum_{i=1}^{j} \bar{h}_i}{\delta - \sum_{i=j+1}^{m} d_i} < 0$$

so j cannot be an optimal index.

2 If $\delta - \sum_{i=k+1}^{m} d_i = d_k$, then $\theta_{k-1} = \infty$ and $\theta_j < 0$ for $1 \leq j \leq k-2$, any index smaller than k cannot be an optimal index.

3 If $\delta - \sum_{i=k+1}^{m} d_i > d_k$, then $\theta_{k-1} \geq \frac{\bar{h}_k}{d_k}$ hence $k - 1$ cannot be an optimal index. To see this, observe the following fact: for constants $A > C > 0$ and $B > D > 0$, $\frac{A}{B} \geq \frac{C}{D}$ implies $\frac{A-C}{B-D} \geq \frac{C}{D}$. Applying it on (10.43) yields

$$\theta_{k-1} \equiv \frac{\sum_{i=1}^{k-1} \bar{h}_i}{\delta - \sum_{i=k}^{m} d_i} \geq \frac{\bar{h}_k}{d_k} \geq \frac{\bar{h}_{k-1}}{d_{k-1}}.$$

From (10.43) and the inequality $\theta_{k-1} \geq \frac{\bar{h}_{k-1}}{d_{k-1}}$ we can continue the argument and conclude that any index smaller than $k - 1$ cannot be optimal.   **QED.**

6.   Concluding Remarks

In very general stochastic ATO systems with a production capacity limit on each component, we have developed a set of tools and results in characterizing the order fill rate, a natural measure of customer service level. We have demonstrated both theoretically and numerically that it is possible to quantify the trade-off between longer leadtimes or higher inventory levels in achieving a target fill rate. Not surprisingly, the trade-off is sharpest in single-component systems. When multiple components are assembled into multiple products, the trade-off depends in part on which components most constrain the product-level fill rate.

We have also developed simple and effective approximate solutions in closed form to the problem of finding the best base-stock policy in an ATO system under high service-level requirements. Our closed-form solution is based on a more stringent, yet asymptotically tight upper bound of the given service-level requirement. A simple refinement to the closed-form solution further reduces the gap to optimality. Numerical studies have shown that the costs of the approximate solutions are typically within a few percentage points of the optimal ones, over a wide range of holding cost parameters and even at mildly high fill rate requirements. The solution and the bounds developed in the analysis have proved to be useful in the direct search of the optimal solution.

Finally, we address a concern that some people may have: the method proposed in this chapter requires high fill rates, which in turn require high inventories, or long leadtimes, or both. These requirements seem to be inconsistent with the philosophy of Just-in-Time (JIT), which advocates low inventory and short leadtime. To this, we note the following: (1) High fill rates are the goal of almost all companies and the common practice of many. (2) The framework of this chapter is about how to achieve high fill rates without changing the demand and production processes. In the presence of uncertainties in demand and production, and with limits in production capacity, holding inventories or quoting appropriate leadtimes is the only means to achieve high fill rates. As a contrast, JIT requires changing the current system — in particular, virtually eliminating the uncertainties. (3) How high a leadtime or inventory has to be in order to achieve a target fill rate is relative and depends on the characteristics of the demand and production processes. When the system has little variability, the values of the parameters α and β will be relatively large. As a result, a moderate leadtime or inventory is high enough to achieve a high fill rate.

Acknowledgments

Much of the work discussed here is based on joint work with Paul Glasserman. I am very grateful to Paul for the many discussions we had and for his encouragement.

References

[1] AGRAWAL, N. AND M.A. COHEN. (1999) Optimal Material Control in an Assembly System with Component Commonality. Working paper, Department of Operations and Information Management, The Wharton School, PA.

[2] ASMUSSEN, S. (1987) *Applied Probability and Queues.* Wiley, New York.

[3] AVIV, Y. AND A. FEDERGRUEN. (1997) Stochastic Inventory Models with Limited Production Capacities and Varying Parameters. *Probability in the Engineering and Information Sciences*, Vol. 11, 107–135.

[4] BESSLER, S.A. AND A.F. VEINOTT. (1965) Optimal Policy for a Dynamic Multiechelon Inventory Model, *Naval Res. Logistics Quarterly*, Vol. 13, 355–389.

[5] BOOKBINDER, J.H. AND J.-Y. TAN. (1988) Strategies for the Probabilistic Lot-Sizing Problem with Service-Level Constraints. *Mgmt. Sci.*, Vol. 34, 1096–1108.

[6] BOYACI T. AND G. GALLEGO. (1998) Serial Production/ Transportation System under Service Constraints. Working Paper, Department of Industrial Engineering and Operations Research, Columbia University.

[7] BUZACOTT, J.A. AND J.G. SHANTHIKUMAR. (1994) Safety Stock versus Safety Time in MRP Controlled Production Systems. *Mgmt. Sci.*, Vol. 40, 1678–1689.

[8] CHENG F., M. ETTL, G. LIN, AND D.D. YAO. (2000) Inventory-Service Optimization in Configure-to-Order Systems: From Machine Type Models to Building Blocks. Technical Report, IBM T.J. Watson Research Center, Yorktown Hights, NY.

[9] CHANG, C.A. (1985) The Interchangeability of Safety Stocks and Safety Time. *J. Oper. Mgmt.*, Vol. 6, 35–42.

[10] CLARK, A.J. AND H. SCARF. (1960) Optimal Policies for a Multi-Echelon Inventory. *Mgmt. Sci.*, Vol. 6, 475–490.

[11] COHEN, M.A., P. KLEINDORFER, AND H. LEE. (1988) Service Constrained (s, S) Inventory Systems with Priority Demand Classes and Lost Sales. *Mgmt. Sci.*, Vol. 34, 482–499.

[12] DEUERMEYER, B.L. AND L.B. SCHWARZ. (1981) A Model for the Analysis of System Service Level in Warehouse-Retailer Distribution Systems: The Identical Retailer Case, in *Multi-Level Production/Inventory Control Systems: Theory and Practice*, L. Swcharz, ed. North-Holland, Amsterdam, 163–193.

[13] EPPEN, G.D. AND L.E. SCHRAGE. (1981) Centralized Ordering Policies in a Multiwarehouse System with Leadtimes and Random Demand, in *Multi-Level Production/Inventory Control Systems: Theory and Practice*, L. Swcharz, ed. North-Holland, Amsterdam, 51–69.

[14] ETTL, M., G.E. FEIGIN, G.Y. LIN, AND D.D. YAO. (1995) A Supply-Chain Model with Base-Stock Control and Service Level Requirements. IBM Research Report, IBM T.J. Watson Research Center, Yorktown Heights, NY.

[15] FEDERGRUEN, A. AND P. ZIPKIN. (1984) Computational Issues in an Infinite-Horizon, Multiechelon Inventory Model. *Oper. Res.*, Vol. 32, 818–836.

[16] FEDERGRUEN, A. AND P. ZIPKIN. (1986a) An Inventory Model with Limited Production Capacity and Uncertain Demands, I: The Average Cost Criterion. *Math. Oper. Res.*, Vol. 11, 193–207.

[17] FEDERGRUEN, A. AND P. ZIPKIN. (1986b) An Inventory Model with Limited Production Capacity and Uncertain Demands, II: The Discounted Cost Criterion. *Math. Oper. Res.*, Vol. 11, 208–215.

[18] FELLER, W. (1971) *An Introduction to Probability Theory and its Applications*, Vol. 2, 2nd Edition. Wiley, New York.

[19] GALLIEN J. AND L.M. WEIN. (1998) A Simple and Effective Component Procurement Policy for Stochastic Assembly Systems. Working paper, MIT Operations Research Center.

[20] GLASSERMAN, P. (1997) Bounds and Asymptotics for Planning Critical Safety Stocks. *Oper. Res.*, Vol. 45, 244–257.

[21] GLASSERMAN, P. AND T.-W. LIU. (1996) Rare-Event Simulation for Multistage Production-Inventory Systems. *Mgmt. Sci.*, Vol. 42, 1292–1307.

[22] GLASSERMAN, P. AND S. TAYUR. (1996) A Simple Approximation for a Multistage Capacitated Production-Inventory System. *Naval Research Logistics*, Vol. 43, 41–58.

[23] GLASSERMAN, P. AND Y. WANG. (1998) Leadtime-Inventory Trade-Offs in Assemble-to-Order Systems. *Oper. Res.*, Vol. 46, 858–871.

[24] GLASSERMAN, P. AND Y. WANG. (1999) Fill-Rate Bottlenecks in Production-Inventory Networks. *Manufacturing and Service Operations Management*, Vol. 1, 62–76.

[25] GRAVES, S. (1985) A Multi-Echelon Inventory Model for a Repairable Item with One-for-One Replenishment. *Mgmt. Sci.*, Vol. 31, 1247–1256.

[26] GUT, A. (1988) *Stopped Random Walks.* Springer, New York.

[27] KAPUCSINZKI, R. AND S. TAYUR (1998) A Capacitated Production-Inventory Model with Periodic Demand. *Oper. Res.*, Vol. 46, 899–911.

[28] KENDALL, M. (1987) *Advanced Theory of Statistics,* Vol. II, 5th Edition. Oxford Pub., New York.

[29] LEE, Y. AND P.H. ZIPKIN. (1992) Tandem Queues with Planned Inventories. *Oper. Res.* Vol. 40, 936–947.

[30] NAHMIAS, S. (1997) *Production and Operations Analysis,* 3rd Edition. Irwin.

[31] PRABHU, N.U. (1980) *Stochastic Storage Processes: Queues, Insurance Risk, and Dams.* Springer, New York.

[32] ROSLING, K. (1989) Optimal Inventory Policies for Assembly Systems Under Random Demands. *Oper. Res.*, Vol. 37, 565–579.

[33] SCHNEIDER, H. AND J.L. RINGUEST. (1990) Power Approximation for Computing (s, S) Policies Using Service Level. *Mgmt. Sci.*, Vol. 36, 822–834.

[34] SCHRANER, E. (1996) Capacity/Inventory Trade-Offs in Assemble-to-Order Systems. Working Paper, IBM T.J. Watson Research Center, Yorktown Heights, New York.

[35] SIEGMUND, D. (1985) *Sequential Analysis: Tests and Confidence Intervals.* Springer, New York.

[36] SONG, J.S. (1998) On the Order Fill Rate in a Multi-Item Inventory System. *Oper. Res.*, Vol. 46, 831–845.

[37] SONG, J.S., S.H. XU, AND B. LIU. (1999) Order-Fulfillment Performance Measures in an Assemble-to-Order System with Stochastic Leadtime. *Oper. Res.*, Vol. 47, 131–149.

[38] SONG, J.S. AND D.D. YAO. (2000) Performance Analysis and Optimization of Assemble-to-Order Systems with Random Leadtimes. *Oper. Res.* to appear.

[39] TAYUR, S. (1993) Computing the Optimal Policy for Capacitated Inventory Models. *Stochastic Models*, Vol. 9, 585–598.

[40] TIJMS, H.C. (1994) *Stochastic Models: an Algorithmic Approach.* Wiley, Chichester.

[41] TIJMS, H.C. AND H. GROENEVELT. (1984) Simple Approximations for the Reorder Point in Periodic and Continuous Review (s, S) Inventory Systems with Service Level Constraints, *Euro. J. Oper. Res.*, Vol. 17, 175–190.

[42] WANG, Y. (2000) Near-Optimal Base-Stock Policies in Assemble-to-Order Systems under Service Level Requirements. Working Paper, MIT Sloan School of Management, MA.

[43] ZHANG, A.X. (1997) Demand Fulfillment Rates in an Assemble-to-Order System with Multiple Products and Dependent Demands, *Prod. and Oper. Mgmt.*, Vol. 6, 309–324.

Chapter 11

DEPENDENCE ANALYSIS OF ASSEMBLE-TO-ORDER SYSTEMS

Susan H. Xu

Department of Management Science and Information Systems
Smeal College of Business Administration
The Pennsylvania State University
University Park, PA 16802
shx@psu.edu

1. Introduction

In this chapter we study the dependence properties of multi-product, assemble-to-order (ATO) systems with either capacitated or uncapacitated suppliers. The aim is to gain insights into the impact of demand correlation on system performance and to develop easily computable, tight bounds for key performance measures.

In recent years, in response to wide-spread adoptions of the ATO strategy in practice, especially by high-tech firms, researchers have shown heightened interests in stochastic assembly systems. An ATO system is a multi-component, multi-product inventory system in which products are designed around interchangeable modules. These modules and major components are typically expensive and need long procurement leadtimes. The company makes and stocks these major components according to forecasted demands, and when a customer order arrives requesting a specific kit of components, the company quickly assembles the components and delivers the end-product to the customer. The ATO strategy helps to postpone the point of commitment of components to specific products and thus increases the probability of meeting a customized demand timely and at a low cost (Lee and Tang [15]).

Since order fulfillment in an ATO system requires simultaneous availability of several different components, the management must jointly control inventories and suppliers across different components. Indeed, perhaps the most significant difference between a classical inventory model and an ATO model is that, in the former case demands occur individually and independently and its analysis is often based on a univariate model, whereas in the latter case demands for different components occur simultaneously and its analysis must be based on a multivariate model. In an ATO system, neglecting statistical dependencies among different components could result in inadequate assessments of system performance. As convincingly argued by Disney et al. [9], "The new areas (of stochastic modelling) are not calling for new ever more complicated inter-event distributions, but rather are calling for new process structures which incorporate dependencies."

An important issue in an ATO system is accurate assessments of *product-based* performance measures, such as, the joint probability that all the items in a customer order are satisfied immediately (the instantaneous product fill rate) and the joint probability that all the requested components in a customer order are filled before a target delivery leadtime. Unfortunately, the presence of demand correlation makes an explicit computation of a joint performance measure either intractable or computationally intensive. For example, the computation of order fill rates in an ATO system with as few as three items is already intensive (Dayanik et al. [8]). Needless to say, effective approximations and heuristic procedures for performance evaluation are in demand. This calls for structural analysis of ATO systems so that a clear understanding of the effects of demand correlation on system performance can not only facilitate the development of easily computable, tight bounds and effective heuristics, but also enhance our ability to make intelligent decisions in product design and joint production/inventory management.

The existing literature on ATO models can be classified into two categories: The work on periodic-review ATO systems and on continuous-review ATO systems. Here we only review the literature on continuous-review ATO systems. The reader is referred to Hausman et al. [13] and references therein, for the work on periodic-review ATO systems.

One line of the literature on continuous-review ATO models considers exact and approximate performance evaluation and optimization issues, under base-stock control. Song [24, 25] assumes a multivariate Poisson demand process in an uncapacitated ATO system with deterministic leadtimes and provides an efficient algorithm to evaluate the order fill rate, defined as the probability of filling an order instantaneously. Glasserman and Wang [12] study several capacitated ATO systems modelled as a set of correlated $M^D/G/1$ and $G^D/G/1$ queues and develop an asymptotic relation between the base-stock level and target delivery leadtime. Built upon Glasserman and Wang's result,

Wang [32] further derives simple and effective base-stock policies for these systems. Song et al. [26] consider a capacitated ATO system modelled as a set of correlated $M/M/1/c$ queues and develop an exact performance analysis based on the matrix-geometric method. Gallien and Wein [11] study a single-product, uncapacitated stochastic assembly system in the setup of correlated $M/G/\infty$ queues, and obtain an approximate base-stock policy with a synchronization constraint (equivalent to the order-for-order policy). Song and Yao [27] investigate a problem similar to Gallien and Wein's, but under the first-come, first-served (FCFS) order fulfillment rule. In addition, Dayanik et al. [8] present a collection of several approaches in approximating and bounding performance measures in an ATO system. Other relevant research includes, but not limited to, Cheung and Hausman [7], Anupindi and Tayur [1] and Zhang [36].

Another line of research that is related to ATO analysis is the literature on correlated queueing models. This body of work stresses the need to develop structural insights via stochastic comparison methods and attempts to establish dependence relations between input parameters and output measures (see, e.g., Nelson and Tantawi [19], Baccelli and Makowski [3], Baccelli et al. [4], Disney et al. [9], Szekli [29], and references therein). In particular, multi-server fork-join queues, which have found prevalent applications in telecommunication systems, closely resemble the features of ATO systems. Based on the notion of associated random variables (Esary et al. [10]; see Section 4 for the definition), Nelson and Tantawi [19], Baccelli and Makowski [3] and Baccelli et al. [4], among others, study multi-server fork-join queues and establish some comparison results and bounds for the order response time (defined as the delay between the fork and join epochs). Unfortunately, association is often a too-strong notion to verify, or even invalid (see Remark 11.16), in a multi-product ATO system with renewal demands. To overcome the difficulty, some authors resort to certain weaker notions of dependencies and show that *supermodular* and *orthant* dependencies are more suitable in dealing with correlated queues with synchronized events. Xu [34] examines stochastic properties of a set of synchronized, $M^D/M/1/c$ queues. Using the so-called h-transform to reduce the degree of correlation in an input process, she establishes supermodular and orthant comparison results for the queue length vector (equivalent to the inventory on-order vector in an ATO system) and waiting time vector (equivalent to the manufacturing leadtime vector in an ATO system). Xu and Li [35] and Li and Xu [16] introduce the notions of *tree majorization orders* and *separable Markov chains with independent generating sequences* and use the notions to study supermodular and orthant dependence properties of several synchronized queueing systems, including the synchronized $GI^D/M/1/c$ queues, $GI^D/G/1$ queues, $GI^D/M/c/c$ loss-systems and $GI^D/M/\infty$ queues. These synchronized queueing systems provide prototypical models to study a variety of ATO systems under different assumptions, such as, capacitated versus uncapacitated suppliers, deterministic versus stochastic manufacturing lead-times, backlogged demands versus lost sales, and unit demand versus batch orders.

The ATO systems studied in this chapter, in the most general setting, can be loosely described as follows (the detailed description is given in Section 3). The inventory system stocks m different items that can be assembled into different end-products. Let $E = \{1, 2, \ldots, m\}$ be the index set of items. The demand process forms a multivariate compound renewal process, characterized by the following generic, independent random vectors: The inter-demand times of different items, $\mathbf{T} = (T_1, \ldots, T_m)$, where T_i is the interarrival time between two type-i batches and $\mathbf{T} = (T_1, \ldots, T_m)$ can be dependent (e.g., if demand is perfectly synchronized, then $T_1 = \cdots = T_m$ almost surely). With probability P^K a demand requests a product-K, $K \subseteq E$, $\sum_{K \subseteq E} P^K = 1$. By a product-$K$ demand it is meant that the demand is comprised of a random quantity of every item in set $K \subseteq E$. One may understand $\{P^K, K \subseteq E\}$ as the joint distribution of an m-dimensional random vector $\mathbf{Z} = (Z_1, \ldots, Z_m)$, where Z_i equals 1 if an item-i batch is requested by a demand and zero otherwise, $i \in E$. We call $\mathbf{Z}$ the *product-type indicator vector*. The random quantities requested by a demand are denoted by the batch size vector $\mathbf{D} = (D_1, \ldots, D_m)$, where D_i is the size of an item-i demand, $i \in E$. We assume that D_i contains independent components. Hence, vector $\mathbf{ZD}$ represents the generic product type and random quantities requested by a demand.

The system is under independent base-stock control. Inventory for each item is initially stocked to its base-stock level, and a demand for an end-product triggers a joint replenishment order for all items constituting the product. Demands are filled on an FCFS basis and unfilled items, if any, are completely backlogged (this assumption is imposed mainly for ease of exposition). Item-i orders are manufactured by facility i with generic service time S_i, $i \in E$. If facility i is *capacitated*, we assume it has a single machine that processes jobs on an FCFS basis, with i.i.d. service times. A capacitated ATO system can be viewed as a set of parallel $GI^D/G/1$ queues with correlated renewal batch demands and general service times. If facility i is *uncapacitated*, we assume that it has infinite machines in the complete backlogging case and finite machines in the partial backlogging or lost-sales case. Manufacturing (or procurement) leadtimes are i.i.d. random variables for each item. An uncapacitated ATO system can be modelled as a set of correlated $GI^D/G/c/c$ queues, where c can be either finite or infinite. A capacitated ATO model is suitable when the company and its suppliers are vertically integrated (e.g., belong to the same organization). An uncapacitated ATO model is more appropriate if the company procures the components from outside suppliers and component orders do not demand a significant portion of the suppliers' production capacities.

Let R_i be the stationary delivery leadtime of batch-i (we assume the stability condition) and $\mathbf{R} = (R_1, \ldots, R_m)$. The performance measure we focus on is the *due-date fill rate* of a product, defined as the joint probability that every batch constituting the product is delivered before its due date:

$$\text{The due-date fill rate of product-}K = P(R_i \leq r_i, i \in K), \quad K \subseteq E,$$
$$(11.1)$$

where r_i is the due-date, or the target delivery leadtime, of batch-i, $i \in K$.

A special case of a multivariate compound renewal process is the multivariate Poisson process, which plays a fundamental role in ATO modelling. It is known that a multivariate Poisson process generated by synchronized interarrival times $(T, \ldots, T)$ and parameter set $\{P^K, K \subset E\}$ is *positively orthant dependent* (in fact, associated), *regardless* of the distribution of $\{P^K, K \subset E\}$ (Xu [34], Li and Xu [16], see Definition 11.2). In addition, Xu [34] shows that the orthant dependence property of a multivariate Poisson process is inherited by the delivery leadtime vector $\mathbf{R}$. As a result, the due-date fill rate given by (11.1) admits a product-form lower bound,

$$P(R_i \leq r_i, i \in K) \geq \prod_{i \in K} P(R_i \leq r_i), \quad (11.2)$$

the bound only requires to evaluate a set of independent single-server queues.

However, if we move away from the Poisson assumption, it is no longer clear whether $\mathbf{R}$ is still positively dependent and, even if so, in what sense. In the study of ATO systems with correlated renewal demands, we would like to understand the advantages and disadvantages brought into the system by positive and negative dependencies of the input process specified by $\{P^K, K \subseteq E\}$ and/or $\mathbf{T}$. For example, as $\{P^K, K \subseteq E\}$ and/or $\mathbf{T}$ change their dependence structure, does the change have a favorable or adverse impact on the due-date fill rates of a product and the system as a whole? With synchronized interarrival times $\mathbf{T} = (T, \ldots, T)$, are the components of $\mathbf{R}$ still associated for any distribution of T, regardless of the properties of $\{P^K, K \subseteq E\}$ (the answer is affirmative if T is exponential)? If not, then under what conditions and in what sense are the components of $\mathbf{R}$ positively and negatively dependent? What are the conditions under which the due-date fill rate still admits the product-form lower bound given in (11.2)? It is conceivable that the product-form bound would degrade as demand becomes more correlated. Then how to develop a better bound that is computationally effective yet incorporates dependencies of the input process? Under what conditions the management can safely ignore demand correlation and rely on the item-based analysis? To answer these questions, dependence analysis of an ATO system that does not rely on specific assumptions of the demand process is desirable.

In this chapter, we demonstrate that the study of dependence structure of ATO systems can enhance our knowledge on both qualitative and quantitative behaviors of ATO systems. We show that supermodular and orthant orders, which compare dependence strengths of two random vectors (e.g., delivery leadtime vectors $\mathbf{R}$ and $\bar{\mathbf{R}}$ in two ATO systems) with identical univariate marginals ($R_i =_{st} \bar{R}_i$, $i \in E$, i.e., item-based performances of the two systems are stochastically identical) are particularly effective in the study of dependence properties of ATO systems. We obtain supermodular and orthant comparison results for several ATO systems based on the interplay between two newly developed notions, *majorization of weighted trees* (Xu and Li [35]) and *separable Markov chains with independent generating sequences* (Li and Xu [16]). The first notion establishes several partial orders on the family of dependent parameter sets $\{P^K, K \subseteq E\}$ and the second notion provides a unified framework to model system dynamics of a variety of ATO systems. More specifically, we study two capacitated ATO systems modelled as a set of correlated $GI^D/M/1$ queues and $GI/G/1$ queues and two uncapacitated ATO systems modelled as a set of correlated $GI^D/M/c/c$ loss-systems and $GI/D/\infty$ queues (our result also holds for $GI/G/\infty$ queues under the order-for-order fulfillment rule, see Section 6.2). We pay special attention to the correlated renewal demand process because its dependence properties and their effects on output measures are not as well understood as that for the correlated Poisson process.

The rest of the chapter is organized as follows. Section 2 outlines the methodology and summarizes the major results. Section 3 describes the model in detail and introduces notation. Section 4 studies dependence structure of the product indicator vector $\mathbf{Z}$ via the notion of tree majorization and their transformations. Sections 5 and 6 study dependence properties of several capacitated and uncapacitated ATO systems, respectively, in the setup of separable Markov chains with independent generating sequences. Section 7 derives upper and lower bounds for the product-based and system-based due-date fill rates and provide numerical examples. Finally, Section 8 concludes the chapter.

2.   The Approach and Main Results

We shall use stochastic comparison as a tool to gain structural insights and to establish bounds for the performances of ATO systems. The idea of stochastic comparisons is to compare the characteristics of two different, yet similar systems and use the result of the simpler system to bound/approximate that of the more complicated one. Because outputs (say $\mathbf{R}$) are determined by inputs (say $\mathbf{T}$ and $\{P^K, K \subseteq E\}$) via transformation process, known as system dynamics (that is, the equations that govern the evolution of a stochastic system), a common approach in stochastic comparisons is to first establish an ordering between the two parameter sets of two compared systems. Then, according to system dynamics, show that a performance measure is a function of parameters and that system dynamics preserve the ordering established for the two parameter sets (Stoyan [28], Chang [5]). Therefore, a well-defined ordering relation between two input parameter sets and ordering-preserving functions that govern system dynamics are two basic ingredients in a stochastic comparison study.

Two distribution functions (either univariate or multivariate) can be compared in different stochastic senses. For example, the usual stochastic order compares the magnitudes of two random variables or two random vectors, whereas the convex order compares the degrees of their dispersion. In this chapter, we focus on the *dependence orders* that compare the dependence strengths of two random vectors with identical univariate marginals. We focus on dependence analysis because the most saliently different characteristic of a product-based ATO model, as compared with an item-based classical inventory model, is its dependence nature caused by synchronized demands.

Many dependence notions are available in the literature that describes the types and strengths of dependencies in different stochastic senses (see, e.g., Shaked and Shanthikumar [22]). We find that *supermodular* and *orthant orders* are particularly effective in the study of ATO systems. These notions are appealing because, firstly, they are weaker notions than association and thus more likely to be valid for a large class of ATO models; secondly, they provide distributional bounds for a performance measure in an ATO system whose explicit expression is unavailable; thirdly, they are easy to verify via parametric characterizations and comparisons; and finally, they enjoy closure properties essential in a stochastic comparison study.

To establish supermodular and orthant comparison results of ATO models, we need two notions introduced by Xu and Li [35] and Li and Xu [16]. The first notion, *majorization of weighted trees*, establishes several partial orders and transforms on the family of parameter sets. The second notion, *separable Markov chains with independent generating sequences*, provides a unified framework to model system dynamics of a variety of ATO systems.

The following describes the idea of tree majorization. In dependence analysis, the parameters of a multivariate model can be considered as univariate parameters and multivariate parameters or dependence parameters (Joe [14]). Loosely speaking, univariate parameters determine univariate marginal distributions and multivariate parameters the joint distribution. Dependence analysis of ATO systems leads naturally to the comparison of multivariate, or dependent, parameters P^K defined on $K \subseteq E$, where E is the index set of items. Let L and K be two subsets of E. We say subset L is a descendent of subset K if $L \subseteq K$. If we regard each subset K as a node and P^K as the weight of node K, $\{P^K, K \subseteq E\}$ can be considered as a weighted tree. A nature question is how should a family of weighted trees be ordered that would characterize different degrees of dependencies, in some stochastic sense, of product-type indicator vectors? To answer this question, Xu and Li propose the notions of tree majorization orders. Specifically, tree majorization orders from the *roots* and *leaves* establish partial orders on the family of parameter trees and provide parametric representations of upper and lower orthant orders of product-type indicator vectors. The authors also introduce the g, $g_\downarrow$ and $g_\uparrow$ transforms, collectively called the inclusion-exclusion (IE) transforms, and show that a tree majorization order can be accomplished via a sequence of IE-transforms. More precisely, a g-transform of $\{P^K, K \subseteq E\}$ will decrease positive dependence strength of $\mathbf{Z}$ in the sense of supermodular order, whereas $g_\downarrow$ and $g_\uparrow$-transforms of $\{P^K, K \subseteq E\}$ will decrease positive dependence strength of $\mathbf{Z}$ in the sense of upper and lower orthant orders, respectively.

The notion of separable Markov chains with independent generating sequences (hereafter, referred to as separable Markov chains) is aimed at providing a unified framework to treat several prototypical parallel queueing systems with correlated arrivals. Simply put, a separable Markov chains is an m-dimensional Markov chain $\{\mathbf{X}_n, n \geq 0\}$, driven by several independent sequences of i.i.d. random vectors $\{\mathbf{Y}_n, n \geq 0\}$, $\{\mathbf{Z}_n, n \geq 0\}$, etc., where the components of vector $\mathbf{Y}_n$ or vector $\mathbf{Z}_n$ in a generating sequence can be dependent. The separability of $\{\mathbf{X}_n, n \geq 0\}$ is meant that its ith univariate chain $\{X_{i,n}, n \geq 0\}$ is driven only by the i component sequences $\{Y_{i,n}, n \geq 0\}$, $\{Z_{i,n}, n \geq 0\}$, etc., $i = 1, \ldots, m$. Li and Xu [16] show that if the transition function that governs the evolution of each univariate chain is monotone in the same direction and increasing, respectively, then the two compared separable Markov chains will preserve the supermodular and orthant orders of their generating sequences, respectively. Separable Markov chains are relevant here because, as we shall see in Sections 5 and 6, a performance vector process in an ATO system can often be treated as a separable Markov chain and the transition function of each univariate chain indeed satisfies the order-preserving conditions of a separable Markov chain. The strength of this "Markov chain" approach is its modelling versatility and generality. Indeed, it allows us to systematically examine dependence structure of capacitated and uncapacitated ATO systems under rather mild conditions. Of course, this method is equally applicable to other possible configurations of ATO systems.

The interplay between tree majorization and separable Markov chains enables us to systematically explore dependence properties of ATO systems. In particular, we study four types of ATO systems whose supply systems are modelled by a set of correlated $GI^D/M/1$ queues, $GI/G/1$ queues, $GI^D/M/c/c$-loss systems and $GI/D/\infty$ systems, respectively (The reader is referred to Remark 11.14 for the dependence properties of correlated $M/G/1$ queues). In each model, we let the marginal performance measures in the two systems be identical in order to expose the pure effects of correlated demands to a joint performance measure. It is worth emphasizing that not only this unified 'Markov chain' approach derives strong comparison results under mild conditions, it is also simpler and more transparent than the methods employed by others.

The motivation to study correlated $GI^D/M/1$ queues, in addition to correlated $GI/G/1$ queues is threefold. Firstly, our result for correlated $GI^D/M/1$ queues is also valid for correlated $GI^D/M/1/c$ queues, where c, the buffer capacity of the production system, can be either finite or infinite. As such, $GI^D/M/1/c$ queues can be used to model a large class of ATO supply systems under various scenarios such as complete or partial backlogging and lost sales. In addition, $GI^D/M/1/c$ queues include, as an important special case, $M/M/1/c$ queues. Song et al. [26] develop the matrix-geometric procedures to evaluate key performance measures in an ATO system modelled as a set of correlated $M/M/1/c$ queues. Secondly, unlike in the correlated $GI/G/1$ queues, where one can only quantify delivery leadtime vector $\mathbf{R}$, in correlated $GI^D/M/1/c$ queues one can also quantity other performance vectors such as inventory on-order, inventory on-hand and backorder vectors. Finally, a $GI/G/1$ queue, in general, defies the closed-form solution. But a $GI^D/M/1/c$ queue can often be evaluated explicitly for a given interarrival time and batch size distributions (e.g., a Gamma interarrival time and a geometric or constant batch size), which allows us to derive the product-form bound effectively.

The major results of the chapter can be summarized as follows.

1 We show that dependence structure of an ATO system is originated from two sources: dependency due to correlated interarrival times of different items, $\mathbf{T} = (T_1, \ldots, T_m)$, and dependency caused by the product-type indicator vector $\mathbf{Z}$ characterized by multivariate parameter set $\{P^K, K \subseteq E\}$. The two types of dependencies are investigated separately and their joint impact on system performance is analyzed.

2 Based on the dependence properties, we derive *setwise* bounds for the product-based and system-based due-date fill rates. Our numerical study shows that the first-order setwise lower bound, which requires only marginal distributions of $\mathbf{R}$, performs extremely well when correlation is low or moderate. We also demonstrate that the first-order setwise bound is very tight when the due-date fill rate is high, regardless of dependence structure of $\mathbf{R}$. This implies that multivariate analysis is necessary only for an ATO system with high demand correlation and high traffic intensities. In such a case, the setwise bound of a higher order can improve the quality of the first-order setwise bound. See Section 7.2 for details.

3 We show that, a more positively lower orthant dependent demand process will *improve* a product-based performance measure. This somewhat counter-intuitive result can be explained by the fact that, as demand becomes more positively lower orthant dependent, the delivery leadtime $\mathbf{R}$ also becomes more positively lower orthant dependent. This, in turn, increases the probability that all the batches in a product meet their respective due-dates. A direct consequence of this result is that a more positively lower orthant dependent $\mathbf{R}$ will stochastically shorten the longest delivery leadtime of a product and thus increase the probability of fulfilling a demand before a common due-date. We caution the reader that this claim is made on a product-to-product comparison basis, assuming a product is served in both systems. In particular, the system with a more positively lower orthant dependent $\mathbf{R}$ does not mean that it will have a higher system-based due-date fill rate

(see (11.10) for definition), because the demand rates of different products in the two systems are different (see Remark 11.10). In fact, our numerical result in Section 7.2 shows that as the demand process becomes more positively lower orthant dependent, the system-based due-date fill rate deteriorates.

4 Since the supermodular dependence implies both the positively upper and lower orthant dependence (see (11.12)), increasing supermodular dependence strength of $\mathbf{Z}$ (via the g-transform) and/or $\mathbf{T}$ will increase supermodular dependence strength of $\mathbf{R}$; this, in turn, reduces the expected sum of the i *longest* delivery leadtimes in $\mathbf{R}$, and increases the expected sum of the i *shortest* delivery leadtimes in $\mathbf{R}$, for any $i \in E$. This further implies that the expected *range* of the delivery leadtimes of a product, defined as the expected difference between the first and the last batch deliveries of a product, becomes shorter as supermodular dependence strength of $\mathbf{R}$ increases. While this indicates that $\mathbf{R}$ become less disperse and is considered as a welcome feature, it also means that the average delivery leadtime of a batch in product K, $K \subset E$, defined as $\frac{1}{|K|} \sum_{i \in K} R_i$, where $|K|$ is the cardinality of product K, becomes more variable in the sense of the convex order (see Remark 11.11). In addition, our numerical examples in Section 7.2 shows that the expected delivery time of the last batch of an arbitrarily chosen product is larger with a more supermodular dependent $\mathbf{R}$.

5 We show that the common perception that simultaneous demands always introduce positive dependencies to a performance vector is false. It is known that if a demand process is a multivariate Poisson process characterized by $\mathbf{T} = (T, \ldots, T)$ and $\{P^K, K \subseteq E\}$, where T is an exponential random variable, then, regardless of the distribution of $\{P^K, K \subseteq E\}$, $\mathbf{R}$ is positively upper and lower orthant dependent. This fact has been used by several authors to derive the product-form lower bound of the due-date fill rate. However, if we allow T to follow an arbitrary distribution, then the dependence structure of $\mathbf{R}$ is no longer independent of $\{P^K, K \subseteq E\}$. It turns out that $\mathbf{R}$ is not necessarily positively dependent, and, in fact, can even be negatively dependent (see Remark 11.16).

3. The Model and Performance Measures

In this section we describe a continuous-review ATO system and identify several key performance measures. We consider an infinite planning horizon and assume that assembly times are negligible.

Throughout the chapter, equality in distribution between two random variables or vectors is denoted by $=_{st}$. Inequalities between vectors are componentwise inequalities. For a m-dimensional vector $\mathbf{a} = (a_1, a_2, \ldots, a_m)$ and $K \subseteq E$, we let $\mathbf{a}^K = (a_i : i \in K)$, $a_{\max}^K = \max_{i \in K}\{a_i\}$ and $a_{\min}^K = \min_{i \in K}\{a_i\}$.

We also take the following convention: An upper-case letter, supplemented by subscripts when necessary, say X_n or $X_{i,n}$, is used for a random variable, a bold upper-case letter, often with subscripts, say $\mathbf{X}_n$, represents an m-dimensional random vector, and a script upper-case letter, say $\mathcal{X}$, is used for a sequence of m-dimensional random vectors. The generic versions of a random variable X_n and a random vector $\mathbf{X}_n$ are denoted by X and $\mathbf{X}$, respectively.

The system stocks m items, where $E = \{1, \ldots, m\}$ denotes the index set. The demand process of the m items forms a multivariate compound renewal process. More specifically, we assume that the inter-arrival times of *potential* batch demands for item-i are i.i.d. random variables $\{T_{i,n}, \ n \geq 1\}$, where $T_{i,n} =_{st} T$, $i \in E, n \geq 1$, follows an arbitrary distribution. The nth demand is comprised of every *potential* item-i batch demand arrived at time $\sum_{j=1}^{n} T_{i,j}$, $i \in E$, $n \geq 1$. Let $\mathbf{T}_n = (T_{1,n}, \ldots, T_{m,n})$ be the interarrival times between demands n and $n+1$. In general, the components of $\mathbf{T} =_{st} \mathbf{T}_n$ are dependent random variables. For example, $\mathbf{T} = (T, T, \ldots, T)$ represents a perfectly synchronized demand pattern, where all the batch requests constituting an order arrive simultaneously. Let $E[T] = 1/\lambda$, where λ is understood as the time-average arrival rate of overall demands.

Let $Z_{i,n} = 1$ if demand n requests an item-i batch, and zero otherwise, $i \in E$. Then $\mathbf{Z}_n = (Z_{1,n}, \ldots, Z_{m,n})$ is the indicator vector of the product type requested by demand n. For each $K \subseteq E$, a demand is call a product-K demand, if $Z_{i,n} = 1$ for $i \in K$ and $Z_{i,n} = 0$ for $i \notin K$. Let $\mathbf{Z} =_{st} \mathbf{Z}_n$ follow the distribution

$$P(\mathbf{Z} = \mathbf{e}^K) = P^K = \frac{\lambda^K}{\lambda}, \quad K \subseteq E, \tag{11.3}$$

where λ^K is understood as the time-average demand rate of product-K, and $\mathbf{e}^K$ is a binary vector with its ith element, $i \in K$, being 1 and zero otherwise. Note that $\mathbf{Z}$, in general, contains dependent components.

The random batch sizes of demand n are denoted by vector $\mathbf{D}_n = (D_{1,n}, \ldots, D_{m,n})$, where $\mathbf{D}_n =_{st} \mathbf{D}$, $n \geq 1$, are i.i.d. random vectors, independent of all other random quantities. The product type and random quantities requested by a demand can be represented by vector $\mathbf{ZD}$, where the product of two vectors is meant componentwise.

In summary, we model a multivariate compound renewal process as an m-dimensional, *correlated vector process*: The inter-demand times are a vector consisting of m random variables *identically distributed* as T, $\mathbf{T} = (T_1, T_2, \ldots, T_m)$. Each demand requests random quantities $\mathbf{D}$ with independent components, but with probability P^K, all the batches not in set K are dummy batches and will be discarded.

Throughout the chapter, we use subscripts to indicate item types and superscripts to denote product types. The probability that a demand requests an item-i batch is

$$P_i = \sum_{K: i \in K} P^K, \quad i \in E. \tag{11.4}$$

Therefore, the (real) batch-i demand process is a thinning of $\{T_{i,n}, n \geq 1\}$, $i \in E$, with thinning probability P_i, $i \in E$. The time-average batch-i demand rate is

$$\lambda_i = \lambda P_i = \lambda \sum_{K: i \in K} P^K, \quad i \in E. \tag{11.5}$$

A batch demand is filled by on-hand inventory whenever possible, with unsatisfied items, if any, completely backlogged. When an item is filled we will either ship the item to its customer if partial shipment is allowed or put aside the item as a "committed" inventory and as soon as all the out-of-stock items of the same product become available the product will be assembled quickly and shipped to its customer as a whole.

We assume that there is no economy of scale in replenishment. Each item is controlled by an independent base-stock policy, with the base-stock level for item-i being s_i. This means that the arrival of a batch-i demand will immediately trigger a batch-i replenishment order of size D_i. Independent base-stock control has been used extensively in ATO modelling, due to its simple structural form, ease of implementation and analytical tractability. Wang [32] shows that with capacitated suppliers, the base-stock policy is near-optimal under high order fill rates, defined as the proportion of orders filled within a target delivery leadtime.

The production system consists of m separate production facilities, where facility i is dedicated to manufacture item-i orders, $i \in E$. A facility can be either capacitated or uncapacitated. If a facility is capacitated, we assume that it has a single machine. Upon arrival, a batch is routed to its dedicated facility and be processed on an FCFS basis. Let $S_{i,n} =_{st} S_i$ be the processing time of the nth item by machine i, with $E(S_i) = 1/\mu_i$. Let $\lambda_i < \mu_i$, $i \in E$, so that the system is stable in equilibrium (Baccelli et al. [4]). Let $\mathbf{S}_n := (S_{1,n}, \ldots, S_{i,n})$ consist of independent components. We assume that a serviced item will be transported to its inventory location individually and the transportation time from the production facility to the inventory location is negligible. If facility i is uncapacitated and unsatisfied demand is completely backlogged, we assume that the facility has infinitely many machines; if facility i is uncapacitated and unsatisfied demand is either partially backlogged or blocked, we assume that the facility has $c_i = s_i + b_i$ machines, where $0 \leq b_i < \infty$ is the capacity of backlog queue i. Upon arrival, each item in a batch will simultaneously seize one machine from its dedicated facility, whenever possible.

Following standard argument of inventory theory, the production system can be viewed as m parallel queues with correlated arrivals. The numbers of occupancies in different facilities correspond to the outstanding orders of different items. In a capacitated production environment, the supply system can be modelled as m correlated $GI^D/G/1$ queues, hereafter, denoted as a $\{GI^{D_i}/G/1, i \in E\}$ system. In an uncapacitated production environment, the supply system can be modelled by m correlated $GI^D/G/c/c$ queues, hereafter denoted as a $\{GI^{D_i}/G/c_i/c_i, i \in E\}$ system, $s_i \leq c_i \leq \infty$. Under either setting, the supply system is driven by the following *independent sequences*:

- A sequence of i.i.d. inter-demand time vectors: $\mathcal{T} := \{\mathbf{T}_n, n \geq 1\}$, where $\mathbf{T}_n =_{st} \mathbf{T} = (T_1, T_2, \dots T_n)$, and $T =_{st} T_1 =_{st} \cdots =_{st} T_m$;

- A sequence of i.i.d. demand type/batch quantity vectors: $\mathcal{ZD} := \{\mathbf{Z}_n \mathbf{D}_n, n \geq 1\}$, where $\mathbf{D}_n =_{st} \mathbf{D}$ contains independent components and $\mathbf{Z}$ and $\mathbf{D}$ are independent;

- A sequence of i.i.d. service time vectors: $\mathcal{S} = \{\mathbf{S}_n, n \geq 1\}$, where $\mathbf{S}_n =_{st} \mathbf{S}$ contains independent components.

We emphasize that only $\mathbf{T}$ and $\mathbf{Z}$ are dependent random vectors, reflecting correlated inter-arrival times and dependent product-type indicator vectors. Collectively, we denote these sequences by $\{\mathcal{T}, \mathcal{ZD}, \mathcal{S}\}$ and call them the independent generating sequences of a $\{GI^{D_i}/G/1, i \in E\}$ or $\{GI^{D_i}/G/c_i/c_i, i \in E\}$ system. A marked feature of these parallel queueing systems is that each queue viewed alone is a prototypical $GI^{D_i}/G/1$ or $GI^{D_i}/G/c_i/c_i$ system extensively studied in the queueing literature, but these queues are statistically dependent and must be treated as a whole.

Next we identify several key performance vectors. An ATO system is characterized by the following random quantities in equilibrium, which we call collectively the *system performance vectors*.

- *Inventory on-order vector* $\mathbf{N} = \{N_1, N_2, \dots, N_m\}$, where N_i is inventory on-order of item i, $i \in E$. Also let $\mathbf{N}^a = \{N_1^a, N_2^a, \dots, N_m^a\}$, where N_i^a is the inventory on-order of item i observed by a batch i demand, $i \in E$;

- *Inventory on-hand vector* $\mathbf{I} = \{I_1, I_2, \ldots, I_m\}$, where I_i is inventory on-hand of item i, $0 \le I_i \le s_i$, $i \in E$. Also let $\mathbf{I}^a = (I_1^a, \ldots, I_m^a)$, where I_i^a is the inventory on-hand of item i observed by a batch i demand;

- *Backorder vector* $\mathbf{B} = \{B_1, B_2, \ldots, B_m\}$, where B_i is the backsliders of item i, $0 \le B_i \le b_i$, $i \in E$. Let $\mathbf{B}^a = (B_1^a, \ldots, B_m^a)$, where B_i^a is the backsliders of item i observed by a batch i demand;

- *Manufacturing leadtime vector* $\mathbf{L} = \{L_1, L_2, \ldots, L_m\}$, where L_i is the manufacturing leadtime (queue time + service time) of a batch i demand;

- *Delivery leadtime vector* $\mathbf{R} = \{R_1, R_2, \ldots, R_m\}$, where R_i is the delivery leadtime of a batch-i demand, defined as the time to fill every item in the batch.

These stationary measures are well defined under the stability conditions $\lambda_i < \mu_i$, $i \in E$.

For ease of exposition, we will choose the *due-date fill rate*, defined by

$$P(\mathbf{R}^K \le \mathbf{r}^K) = P(R_i \le r_i, i \in K), \quad K \subseteq E, \qquad (11.6)$$

as our sole product-based performance measure, where $r_i \ge 0$ is the due-date, or the target delivery leadtime, of a batch-i demand and r_i may take on different values for different items. Thus, (11.6) gives the probability that each batch of a product is filled with respect to its individual due-date. In an ATO environment at the distribution level, where serviced batches are finished goods and can be shipped to its customer independently, it is appropriate to let r_i take on different values. In an ATO environment at the manufacturing level, where serviced batches represent semi-finished goods or subassemblies and the final assembly is required, it is more suitable to let all batches have a common due date, $r_i = r_{\max}$. In this case, (11.6) becomes

$$P(R_{\max}^K \le r_{\max}) := P\left(\max_{i \in K}\{R_i\} \le r_{\max}\right), \quad K \subseteq E. \qquad (11.7)$$

Another measure that might be of interest is the probability of filling the first batch of a product before a given due-date:

$$P(R_{\min}^K \le r_{\min}) := P\left(\min_{i \in K}\{R_i\} \le r_{\min}\right), \quad K \subseteq E, \qquad (11.8)$$

where $r_{\min}$ is the target delivery leadtime of the first batch of a product. Of course, $r_{\max}$ and $r_{\min}$ can be product dependent. The quantity $E[R_{\max}^K - R_{\min}^K]$ then represents the expected *delivery leadtime range* to fill a product. A smaller delivery leadtime range means delivery leadtime vector becomes less disperse.

The due-date fill rate also allows us to gain important information of other product-based performance measures of interest. For example, the *instantaneous* product fill rate, defined as the probability that a given product is filled immediately by on-hand inventory, can be obtained from (11.6) by letting $r_i = 0$,

$$F^K := P(R_i = 0, \ i \in K) = P\big(R_{\max}^K = 0\big)$$
$$= P\big(s_i - D_i \geq N_i^a, \ i \in K\big) \quad K \subseteq E. \tag{11.9}$$

Indeed, in order to fill a product-K demand instantaneously, one must have at least D_i items on hand when the batch-i demand of the product arrives, for each item $i \in K$.

A *system-based*, time-average performance measure can be obtained from its corresponding *product-based* performance measure. For example, the *system-based* due-date fill rate is given by

$$\sum_{K \in E} \frac{\lambda^K}{\lambda} P(\mathbf{R}^K \leq \mathbf{r}^K) = \sum_{K \in E} P^K P(\mathbf{R}^K \leq \mathbf{r}^K). \tag{11.10}$$

Therefore, it is sufficient to focus on product-based performance measures.

4.   Dependence Orders of Product-Type Indicator Vectors

In this section we first review several relevant dependence concepts. We then provide parametric characterizations of dependence orders of **Z**, using the notion of *majorization of weighted trees*. We also introduce several classes of transformations, termed *inclusion-exclusion* (IE) transforms and show that tree majorization orders can be achieved via IE-transforms. More information on tree majorizations and IE-transforms can be found in Xu and Li [35].

4.1.   Notions of dependence

Many different notions of dependence have been introduced and studied extensively in the literature (see, for example, Tong [31], Shaked and Shanthikumar [22], and Szekli [29]). Here we only discuss dependence orders *between* two random vectors and dependence relations *among* the components of a random vector that are most relevant for our purpose.

To compare dependence structure between two random vectors $\mathbf{X} = (X_1, \ldots, X_m)$ and $\bar{\mathbf{X}} = (\bar{X}_1, \ldots, \bar{X}_m)$ in the weakest sense, one may compare their covariance matrices element-wise. However, some stronger notions of dependence orders have been developed in the literature, which are summarized in the following definition. Here and in the sequel, inequalities in two separate case such as $A > a$ $(A \leq a)$ are written in the compact form $A > (\leq)a$.

Definition 11.1 Let $\mathbf{X} = (X_1, \ldots, X_m)$ and $\bar{\mathbf{X}} = (\bar{X}_1, \ldots, \bar{X}_m)$ be two $\mathcal{R}^m$-valued random vectors.

1. $\mathbf{X}$ is said to be larger (smaller) than $\bar{\mathbf{X}}$ in the upper (lower) orthant order, denoted by $\mathbf{X} \geq_{uo} (\leq_{lo}) \bar{\mathbf{X}}$, if $P(\mathbf{X} \geq (\leq) \mathbf{x}) \geq P(\bar{\mathbf{X}} \geq (\leq) \mathbf{x})$, for all $\mathbf{x} \in \mathcal{R}^m$. If, in addition, $X_i = \bar{X}_i$ for all $i \in E$, then $\mathbf{X}$ is said to be more positively upper (lower) orthant dependent than $\bar{\mathbf{X}}$.

2. $\mathbf{X}$ is said to be larger than $\bar{\mathbf{X}}$ in the supermodular order, denoted by $\mathbf{X} \geq_{sm} \bar{\mathbf{X}}$, if $E\phi(\mathbf{X}) \geq E\phi(\bar{\mathbf{X}})$ whenever ϕ is supermodular (a function ϕ is supermodular if for any $\mathbf{u}, \mathbf{v} \in \mathcal{R}^m$, $\phi(\mathbf{u} \vee \mathbf{v}) + \phi(\mathbf{u} \wedge \mathbf{v}) \geq \phi(\mathbf{u}) + \phi(\mathbf{v})$, where $\mathbf{u} \wedge \mathbf{v} = (\min(u_1, v_1), \ldots, \min(u_m, v_m))$ and $\mathbf{u} \vee \mathbf{v} = (\max(u_1, v_1), \ldots, \max(u_m, v_m)))$.

Intuitively, Definition 11.1 (1) means that, if $\mathbf{X}$ dominates $\bar{\mathbf{X}}$ in the upper (lower) orthant sense, then the components of $\mathbf{X}$ are more likely to simultaneously take on large (small) values, compared with the components of $\bar{\mathbf{X}}$ that have the same univariate marginals. Definition 11.1 (2) means that, if $\mathbf{X}$ dominates $\bar{\mathbf{X}}$ in the supermodular sense, then the components of $\mathbf{X}$ are more likely to simultaneously take on large values *or* simultaneously take on small values, compared with the components of $\bar{\mathbf{X}}$.

In the context of our problem, if $\mathbf{Z}$ and $\bar{\mathbf{Z}}$ are two product-type indicator vectors with identical marginals, then $\mathbf{Z} \geq_{uo} (\leq_{lo})\bar{\mathbf{Z}}$ means that demand specified by $\mathbf{Z}$ is more likely to request *at least* (*at most*) all the items in set K, compared with demand specified by $\bar{\mathbf{Z}}$. Also, $\mathbf{Z} \geq_{sm} \bar{\mathbf{Z}}$ implies that demand specified by $\mathbf{Z}$ is more likely to request *at least* all items in set $K_1 \subseteq E$ *or at most* all items in $K_2 \subseteq E$, compared with demand specified by $\bar{\mathbf{Z}}$.

The following facts are easy to verify (see, for example, Tong [31] and Szekli [29]):

$$\mathbf{X} \geq_{st} \bar{\mathbf{X}} \Longrightarrow \mathbf{X} \geq_{uo} \bar{\mathbf{X}} \quad \text{and} \quad \mathbf{X} \geq_{lo} \bar{\mathbf{X}}, \tag{11.11}$$

$$\mathbf{X} \geq_{sm} \bar{\mathbf{X}} \Longrightarrow \mathbf{X} \geq_{uo} \bar{\mathbf{X}}, \quad \mathbf{X} \leq_{lo} \bar{\mathbf{X}} \tag{11.12}$$

$$\text{and } X_i =_{st} \bar{X}_i, \quad \text{for } i \in E.$$

Here and in the sequel, $\mathbf{X} \geq_{st} \mathbf{Y}$ means $Ef(\mathbf{X}) \geq Ef(\bar{\mathbf{X}})$ for any increasing function f. Note that if $\mathbf{X} \geq_{sm} \bar{\mathbf{X}}$, then $X_i =_{st} \bar{X}_i$ for each i.

Thus, supermodular dependence orders and orthant orders coupled with identical marginals emphasize the comparisons of dependence strengths of two vectors by separating the marginals from consideration. It is also worth mentioning that if $\mathbf{X}$ and $\bar{\mathbf{X}}$ are two-dimensional random vectors, then $\mathbf{X} \geq_{sm} \bar{\mathbf{X}}$ was shown to be equivalent to $\mathbf{X} \geq_{uo} \bar{\mathbf{X}}$, $\mathbf{X} \leq_{lo} \bar{\mathbf{X}}$ and $X_i =_{st} \bar{X}_i$, $i = 1, 2$ (see, for example, Rüschendorf [21]).

To express dependence relations among the components of a random vector, one can of course use its covariance matrix. However, the following notions of dependence are stronger and frequently used in the literature.

Definition 11.2 Let $\mathbf{X}$ and $\tilde{\mathbf{X}}$ be two $\mathcal{R}^m$-valued random vectors such that $X_i =_{st} \tilde{X}_i$ for each i and $\tilde{X}_1, \ldots, \tilde{X}_m$ are independent. We call $\tilde{\mathbf{X}}$ an *independent copy* of $\mathbf{X}$.

1 $\mathbf{X}$ is said to be associated if $\mathrm{Cov}(f(\mathbf{X}), g(\mathbf{X})) \geq 0$ whenever f and g are non-decreasing.

2 $\mathbf{X}$ is said to be positively upper (lower) orthant dependent (PUOD, PLOD) if $\mathbf{X} \geq_{uo} (\leq_{lo})\tilde{\mathbf{X}}$. $\mathbf{X}$ is said to be negatively upper (lower) orthant dependent (NUOD, NLOD) if $\mathbf{X} \leq_{uo} (\leq_{lo})\tilde{\mathbf{X}}$.

It is known that (Esary et al. [10], Szekli [29])

$$\mathbf{X} \text{ is associated} \Longrightarrow \mathbf{X} \text{ is PUOD and PLOD}$$
$$\Longrightarrow \mathrm{Cov}(\mathbf{X}) \geq \mathbf{0}. \qquad (11.13)$$

Association is a positive dependence notion because it requires that two increasing functions of $\mathbf{X}$ be positively correlated. Association is a stronger notion than positive pairwise correlation because $f(\mathbf{X}) = X_i$ and $g(\mathbf{X}) = X_j$, $i, j \in E$, are non-decreasing functions, thus

$$\mathrm{Cov}(f(\mathbf{X}), g(\mathbf{X})) \geq 0 \text{ for } f \text{ and } g \text{ non-decreasing}$$
$$\Longrightarrow \mathrm{Cov}(X_i, X_j) \geq 0, \quad i, j \in E.$$

Similarly, PUOD (PLOD) is a positive dependence concept because it means that the components of $\mathbf{X}$ are more likely to simultaneously take on large (small) values, compared with its independent copy. By the same token, NUOD and NLOD are negative dependence concepts.

The orthant dependence property of a random vector means that its joint distribution or survival functions can be bounded below or above by the product of its marginal distribution or survival functions. In general, we found that orthant and supermodular dependencies are the most useful notions in characterizing dependence behaviors of ATO systems.

Pairwise-correlation is often too weak to describe dependence nature of an ATO system and unable to generate distributional bounds for a performance measure, yet association is often difficult to verify, sometimes even invalid, when the input process is no longer Poisson.

It is worth mentioning that orthant dependence, while a weaker notion than association and supermodular dependence, allows the separation of upper and lower orthant dependencies. In other words, it is possible that a random vector is PUOD and NLOD (of course, other three combinations are also possible). As such, orthant orders offer more flexibility than the supermodular order in specifying dependence properties of a random vector. The following example illustrates that system performance can be improved by properly combining positively and negatively orthant dependencies of its input process.

Example 11.3 (a) Consider two possible product-type indicator vectors $\mathbf{Z}$ and $\bar{\mathbf{Z}}$ in a three-item system, with distributions

$$P^{123} = P^1 = P^2 = P^3 = \frac{1}{4}, \quad P^{12} = P^{13} = P^{23} = P^{\emptyset},$$

$$\bar{P}^{123} = \bar{P}^1 = \bar{P}^2 = \bar{P}^3 = 0, \quad \bar{P}^{12} = \bar{P}^{13} = \bar{P}^{23} = \bar{P}^{\emptyset} = \frac{1}{4},$$

where product-$\emptyset$ is a dummy request and $\emptyset \subseteq E$. The dummy product is introduced in order to normalize the input processes of compared systems. Because $P_i = \bar{P}_i = 1/2$, $i = 1, 2, 3$, the two random vectors have identical univariate marginals. However, the number of items requested under $\mathbf{Z}$ is more variable than that under $\bar{\mathbf{Z}}$. The independent copy of $\mathbf{Z}$ (or $\bar{\mathbf{Z}}$), denoted by $\tilde{\mathbf{Z}}$, has the distribution $\tilde{P}^K = 1/8$, $K \subseteq E$. One can easily check that $\mathbf{Z}$ is PUOD and NLOD and $\bar{\mathbf{Z}}$ is NUOD and PLOD. In fact, they are ordered as

$$\mathbf{Z} \geq_{uo} \tilde{\mathbf{Z}} \geq_{uo} \bar{\mathbf{Z}} \text{ and } \bar{\mathbf{Z}} \leq_{lo} \tilde{\mathbf{Z}} \leq_{lo} \mathbf{Z}.$$

We shall revisit this system in Example 11.3 (b)-(d) in the following sections, where we demonstrate that, with other things being equal, the system with $\bar{\mathbf{Z}}$ outperforms the system with $\mathbf{Z}$ in the sense that the former system stochastically improves the best and the worst components in a performance vector, compared with their counterparts in the latter system, even though their item-based performances are stochastically identical. It is precisely these desirable properties motivate us to focus on orthant dependence, rather than other stronger dependence concepts. ■

4.2.   Parametric characterizations of dependence orders via tree majorization

It is often difficult to verify supermodular and orthant orders of two random vectors via Definition 11.1. Therefore, parametric characterizations of supermodular and orthant orders between two random vectors are desirable. Toward this end, Xu and Li [35] introduce the notion of *majorization of weighted trees* and use it to analyzed the dependence structure of **Z**. Next, we briefly review their notion.

Consider a set of parameter values $\Lambda = \{\lambda^K, K \subseteq E\}$ defined on a partially ordered index set $S(E)$ (define the partial order $L < K$ if $L \subseteq K$) where $S(E)$ is the collection of all subsets of E, including the empty set $\emptyset$. We treat each subset of E as a node. If $L \subseteq K \subseteq E$, then we say L is a descendant of K and K is an ancestor of L. A node is considered its own descendant and ancestor and the root E (node $\emptyset$) is the ancestor (descendant) of everyone. For each node K, we assign a real number λ^K, and call it the *weight* of K (in the context of our system, λ^K is the time-average arrival rate of product-K demands). As such, Λ can be thought as a weighted tree. If each weight in a tree is between 0 and 1 and the total weight of the tree equals 1, then the tree is called a *probability tree*. Any positively weighted tree (i.e., all weights are nonnegative and at least one weight is positive) can always be normalized to a probability tree.

Definition 11.4 (Xu and Li) Let Λ and $\bar{\Lambda}$ be two weighted trees on the same index set $S(E)$, with total weights $\lambda = \sum_{K \subseteq E} \lambda^K$ and $\bar{\lambda} = \sum_{K \subseteq E} \bar{\lambda}^K$, respectively. Λ is said to *majorize* $\bar{\Lambda}$ from *roots* (*leaves*), denoted by $\Lambda \geq_{\mathcal{T}_r(\mathcal{T}_l)} \bar{\Lambda}$, if

$$\lambda = \bar{\lambda}, \tag{11.14}$$

$$\sum_{K \subseteq (\supseteq) L} \lambda^L \geq \sum_{K \subseteq (\supseteq) L} \bar{\lambda}^L, \qquad K \subseteq E. \tag{11.15}$$

Definition 11.4 states that Λ majorizes $\bar{\Lambda}$ from roots (leaves) if the total weights of the two trees, λ and $\bar{\lambda}$, are the same, and, for each node K, $K \subseteq E$, the total weight of its ancestors (descendants) in Λ is no less than its counterpart in $\bar{\Lambda}$. In other words, if Λ majorizes $\bar{\Lambda}$ from roots (leaves), then the weights of Λ is more concentrated on the nodes in the earlier (later) generations than $\bar{\Lambda}$ does.

Due to (11.3), the distribution of **Z** can be described by either $\{P^K, K \subseteq E\}$ or $\{\lambda^K, K \subseteq E\}$. We find that it is more convenient to work with $\{\lambda^K, K \subseteq E\}$, especially when the demand process is a multivariate compound Poisson process. Here and in the sequel, let $\mathbf{Z}$ and $\bar{\mathbf{Z}}$ be the product-type indicator vectors represented by trees $\Lambda = \{\lambda^K, K \subseteq E\}$ and $\bar{\Lambda} = \{\bar{\lambda}^K, K \subseteq E\}$, respectively. Since, by Definition 11.4,

$$P(\mathbf{Z} \geq (\leq) \, \mathbf{e}^K) - P(\bar{\mathbf{Z}} \geq (\leq) \, \mathbf{e}^K)$$

$$= \sum_{K \subseteq (\supseteq) L} \frac{\lambda^L}{\lambda} - \sum_{K \subseteq (\supseteq) L} \frac{\bar{\lambda}^L}{\bar{\lambda}}$$

$$= \frac{1}{\lambda} \left[\sum_{K \subseteq (\supseteq) L} \lambda^L - \sum_{K \subseteq (\supseteq) L} \bar{\lambda}^L \right] \geq 0, \qquad (11.16)$$

it implies

$$\Lambda \geq_{T_r(T_l)} \bar{\Lambda} \iff \mathbf{Z} \geq_{uo} (\leq_{lo}) \bar{\mathbf{Z}}. \qquad (11.17)$$

In words, it states that root and leaf majorization orders between two trees are the parametric representations of upper and lower orthant orders between two product-type indicator vectors.

Now let the two trees have the identical marginal weights, $\lambda_i - \bar{\lambda}_i$, $i \in E$. This implies $P_i = \bar{P}_i$, $i \in E$. Since Z_i and $\bar{Z}_i$ are indicator functions, it further implies that $Z_i =_{st} \bar{Z}_i$. Thus

$$\Lambda \geq_{T_r(T_l)} \bar{\Lambda} \text{ and } \lambda_i = \bar{\lambda}_i, \quad i \in E$$
$$\iff \mathbf{Z} \geq_{uo} (\leq_{lo}) \bar{\mathbf{Z}} \text{ and } Z_i =_{st} \bar{Z}_i, \quad i \in E. \qquad (11.18)$$

4.3. The pairwise g-transform and inclusion-exclusion transforms

While equations (11.17) and (11.17) allow us to compare the orthant orders between two product-type indicator vectors via tree majorization orders, verifying tree majorization orders via Definition 11.4 is not only cumbersome for large trees, but also reveals little structural differences between two indicator vectors. Xu and Li propose the inclusion-exclusion (IE) transforms and show that tree majorization orders can be achieved via IE-transforms.

The Pairwise g-Transform   To gain insight, we start with a simple form of an IE-transform called the pairwise g-transform. It turns out that a pairwise g-transform order between two trees implies the supermodular order between their corresponding product-type indicator vectors.

Here and in the sequel, scalar multiplication and summation of trees defined on the same index set $S(E)$ are operated node-wise. For any $K \subseteq E$, Let ϵ^K be the tree with the weight of node K being 1 and others zero.

Definition 11.5 Let K_1 and K_2 be two nodes in $S(E)$ and let $K_1 \cup K_2$ and $K_1 \cap K_2$ be the common parent and common child of K_1 and K_2, respectively. For any $\gamma \geq 0$, we define $g^{\gamma,(K_1,K_2)} : \mathcal{R}^{2^m} \to \mathcal{R}^{2^m}$ as

$$g^{\gamma,(K_1,K_2)}(\Lambda) = \Lambda - \gamma(\epsilon^{K_1 \cup K_2} - \epsilon^{K_1} - \epsilon^{K_2} + \epsilon^{K_1 \cap K_2}). \qquad (11.19)$$

A transform satisfying (11.19) is called a pairwise g-transform and is denoted by g^2. We say Λ is larger than $\bar{\Lambda}$ in the g^2-transform order, written $\Lambda \geq_{g^2} \bar{\Lambda}$, if $\bar{\Lambda}$ can be obtained from Λ through successive g^2-transforms.

In words, $g^{\gamma,(K_1,K_2)}$ is a transform that simultaneously decreases the demand rates of products $(K_1 \cup K_2)$ and $(K_1 \cap K_2)$ by γ, and simultaneously increases the demand rates of products K_1 and K_2 by γ. Thus the cardinality of a product becomes less variable after a g^2-transform. It is easily seen that a g^2-transform does not change the total weight and marginal weights of a tree.

What is the implication of a g^2-transform order between Λ and $\bar{\Lambda}$ to the dependence order between $\mathbf{Z}$ and $\bar{\mathbf{Z}}$? Li and Xu [16] show that a g^2-transform order between two trees leads to the supermodular order between their product-type indicator vectors:

$$\Lambda \geq_{g^2} \bar{\Lambda} \Longrightarrow \mathbf{Z} \geq_{sm} \bar{\mathbf{Z}}$$
$$\Longrightarrow \mathbf{Z} \geq_{uo} \bar{\mathbf{Z}}, \ \mathbf{Z} \leq_{lo} \bar{\mathbf{Z}} \quad \text{and} \quad Z_i = \bar{Z}_i, \ i \in E. \quad (11.20)$$

Example 11.6 (a)   Consider a two-item system with the demand rate tree

$$\Lambda = (\lambda^{12}, \lambda^1, \lambda^2, \lambda^0) = (2, 0, 0, 2).$$

The total demand rate is 4 and the marginal demand rate of each item is 2. Then, by (11.19),

$$g^{1,(1,2)}(\Lambda) = (1, 1, 1, 1) := \tilde{\Lambda}.$$

Clearly, the total and marginal demand rates of the tree after the transform are unchanged. Observe that $\tilde{\Lambda}$ corresponds to the independent product-type indicator vector. Let us apply transform $g^{1,(1,2)}$ to $\tilde{\Lambda}$, then

$$g^{1,(1,2)}(\tilde{\Lambda}) = (0, 2, 2, 0) := \bar{\Lambda}.$$

Then (11.19) implies that the product-type indicator vectors corresponding to these trees satisfy the supermodular order: $\mathbf{Z} \geq_{sm} \tilde{\mathbf{Z}} \geq_{sm} \bar{\mathbf{Z}}$. It is easily verified that $\mathbf{Z}$ is PUOD and PLOD and $\bar{\mathbf{Z}}$ is NUOD and NLOD.

In fact, **Z** and **Z̄** are more PUOD and PLOD and NUOD and NLOD, respectively, than any other two-item product indicator vector of identical marginals. We shall further examine the properties associated with these systems in Example 11.6 (b)–(c). ■

Inclusion-Exclusion (IE) Transforms   IE-transforms are extensions of the g^2-transform. Intuitively, an IE-transform is a scheme to systematically redistribute the weights on a tree that leads to either the upper or lower orthant order between two product-type indicator vectors.

Definition 11.7 Let $\mathbf{K} = (K_1, K_2, \ldots, K_\nu)$ be ν different nodes in $S(E)$. Let $\gamma \geq 0$.

1 Let $g_\downarrow^{\gamma, \mathbf{K}} : \mathcal{R}^{2^m} \to \mathcal{R}^{2^m}$ be defined by

$$g_\downarrow^{\gamma, \mathbf{K}}(\Lambda)$$
$$= \Lambda - \gamma \left[\epsilon^{\cup_{i=1}^\nu K_i} - \sum_{i=1}^\nu \epsilon^{K_i} + \sum_{i<j} \epsilon^{K_i \cap K_m} - \cdots + (-1)^\nu \epsilon^{\cap_{i=1}^\nu K_i} \right].$$
(11.21)

Such a transform is called a $g_\downarrow$-transform. We say Λ is larger than $\bar{\Lambda}$ in the $g_\downarrow$-transform order, written $\Lambda \geq_{g_\downarrow} \bar{\Lambda}$, if $\bar{\Lambda}$ can be obtained from Λ through successive $g_\downarrow$-transforms.

2 Let $g_\uparrow^{\gamma, \mathbf{K}} : \mathcal{R}^{2^m} \to \mathcal{R}^{2^m}$ be defined by

$$g_\uparrow^{\gamma, \mathbf{K}}(\Lambda)$$
$$= \Lambda - \gamma \left[\epsilon^{\cap_{i=1}^\nu K_i} - \sum_{i=1}^\nu \epsilon^{K_i} + \sum_{i<j} \epsilon^{K_i \cup K_m} - \cdots + (-1)^\nu \epsilon^{\cup_{i=1}^\nu K_i} \right].$$
(11.22)

Such a transform is called a $g_\uparrow$-transform. We say Λ is larger than $\bar{\Lambda}$ in the $g_\uparrow$-transform order, written $\Lambda \geq_{g_\uparrow} \bar{\Lambda}$, if $\bar{\Lambda}$ can be obtained from Λ through successive $g_\uparrow$-transforms.

We also call a $g_\downarrow$ or $g_\uparrow$-transform an *inclusion-exclusion* (IE) transform, for the reason that is evident from (11.20) and (11.21). Clearly, for $\nu = 2$, $g_\downarrow^{\gamma, (K_1, K_2)} = g_\uparrow^{\gamma, (K_1, K_2)} = g^{\gamma, (K_1, K_2)}$.

Loosely speaking, a $g_\downarrow$-transform shifts a partial weight γ from some ancestor nodes to their descendant nodes in a "top-down" fashion; consequently, the total weight of the ancestors of node K becomes smaller, for any $K \subseteq E$. On the other hand, a $g_\uparrow$-transform shifts a partial weight γ from some descendant nodes to their ancestor nodes in a "bottom-up" fashion; as a result, the total weight of descendants of node K becomes smaller, for any $K \subseteq E$. If a $g_\downarrow$-transform is also a $g_\uparrow$-transform (e.g., a pairwise g-transform), the weights of the tree after the transform will be more concentrated among the nodes of middle generations.

The implication of an IE-transform is the following. First, it can be shown that $g_\downarrow$ and $g_\uparrow$-transforms preserve the total weight λ and marginal weights λ_i, $i \in E$, of Λ. This means that an IE transform will not alter the marginal distributions of its product-type indicator vector. Second, a $g_\downarrow$ ($g_\uparrow$)-transform order between two trees is a *necessary and sufficient* condition for the upper (lower) orthant order between their product-type indicator vectors, as stated in the following (see Xu and Li for proof):

$$\Lambda \geq_{g_\downarrow} (\geq_{g_\uparrow}) \bar{\Lambda} \iff \mathbf{Z} \geq_{uo} (\leq_{lo}) \bar{\mathbf{Z}} \quad \text{and} \quad Z_i =_{st} \bar{Z}_i, \quad i \in E. \quad (11.23)$$

From (11.17), (11.17) and (11.23), it is clear that the tree majorization order from roots (leaves) characterizes the upper (lower) orthant order between two indicator vectors that may have different marginals, and the $g_\downarrow$ ($g_\uparrow$)-transform order characterizes the upper (lower) orthant order between two indicator vectors with identical marginals. Since we are mainly interested in the dependence orders rather than the usual stochastic order, in the remainder of the chapter we will use IE-transform orders, rather than tree majorization orders, to compare product-type indicator vectors.

An IE-transform can simultaneously decrease (or increase) upper orthant dependence strength and increase (or decrease) lower orthant dependence strength of $\mathbf{Z}$. This is illustrated by the following example.

Example 11.3 (b)   Consider the three-item ATO system discussed in Example 11.3 (a), where $\tilde{\mathbf{Z}}$, the independent copy of $\mathbf{Z}$ (and $\bar{\mathbf{Z}}$), has demand rates $\lambda^K = 1$, $K \subseteq E$. In symbols,

$$\tilde{\Lambda} = \sum_{K \subseteq 123} \epsilon^K.$$

To illustrate the $g_\uparrow$-transform, we select $\gamma = 1$, $\mathbf{K} = (12, 13, 23)$ and apply (11.21),

$$g_\uparrow^{1,(12,13,23)}(\tilde{\Lambda}) = \tilde{\Lambda} - (\epsilon^0 - \epsilon^{12} - \epsilon^{13} - \epsilon^{23} + 2\epsilon^{123})$$
$$= 2[\epsilon^{123} + \epsilon^1 + \epsilon^2 + \epsilon^3] := \Lambda. \quad (11.24)$$

As observed in Example 11.3(a), $\mathbf{Z}$ is PUOD and NLOD. This is not surprising, as it can be shown that, if $\mathbf{K}$ consists of every node in the same generation and has an odd (even) number of nodes, then a $g_\uparrow^{\gamma,\mathbf{K}}$-transform will generate a vector that is more PUOD and NLOD (less PUOD and PLOD) than itself. To illustrate a $g_\downarrow$-transform, we let $\gamma = 1$, $\mathbf{K} = (1,2,3)$ and apply (11.20),

$$g_\downarrow^{1,(1,2,3)}(\tilde{\Lambda}) = 2[\epsilon^{12} + \epsilon^{13} + \epsilon^{23} + \epsilon^0] := \bar{\Lambda}$$

Now $\bar{\mathbf{Z}}$ is NUOD and PLOD. Again, this follows from the fact that if $\mathbf{K}$ consists of every node in the same generation and contains an odd (even) number of nodes, then a $g_\downarrow^{\gamma,\mathbf{K}}$-transform will result in a vector that is more NUOD and PLOD (less PUOD and PLOD) than itself. More properties of IE-transforms have been developed by Xu and Li. ■

5. Capacitated Assemble-to-Order Systems

This section studies dependence structure of two capacitated ATO systems, modelled as the $\{GI^{D_i}/M/1, i \in E\}$ system and the $\{GI/G/1, i \in E\}$ system.

As discussed in Section 2, in addition to a well-defined order between two input parameter sets, another important ingredient in a stochastic comparison study is system dynamics that preserve the ordering established for the compared parameter sets. It is known (see Section 9.1) that the supermodular order is preserved under *monotone* (either increasing or decreasing) transformations, whereas orthant orders are preserved only under *increasing* transformations. We show that for either capacitated ATO model, its performance vector process (e.g., the m-dimensional inventory on-order process) can be modelled as a separable Markov chain generated by independent sequences $\mathcal{T}$, $\mathcal{ZD}$ and $\mathcal{S}$. We also show that the function that governs the evolution of each marginal process of the performance vector process (e.g., the inventory on-order process of a given item) is increasing. As a direct consequence, the two compared performance vector processes, generated by sequences $(\mathcal{T}, \mathcal{ZD}, \mathcal{S})$ and $(\bar{\mathcal{T}}, \bar{\mathcal{ZD}}, \mathcal{S})$, respectively, preserve the orthant and supermodular orders of their generating sequences.

5.1. Separable Markov chains with independent generating sequences

We first provide the definition of separable Markov chains generated by independent generating sequences and then state its closure property.

Let $\{\mathbf{Y}_n, n \geq 1\}$ be a sequence of m-dimensional i.i.d. random vectors. An m-dimensional Markov chain $\{\mathbf{X}_n, n \geq 0\}$ is called a separable Markov chain generated by the sequence $\{\mathbf{Y}_n, n \geq 1\}$, if $\mathbf{X}_0 = (0, \ldots, 0)$ and

$$
\begin{aligned}
\mathbf{X}_{n+1} &= (X_{1,n+1}, \ldots, X_{J,n+1}) \\
&= (g_1(X_{1,n}, Y_{1,n}), \ldots, g_m(X_{m,n}, Y_{m,n})), \quad n \geq 0. \quad (11.25)
\end{aligned}
$$

In words, a separable Markov chains $\{\mathbf{X}_n, n \geq 0\}$ is an m-dimensional Markov chain driven by sequence $\{\mathbf{Y}_n, n \geq 0\}$, where the components of vector $\mathbf{Y}_n$ can be dependent. The following theorem provides a sufficient condition under which two separable Markov chains preserve the orthant and supermodular orders of their generating sequences.

Proposition 11.8 (Li and Xu [16]) Let $\{\mathbf{X}_n, n \geq 0\}$ and $\{\bar{\mathbf{X}}_n, n \geq 0\}$ be two $\mathcal{R}^m$-valued separable Markov chains generated by sequences of i.i.d. random vectors $\{\mathbf{Y}_n, n \geq 1\}$ and $\{\bar{\mathbf{Y}}_n, n \geq 1\}$, respectively. Let $g_i : \mathcal{R}^2 \to \mathcal{R}$, $i \in E$, be monotone in the same direction (increasing). If $\mathbf{Y}_n \geq_{sm} (\geq_{uo}, \leq_{lo}) \bar{\mathbf{Y}}_n$, then for any $n \geq 0$,

$$
\{\mathbf{X}_n, n \geq 0\} \geq_{sm} (\geq_{uo}, \leq_{lo}) \{\bar{\mathbf{X}}_n, n \geq 0\}, \quad (11.26)
$$

where the supermodular or orthant order of two Markov chains is meant that for any $j \geq 1$ and any choice of $n_1, \ldots, n_j$,

$$
(\mathbf{X}_{n_1}, \ldots, \mathbf{X}_{n_j}) \geq_{sm} (\geq_{uo}, \leq_{lo})(\bar{\mathbf{X}}_{n_1}, \ldots, \bar{\mathbf{X}}_{n_j}).
$$

For notation simplicity, here we only consider a separable Markov chain with a single generating sequence. The definition can be easily extended to a separable Markov chain generated by several *independent* sequences (such as the inventory on-order process generated by independent sequences $\mathcal{T}$, $\mathcal{ZD}$ and $\mathcal{S}$), and Proposition 11.8 still holds under this generalization.

5.2.   A capacitated ATO system with a $\{GI^{D_i}/M/1, i \in E\}$ supply system

In this model, the service time of item i, S_i, are exponentially distributed with rate μ_i, $i \in E$, and other random vectors, $\mathbf{T}$, $\mathbf{Z}$, and $\mathbf{D}$, follow general distributions. The following identities, which connect inventory on-order to inventory on-hand and backsliders, are well-known:

$$
I_i = (s_i - N_i)^+, \quad I_i^a = (s_i - N_i^a)^+, \quad i \in E, \quad (11.27)
$$

$$
B_i = (N_i - s_i)^+, \quad B_i^a = (N_i^a - s_i)^+, \quad i \in E. \quad (11.28)
$$

The delivery leadtime of a batch-i demand is related to N_i^a via the following identity:

$$R_i = \sum_{n=1}^{[N_i^a + D_i - s_i]^+} S_{i,n}, \quad i \in E, \tag{11.29}$$

where $S_{i,n}$ are i.i.d. exponential random variables with rate μ_i. Here, if $[N_i^a + D_i - s_i]^+ = 0$, then the batch is filled immediately and hence $R_i = 0$; otherwise the last item in the batch takes on the $(N_i^a + D_i - s_i)$th position in backlog queue i and will be filled when machine i releases the $(N_i^a + D_i - s_i)$th job.

Equations (11.27)–(11.29) show that inventory on-order vector observed by a demand, $\mathbf{N}^a$, plays a pivotal role in determining the distributions of other performance measures, such as $\mathbf{B}^a$, $\mathbf{I}^a$ and $\mathbf{R}$. Therefore, in this section we will mainly consider dependence properties of $\mathbf{N}^a$ and $\mathbf{R}$.

We first derive a sample path representation of the inventory on-order process embedded at arrival epochs. For $n \geq 1$, let $N_{i,n}^a$ be the inventory on-order of item i at time $\sum_{j=1}^{n} T_{i,j}$, $i \in E$, and let $\mathbf{N}_n^a = (N_{1,n}^a, \ldots, N_{m,n}^a)$, $n \geq 1$. Initially, each item is stocked to base-stock level s_i, $i \in E$, and there is no stock on order for any item, i.e., $\mathbf{N}_1^a = (0, \ldots, 0)$. Let

$$Y_{i,n} = \max \left\{ k : \sum_{j=1}^{k} S_{i,j} \leq T_{i,n+1} \right\}, \tag{11.30}$$

where $S_{i,j}$, $i \in E$, $j \geq 1$, are i.i.d. exponential random variables with rate μ_i. Here, provided that machine i is busy, $Y_{i,n}$ is the number of type-i jobs completed during $T_{i,n+1}$, $i \in E$. Note that $\mathbf{Y}_n := (Y_{1,n}, \ldots, Y_{m,n})$ contains dependent components, because the numbers of job completions by different machines depend on correlated interarrival times $T_{i,n+1}$. $i \in E$. However, $\mathcal{Y} := \{\mathbf{Y}_n, n \geq 1\}$ is a sequence of i.i.d. random vectors and is also independent of $\mathcal{Z}\mathcal{D}$. Now, the number of inventory on-order of item i observed at time $\sum_{j=1}^{n+1} T_{i,j}$ satisfies the recursion:

$$N_{i,n+1}^a = [N_{i,n}^a + Z_{i,n} D_{i,n} - Y_{i,n}]^+, \quad i \in E, \ n \geq 1. \tag{11.31}$$

It means that if the nth batch of item i finds $N_{i,n}^a$ items on order, then the next batch will find $N_{i,n} + Z_{i,n} D_{i,n}$ items minus the items served during $T_{i,n+1}$. Therefore, $\{\mathbf{N}_n^a, n \geq 0\}$ is a separable Markov chain generated by independent sequences $\mathcal{Z}\mathcal{D}$ and $-\mathcal{Y}$.

Consider another $\{GI^{D_i}/M/1, i \in E\}$ system, driven by the independent sequences $\{\bar{T}, \bar{\mathcal{Z}}\mathcal{D}, \mathcal{S}\}$, where $\bar{T} = \{\bar{\mathbf{T}}_n, n \geq 1\}$ and $\bar{\mathcal{Z}} = \{\bar{\mathbf{Z}}_n, n \geq 1\}$.

Here and in the sequel, without explicit mention, we always assumed that $\bar{\mathbf{T}}$ and $\mathbf{T}$ ($\bar{\mathbf{Z}}$ and $\mathbf{Z}$) have identical marginal distributions, but may have different joint distributions. Then, following a similar argument, $\{\bar{\mathbf{N}}_n^a, n \geq 0\}$ is a separable Markov chain generated by independent sequences $\bar{\mathcal{Z}}\mathcal{D}$ and $-\bar{\mathcal{Y}}$.

The next theorem states the dependence comparison results of $\mathbf{N}^a$ and $\mathbf{R}$. Its proof can be found in Appendix B (Section 9.2).

Theorem 11.9 Consider two $\{GI^{D_i}/M/1, i \in E\}$ supply systems, driven by independent sequences $\{\mathcal{T}, \mathcal{Z}\mathcal{D}, \mathcal{S}\}$ and $\{\bar{\mathcal{T}}, \bar{\mathcal{Z}}\mathcal{D}, \mathcal{S}\}$, respectively. If $\Lambda \geq_{g^2} (\geq_{g_\downarrow}, \geq_{g_\uparrow})\bar{\Lambda}$, and $\mathbf{T} \geq_{sm} (\leq_{lo}, \geq_{uo})\bar{\mathbf{T}}$, then

$$\{\mathbf{N}_n^a, n \geq 1\} \geq_{sm} (\geq_{uo}, \leq_{lo})\{\bar{\mathbf{N}}_n^a, n \geq 1\} \qquad (11.32)$$

and $\mathbf{N}_n^a$ and $\bar{\mathbf{N}}_n^a$ have identical marginal distributions for any $n = 1, 2, \ldots$. Under stability conditions $\lambda_i < \mu_i$, $i \in E$,

$$\mathbf{N}^a \geq_{sm} (\geq_{uo}, \leq_{lo}) \bar{\mathbf{N}}^a, \text{ and } N_i^a =_{st} \bar{N}_i^a, \ i \in E. \qquad (11.33)$$

Furthermore, the stationary delivery leadtime vectors in the two systems satisfy

$$\mathbf{R} \geq_{sm} (\geq_{uo}, \leq_{lo})\bar{\mathbf{R}} \quad \text{and} \quad R_i =_{st} \bar{R}_i, \ i \in E. \qquad (11.34)$$

In the remainder of the section, we discuss the implications of Theorems 11.9 and provide examples. Observe that, because supermodular and orthant orders are closed under marginalization, (11.34) extends trivially to

$$\mathbf{R}^K \geq_{sm} (\geq_{uo}, \leq_{lo})\bar{\mathbf{R}}^K \quad \text{and} \quad R_i =_{st} \bar{R}_i, \ i \in K \subseteq E. \qquad (11.35)$$

Let us first consider the implications of the supermodular order between $\mathbf{R}^K$ and $\bar{\mathbf{R}}^K$. Since the supermodular order implies positively upper and lower orthant dependence orders (see (11.12)), for each $K \subseteq E$,

$$\mathbf{R}^K \geq_{sm} \bar{\mathbf{R}}^K \implies P(\mathbf{R}^K \geq \mathbf{r}^K) \geq P(\bar{\mathbf{R}}^K \geq \mathbf{r}^K)$$
$$\implies P(R_{\min}^K \geq r_{\min}) \geq P(\bar{R}_{\min}^K \geq r_{\min}),$$

$$(11.36)$$

$$\mathbf{R}^K \geq_{sm} \bar{\mathbf{R}}^K \implies P(\mathbf{R}^K \leq \mathbf{r}^K) \geq P(\bar{\mathbf{R}}^K \leq \mathbf{r}^K)$$
$$\implies P(R_{\max}^K \leq r_{\max}) \geq P(\bar{R}_{\max}^K \leq r_{\max}).$$

One understands the above result as follows. If product-K demands arrive in both systems, i.e., $\lambda^K > 0$ and $\bar{\lambda}^K > 0$, then a more supermodular dependent $\mathbf{R}^K$ will have a higher due-date fill rate of product-K, but also a higher probability that every batch of the product is late-due. If product-K demands only arrive to one of the systems, say $\lambda^K > 0$ but $\bar{\lambda}^K = 0$, then the probability such as $P(\bar{\mathbf{R}}^K \leq \mathbf{r}^K)$ is still well defined and can serve as a lower bound of $P(\mathbf{R}^K \leq \mathbf{r}^K)$, though it cannot be interpreted as the due-date fill rate of product-K. Next consider the implication of the orthant orders between $\mathbf{R}^K$ and $\bar{\mathbf{R}}^K$. Equation (11.35) implies that

$$\mathbf{R}^K \geq_{uo} \bar{\mathbf{R}}^K \Longrightarrow P(\mathbf{R}^K \geq \mathbf{r}^K) \geq P(\bar{\mathbf{R}}^K \geq \mathbf{r}^K)$$
$$\Longrightarrow P(R_{\min}^K \geq r_{\min}) \geq P(\bar{R}_{\min}^K \geq r_{\min}), \quad (11.37)$$

$$\mathbf{R}^K \leq_{lo} \bar{\mathbf{R}}^K \Longrightarrow P(\mathbf{R}^K \leq \mathbf{r}^K) \geq P(\bar{\mathbf{R}}^K \leq \mathbf{r}^K)$$
$$\Longrightarrow P(R_{\max}^K \leq r_{\max}) \geq P(\bar{R}_{\max}^K \leq r_{\max}). \quad (11.38)$$

The interpretation of the above result is similar to that of (11.35).

Remark 11.10 A word of caution is warranted here. We emphasize that our dependence comparison of due-date fill rates in two systems is on a product-to-product basis. In particular, it is not true that a more positively lower orthant dependent $\mathbf{R}$ will yield a higher system-based due-date fill rate. That is

$$\mathbf{R} \leq_{lo} \bar{\mathbf{R}} \nRightarrow \sum_{K \subseteq E} \frac{\lambda^K}{\lambda} P(\mathbf{R}^K \leq \mathbf{r}^K) \geq \sum_{K \subseteq E} \frac{\bar{\lambda}^K}{\lambda} P(\bar{\mathbf{R}}^K \leq \mathbf{r}^K), \quad (11.39)$$

because the weights of the two trees are distributed differently. In fact, our numerical examples, given in Section 7.2, show that the opposite is true. Nevertheless, each of the above probabilities is well defined and we can obtain the following inequality:

$$\mathbf{R} \leq_{lo} \bar{\mathbf{R}} \Longrightarrow \sum_{K \subseteq E} \frac{\lambda^K}{\lambda} P(\mathbf{R}^K \leq \mathbf{r}^K) \geq \sum_{K \subseteq E} \frac{\lambda^K}{\lambda} P(\bar{\mathbf{R}}^K \leq \mathbf{r}^K). \quad (11.40)$$

In Section 7, we will derive the bounds for the system-based due-date fill rate. ∎

Note that (11.36) and (11.37) also imply the following, for $K \subseteq E$:

$$\mathbf{R}^K \geq_{uo} \bar{\mathbf{R}}^K \Longrightarrow R_{\min}^K \geq_{st} \bar{R}_{\min}^K \Longrightarrow E[R_{\min}^K] \geq E[\bar{R}_{\min}^K], \quad (11.41)$$

$$\mathbf{R}^K \leq_{lo} \bar{\mathbf{R}}^K \Longrightarrow R_{\max}^K \leq_{st} \bar{R}_{\max}^K \Longrightarrow E[R_{\max}^K] \leq E[\bar{R}_{\max}^K]. \quad (11.42)$$

This further implies that decreasing both upper and lower dependencies of $\mathbf{R}$ increases the expected *range* of delivery leadtimes of product K:

$$\mathbf{R}^K \geq_{uo} \bar{\mathbf{R}}^K \text{and} \mathbf{R}^K \leq_{lo} \bar{\mathbf{R}}^K$$
$$\Longrightarrow E[R_{\max}^K] - E[R_{\min}^K] \leq E[\bar{R}_{\max}^K] - E[\bar{R}_{\min}^K]. \quad (11.43)$$

Remark 11.11 Although increasing both upper and lower orthant dependencies of $\mathbf{R}$ makes it less disperse, it also causes the average batch delivery leadtime of a product becomes more volatile. To see this, observe that $\phi(\mathbf{x}) = f(\frac{1}{n} \sum_{i=1}^{n} x_i)$ is supermodular for any convex function f. Thus,

$$Ef\left(\frac{1}{|K|} \sum_{i \in K} R_i\right) \geq Ef\left(\frac{1}{|K|} \sum_{i \in K} \bar{R}_i\right), \qquad K \subseteq E. \qquad (11.44)$$

This relation is known as convex ordering (see, e.g., Shaked and Shanthikumar [22]).

To illustrate that the differences given by (11.42) and (11.44) can be significant, consider a perfectly synchronized ATO system with a single end-product. Service times of different items are all equal to a constant. Compared with the system with independent demands, it is patently obvious that the perfectly synchronized system decreases the range of $\mathbf{R}^K$ to zero, and increases the variance of the average batch delivery leadtime by a factor of m. The phenomena that correlation can improve one performance measure yet hurt another is also observed by Disney et al. [9], Xu [34], and Li and Xu [16]. ∎

If $\mathbf{R}$ is more supermodular dependent than $\bar{\mathbf{R}}$, we can obtain an even stronger result. It was shown that (Chang [5], Lemma 2.4 (ii))

$$f_i(\mathbf{x}) = \sum_{j=1}^{i} x_{[j]}, \quad \text{and} \quad g_i(\mathbf{x}) = -\sum_{j=i}^{m} x_{[j]}, \quad i \in E,$$

are *submodular* functions, where $x_{[j]}$ is the j-th largest component in vector $\mathbf{x}$. Since if f is submodular then $-f$ is supermodular, (11.35) implies that

$$E \sum_{j=1}^{i} R_{[j]}^K \leq E \sum_{j=1}^{i} \bar{R}_{[j]}^K \quad \text{and} \quad E \sum_{j=i}^{|K|} R_{[j]}^K \geq E \sum_{j=i}^{|K|} \bar{R}_{[j]}^K, \quad i \leq |K|,$$

where $R_{[j]}^K$ is the j-th largest component in $\mathbf{R}^K$. In words, this result states that, as the product-type indicator vector and/or interarrival time vector become less supermodular dependent, the expected sum of the i *longest* delivery leadtimes becomes *larger*, whereas the expected sum of the i *shortest* delivery leadtimes becomes *smaller*, for any $i \leq |K|$.

We illustrate our results using the input parameters given by Example 11.6 (a) and Example 11.3 (a)–(b).

Example 11.6 (b)   We compare the performance of two $\{M/M/1, i \in \{1,2\}\}$ supply systems whose demand rate trees are given by Example 11.6 (a). For simplicity, we call the system induced by $\mathbf{Z}$ System-$\mathbf{Z}$. System-$\tilde{\mathbf{Z}}$ is defined similarly. Let System-$\mathbf{Z}$ and System-$\tilde{\mathbf{Z}}$ have inter-arrival time vectors $\mathbf{T} = (T, T)$ and $\tilde{\mathbf{T}} = (T_1, T_2)$, respectively, where T, T_1, and T_2 are i.i.d exponential random variables with rate 4. Because both $\tilde{\mathbf{Z}}$ and $\tilde{\mathbf{T}}$ contain independent components, $\tilde{\mathbf{R}}$ also contains two independent components. In addition, $\mathbf{R} \geq_{sm} \tilde{\mathbf{R}}$. By (11.35), the product-form lower bound of the due-date fill rate of product-12 satisfies

$$P(\mathbf{R}^{12} \leq \mathbf{r}^{12}) \geq \prod_{i=1}^{2} \left(1 - \left(\frac{\lambda_i}{\mu_i} \right)^{s_i} e^{-(\mu_i - \lambda_i) r_i} \right),$$

where $\lambda_i = \lambda^i + \lambda^{12}$, $i = 1, 2$ and μ_i is the service rate of machine i. We can also obtain the lower bound for the system-based due-date fill rate as follows:

$$\sum_{K \in \{1,2\}} \frac{\lambda^K}{\lambda} P(\mathbf{R}^K \leq \mathbf{r}^K) \geq \frac{\lambda^1}{\lambda} \left(1 - \left(\frac{\lambda_1}{\mu_1} \right)^{s_1} e^{-(\mu_1 - \lambda_1) r_1} \right)$$
$$+ \frac{\lambda^2}{\lambda} \left(1 - \left(\frac{\lambda_2}{\mu_2} \right)^{s_2} e^{-(\mu_2 - \lambda_2) r_2} \right)$$
$$+ \frac{\lambda^{12}}{\lambda} \prod_{i=1}^{2} \left(1 - \left(\frac{\lambda_i}{\mu_i} \right)^{s_i} e^{-(\mu_i - \lambda_i) r_i} \right).$$

In Example 11.6 (c), we will illustrate how to compute the bounds of the expected product delivery leadtime and the expected range of product delivery leadtimes.   ■

Example 11.3 (c)   Let $\mathbf{Z}$ and $\bar{\mathbf{Z}}$ be given in Example 11.3 (b) and let $\tilde{\mathbf{Z}}$ be an independent copy of $\mathbf{Z}$. The overall demand process to each system is Poisson with rate 1. Let $\mathbf{T} =_{st} \tilde{\mathbf{T}} =_{st} \bar{\mathbf{T}} = (T, T, T)$, where T is an exponential random variable with rate 1. Because the three systems have the same interarrival time vectors, their structural differences are determined by their product-type indicator vectors. For simplicity, we call the system induced by $\mathbf{Z}$ System-$\mathbf{Z}$. System-$\bar{\mathbf{Z}}$ and System-$\tilde{\mathbf{Z}}$ are similarly defined.

We have shown that $\mathbf{Z}$ is PUOD and NOLD and $\bar{\mathbf{Z}}$ is NUOD and PLOD. The item-based performance measures in these systems are stochastically identical. Let us rank order these systems from the viewpoint of a *system-based* performance measure. To illustrate, we rank order their inventory on-order vectors. By the Poisson Arrivals See Time

Average (PASTA) property, $\mathbf{N}^a = \mathbf{N}$. By (11.33),

$$\mathbf{N} \geq_{uo} \tilde{\mathbf{N}} \geq_{uo} \bar{\mathbf{N}} \text{ and } \mathbf{N} \geq_{lo} \tilde{\mathbf{N}} \geq_{lo} \bar{\mathbf{N}}.$$

Then it immediately follows

$$N_{\min} \geq_{st} \tilde{N}_{\min} \geq_{st} \bar{N}_{\min} \text{ and } N_{\max} \geq_{st} \tilde{N}_{\max} \geq_{st} \bar{N}_{\max}.$$

This means that System-$\mathbf{Z}$ underperforms System-$\tilde{\mathbf{Z}}$ in the sense that the smallest and largest outstanding orders in the former system are both stochastically greater than their counterparts in the latter system. Similarly, System-$\tilde{\mathbf{Z}}$ underperforms System-$\bar{\mathbf{Z}}$ in the same sense. This implies that the system with its demand cardinality less variable can simultaneously improve the best and worse components in a system-based performance measure. This result illustrates that an appropriate combination of positive and negative dependencies can simultaneously improve the best and the worst components of a performance vector.

Finally, we consider a *product-based* performance measure, say the due-date fill rate, in these systems. Since Systems $\mathbf{Z}$ and $\bar{\mathbf{Z}}$ serve different product-types, they cannot be compared directly on a product-to-product basis. However, the performance of a product in either System-$\mathbf{Z}$ or $\bar{\mathbf{Z}}$ can be compared with its counterpart in System-$\tilde{\mathbf{Z}}$, who serves every product-type. From (11.35),

$$P(\mathbf{R}^K \leq \mathbf{r}^K) \leq P(\tilde{\mathbf{R}}^K \leq \mathbf{r}^K) \leq P(\bar{\mathbf{R}}^K \leq \mathbf{r}^K), \ K \subseteq E.$$

Again, we observe that System-$\mathbf{Z}$ underperforms system $\tilde{\mathbf{Z}}$ in terms of the due-date fill rate of a product, as long as the product is served by both systems. Otherwise, one probability is understood as a stochastic bound of another. Similarly, System-$\tilde{\mathbf{Z}}$ underperforms System-$\bar{\mathbf{Z}}$ in the same sense. In Section 7, we use these inequalities to derive stochastic bounds of the due date fill-rate. ∎

Carefully examining the proof of Theorem 11.9 shows that if batch size vectors $\mathbf{D}$ and $\bar{\mathbf{D}}$ are correlated but satisfy the condition $\mathbf{D} \geq_{sm} (\geq_{uo}, \leq_{lo})\bar{\mathbf{D}}$ and $D_i =_{st} \bar{D}_i$, $i \in E$, then Theorem 11.9 remains true.

5.3. A capacitated ATO system with a $\{GI/G/1, i \in E\}$ supply system

In this model, we assume that each batch contains a single unit, $D_i \equiv 1$, $i \in E$. Then the demand type/batch quantity vector becomes $\mathbf{ZD} = \mathbf{Z}$. The distribution of other random variables, $\mathbf{T}$ and $\mathbf{S}$, are arbitrarily distributed.

For this system, it is no longer easy to express $\mathbf{R}$ in terms of $\mathbf{N}^a$ as in (11.29). Instead, we will express $\mathbf{R}$ in terms of the manufacturing leadtime vector **L**. Since the system implements a base-stock policy and orders are received in the same sequence as they are placed, we recognize that *any item i demand is filled by the previous s_ith replenishment order before the current item-i demand,* hereafter, referred to as *its order* (see, for example, Asxäter [2]). Let us assume that an item-i demand has just arrived. Let the previous J_ith order contain *its order* (recall that not every replenishment order contains an item i order). Then

$$J_i = \min \left\{ j : \sum_{n=1}^{j} Z_{i,n} = s_i \right\}. \qquad (11.45)$$

The elapsed time since *its order* is placed is $\sum_{n=1}^{J_i} T_{i,n}$, which will be serviced after leadtime L_i. Thus the delivery leadtime of the item i demand is

$$R_i = \left[L_i - \sum_{n=1}^{J_i} T_{i,n} \right]^+, \qquad i \in E. \qquad (11.46)$$

Equations (11.45)–(11.46) show that **R** is determined by **L**. Thus our study of the $\{GI/G/1, i \in E\}$ system will be focused on the stochastic properties of random vectors **L** and **R**.

Let demand n contain an item-i demand and let $L_{i,n}$ be the manufacturing leadtime (queue time plus service time) of the item. Let $\mathbf{L}_n = (L_{1,n}, \ldots, L_{m,n})$, $n \geq 1$. The proof of the following theorem is given in Appendix C (Section 9.3).

Theorem 11.12 Let $\{\mathbf{L}_n, n \geq 1\}$ and $\{\bar{\mathbf{L}}_n, n \geq 1\}$ be the manufacturing leadtime processes in two $\{GI/G/1, i \in E\}$ supply systems, driven by independent sequences $\{\mathcal{T}, \mathcal{ZD}, \mathcal{S}\}$ and $\{\bar{\mathcal{T}}, \bar{\mathcal{ZD}}, \mathcal{S}\}$, respectively. If $\Lambda \geq_{g2} (\geq_{g_\downarrow}, \geq_{g_\uparrow})\bar{\Lambda}$, and $\mathbf{T}_n \geq_{sm} (\leq_{lo}, \geq_{uo})\bar{\mathbf{T}}_n$, then, for $i \in E$ and $n \geq 1$,

$$\{\mathbf{L}_n, n \geq 1\} \geq_{sm} (\geq_{uo}, \leq_{lo})\{\bar{\mathbf{L}}_n, n \geq 1\} \quad \text{and} \quad L_{i,n} =_{st} \bar{L}_{i,n}, \quad (11.47)$$

and, in equilibrium,

$$\mathbf{L} \geq_{sm} (\geq_{uo}, \leq_{lo})\bar{\mathbf{L}} \quad \text{and} \quad L_i =_{st} \bar{L}_i, \ i \in E. \qquad (11.48)$$

In addition,

$$\mathbf{R} \geq_{uo} (\leq_{lo})\bar{\mathbf{R}} \quad \text{and} \quad R_i =_{st} \bar{R}_i, i \in E. \qquad (11.49)$$

We can also obtain dependence properties of **R** for a $\{GI^{D_i}/G/1, i \in E\}$ supply system, using an approach similar to that of Glasserman and Wang [12], who derive a sample path representation of **R** for a single-product ATO system with a $\{GI^{D_i}/G/1, i \in E\}$ supply system. To limit our exposure, we shall not pursue the general case in this chapter.

6.   Uncapacitated Assemble-to-Order Systems

In this section we study two uncapacitated ATO systems, namely, the $\{GI^{D_i}/M/c_i/c_i, \ i \in E\}$ system and the $\{GI/D/\infty, \ i \in E\}$ system. We show that a performance vector process in each system can also be viewed as a separate Markov chain. We provide the sufficient conditions under which $\mathbf{N}^a$ and $\mathbf{R}$ in two systems follow supermodular and orthant orders.

6.1.   An uncapacitated ATO system with a $\{GI^{D_i}/M/c_i/c_i, \ i \in E\}$ supply system

In this partial lost-sales model, it is assumed that manufacturing lead-times of item i are i.i.d. exponential random variables with rate μ_i, $i \in E$. This is equivalent to assuming that the supplier is uncapacitated so that outstanding orders do not interfere with one another (i.e., no queue is formed in front of a production facility).

A batch-i arrival will be filled using on-hand inventory whenever possible. An item-i has a backlog queue with capacity b_i, $0 \le b_i \le \infty$. When $b_i = \infty$, all unfilled items in a batch i demand are completely backlogged. When $b_i = 0$, all unfilled items in a batch i demand are lost. When $0 < b_i < \infty$, backlog queue i will accept as many unfilled items as possible and reject the rest after queue i reaches its capacity b_i. This service scheme is termed the *partial service scheme* because an order can be serviced partially and the acceptance/rejection decision made for each batch is independent of the acceptance/rejection decisions made for other batches of the product. The partial service scheme fits the ATO environment at the distribution level, where customers often accept partial shipments of finished goods.

Let $s_i + b_i = c_i$, $i \in E$. It is easily seen that the production system of this ATO system can be modelled by a $\{GI^{D_i}/M/c_i/c_i, i \in E\}$ loss-system. Upon arrival, each item in a batch will be assigned to an available machine, if possible, with the rest blocked and lost. A serviced item will be shipped to its inventory location individually, where the shipping time is negligible.

If backlogged demand is served on an FCFS basis, then it appears formidable to derive the exact expression of R_i, because stochastic lead-times may cause orders to cross over, that is, orders may be received in a different sequence as they are placed. However, the inventory on-order process embedded at arrival epochs $\{\mathbf{N}_n^a, n \ge 1\}$ still has an explicit sample path representation, so we shall focus on the dependence structure of this process.

Let $\mathbf{N}_n^a = (N_{1,n}^a, \ldots, N_{m,n}^a)$, $n \geq 1$, be defined as before. Let 1_A be the indicator function of event A. It is easily seen that the following recursion holds for $N_{i,n+1}^a$:

$$N_{i,n+1}^a = \sum_{j=1}^{(N_{i,n}^a + Z_{i,n} D_{i,n}) \wedge c_i} 1_{\{S_{i,j} \geq T_{i,n+1}\}}, \qquad i \in E, \qquad (11.50)$$

where $S_{i,j}$ are i.i.d. exponential random variables with rate μ_i. This recursion is understood as: Inventory on-order of item-i immediately after the nth batch demand of item-i is $(N_{i,n}^a + Z_{i,n} D_{i,n}) \wedge c_i$. The inventory on-order observed by the next batch demand are those whose service times are longer than the interarrival time $T_{i,n+1}$.

It is not difficult to see that $\{\mathbf{N}_n^a, n \geq 1\}$ and $\{\bar{\mathbf{N}}_n^a, n \geq 1\}$ are separable Markov chains with independent generating sequences $(\mathcal{T}, \mathcal{ZD}, \mathcal{S})$ and $(\bar{\mathcal{T}}, \bar{\mathcal{ZD}}, S)$, respectively. Using the same approach as in the previous section, we can prove the following (see Li and Xu [16] for details).

Theorem 11.13 Consider two $\{GI^{D_i}/M/c_i/c_i, i \in E\}$ supply systems, driven by independent sequences $\{\mathcal{T}, \mathcal{ZD}, \mathcal{S}\}$ and $\{\bar{\mathcal{T}}, \bar{\mathcal{ZD}}, \mathcal{S}\}$, respectively. If $\Lambda \geq_{g^2} (\geq_{g_\downarrow}, \geq_{g_\uparrow}) \bar{\Lambda}$, and $\mathbf{T}_n \geq_{sm} (\leq_{lo}, \geq_{uo}) \bar{\mathbf{T}}_n$, then for $i \in E$ and $n \geq 1$,

$$\{\mathbf{N}_n^a, n \geq 1\} \geq_{sm} (\geq_{uo}, \leq_{lo})\{\bar{\mathbf{N}}_n^a, n \geq 1\} \quad \text{and} \quad N_{i,n}^a =_{st} \bar{N}_{i,n}^a, \tag{11.51}$$

and in equilibrium,

$$\mathbf{N}^a \geq_{sm} (\geq_{uo}, \leq_{lo})\bar{\mathbf{N}}^a \quad \text{and} \quad N_i^a =_{st} \bar{N}_i^a, \quad i \in E. \tag{11.52}$$

For the lost sales case ($b_i = 0$, $i \in E$), under the conditions of Theorem 11.13, the instantaneous fill rates in the two systems satisfy, for $K \subseteq E$,

$$F^K = P(N_i^a \leq s_i - D_i, i \in K) \geq P(\bar{N}_i^a \leq s_i - D_i, i \in K) = \bar{F}^K. \tag{11.53}$$

That is, the instantaneous fill rate of any product-type will be higher if demand is more likely to contain a smaller subset of items and interarrival times are more likely to jointly assume large values.

If $b_i > 0$ and backlogged demand is served on an FCFS basis, then (11.53) still holds. But it appears formidable to derive the exact expression of R_i, due to the possibility of order crossovers.

6.2.   An uncapacitated ATO system with a $\{GI/D/\infty, \ i \in E\}$ supply system

As we noted in the previous section, the major difficulty in deriving an exact expression of $\mathbf{R}$ is caused by the crossovers of replenishment orders. In this model, we assume that manufacturing leadtimes are constants, so that orders will never cross. Our derivation, however, is also valid for a $\{GI/G/\infty, \ i \in E\}$ supply system under the *order-for-order* fulfillment rule. Under this rule, each item-i order is placed specifically for the s_ith demand following its order, bypassing the difficulty caused by possible order overtakes. This order fulfillment rule has been used extensively in the classical inventory models (Asxäter [2], Zipkin [37]). Gallien and Wein [11] impose a synchronization assumption, which is equivalent to the order-for-order fulfillment rule, in their study of a single-product stochastic assembly system modelled as a set of $M/G/\infty$ queues.

However, observe that with deterministic manufacturing leadtimes, $L_i, \ i \in E$, (11.45) and (11.46) we derived for $\mathbf{R}$ in a $\{G/G/1, i \in E\}$ supply system are also applicable here. Therefore, the result of Theorem 11.12 is trivially extended to the $\{GI/D/\infty, \ i \in E\}$ system.

Clearly, R_i cannot exceed $L_i, \ i \in E$. From (11.46), it is easily seen that

$$R_i \le r_i \Longleftrightarrow \sum_{n=1}^{s_i} T_n \ge L_i - r_i, \quad i \in E. \tag{11.54}$$

In words, it says that in order to deliver an item-i demand before its target due-date, *its order* must arrive before $L_i - r_i$. This, in turn, means that at most s_i units of item-i can arrive during interval $L_i - r_i$. Thus

$$P(\mathbf{R}^K \le \mathbf{r}^K, \ i \in K) = P\left(\sum_{n=1}^{s_i} T_n \ge L_i - r_i, \ i \in K \right)$$
$$= P(A_i(L_i - r_i) \le s_i, \ i \in K). \tag{11.55}$$

The above expression also implies that, the orthant orders of interarrival time vectors and product-type indicator vectors cause the demand processes in the two system to be orthant dependent:

$$\mathbf{Z} \ge_{uo} (\le_{lo})\bar{\mathbf{Z}} \quad \text{and} \quad \mathbf{T} \le_{lo} (\ge_{uo})\bar{\mathbf{T}} \Longrightarrow \mathbf{A}(\mathbf{t}) \ge_{uo} (\le_{lo})\bar{\mathbf{A}}(\mathbf{t}),$$

where the multivariate renewal arrival process $\mathbf{A}(\mathbf{t})$ is given by $\mathbf{A}(\mathbf{t}) = \{A_i(t_i), t_i \ge 0, i \in E\}$.

Remark 11.14 An $\{M/G/\infty, i \in E\}$ supply system plays an important role in the study of uncapacitated ATO systems. A single-product, synchronized $\{M/G/\infty, i \in E\}$ system has been discussed by several authors (Gallien and Wein [11] and Song and Yao [27]).

An $\{M/G/\infty, i \in E\}$ system with simultaneous arrivals enjoys certain special properties and its dependence structure is well understood, due to the much celebrated Palm Theorem (Wolff [33]). More specifically, consider a multivariate Poisson process generated by synchronized interarrival times $\mathbf{T} = (T, \ldots, T)$ and the parameter set $\{P^K, K \subseteq E\}$, where T is an exponential random variable with rate λ. For each $K \subseteq E$, let L be a subset of K. We define N_L^K as the stationary number of the unsatisfied product-K demands that are waiting for the items in set L to be filled. The Palm Theorem states that the random variables N_K^L, $L \subseteq K \subseteq E$, are *independent* Poisson random variables with rates

$$\lambda P^K \int_0^\infty \left[\prod_{i \in K-L} G_i(y) \prod_{i \in L} \bar{G}_i(y) \right] dy, \quad L \subseteq K \subseteq E,$$

where G_i is the service time distribution of a machine in facility i and $\bar{G}_i = 1 - G_i$, $i \in E$. The stationary number of inventory on-order of item i has the explicit stochastic representation

$$N_i = \sum_{i \in L \subseteq K \subseteq E} N_L^K, \quad i \in E. \tag{11.56}$$

For example, in a two-item, three-product system, (11.56) becomes

$$N_1 = N_1^1 + N_1^{12} + N_{12}^{12},$$
$$N_2 = N_2^2 + N_2^{12} + N_{12}^{12}.$$

Due to independence of N_L^K, $i \in L \subseteq K \subseteq E$, each N_i is a Poisson random variable with rate λ_i/μ_i, where μ_i is the time-average service rate of facility i, $i \in E$. It is well-known that Poisson random variables $\mathbf{N} = (N_1, \ldots, N_m)$ given by (11.56) are associated (Joe [14], Song [24]), regardless of the distribution of parameter set $\{P^K, K \subseteq E\}$. Therefore, if (K_1, K_2) is a partition of K, then we immediately have

$$F^K = P(\mathbf{N}^K < \mathbf{s}^K) \geq P(\mathbf{N}^{K_1} < \mathbf{s}^{K_1}) P(\mathbf{N}^{K_2} < \mathbf{s}^{K_2})$$

$$\geq \cdots \geq \prod_{i \in K} \left[\sum_{n < s_i} \frac{e^{-\rho_i} \rho_i^n}{n!} \right]$$

Song and Yao [27] use this idea to derive the product-form lower bound of F^K. ∎

7.   Performance Bounds

So far we have discussed the structural and qualitative properties of capacitated and uncapacitated ATO systems. In this section we show that the results obtained in the previous sections can be used to develop *quantitative* bounds for the statistics of a performance measure. We focus on the classes of *setwise* bounds. Other types of bounds can be found in Dayanik et al. [8]. We also provide several numerical examples to illustrate the tightness of our bounds.

7.1.   Setwise bounds

Recall that the dependence nature of a performance measure is inherited from two sources: The correlated interarrival times $\mathbf{T}$ and the product-type indicator $\mathbf{Z}$. Let $\mathbf{R}$ be the delivery leadtime of *any* system studied in Sections 5 and 6, whenever it is well-defined. To indicate that $\mathbf{R}$ depends on $\mathbf{T}$ and $\mathbf{Z}$, we write

$$\mathbf{R} \equiv \mathbf{R}(\mathbf{T}, \mathbf{Z}).$$

We first give a sufficient condition under which the distribution of $\mathbf{R}(\mathbf{T}, \mathbf{Z})$ is bounded below by the product of its univariate marginals. To this end, Let $\tilde{\mathbf{Z}}$ be the independent copy of $\mathbf{Z}$, with distribution

$$P(\tilde{\mathbf{Z}} = \mathbf{e}^K) = \prod_{i \in K} P_i \prod_{i \in E-K} (1 - P_i), \quad K \subseteq E. \qquad (11.57)$$

Since $P_i = \lambda_i/\lambda$, $i \in E$, we can define the demand rate tree $\tilde{\Lambda}$ of $\tilde{\mathbf{Z}}$ by

$$\tilde{\lambda}^K = \prod_{i \in K} \lambda_i \prod_{i \in E-K} (\lambda - \lambda_i), \quad K \subseteq E. \qquad (11.58)$$

From the theorems derived in the previous two sections,

$$\Lambda \geq_{g_\downarrow} (\geq_{g_\uparrow})\tilde{\Lambda} \Longrightarrow \mathbf{R}(\mathbf{T}, \mathbf{Z}) \geq_{uo} (\leq_{lo})\mathbf{R}(\mathbf{T}, \tilde{\mathbf{Z}})$$
$$\text{and} \quad R_i(T_i, Z_i) =_{st} R_i(T_i, \tilde{Z}_i), \ i \in E. \qquad (11.59)$$

Similarly, if $\tilde{\mathbf{T}}$ is an independent copy of $\mathbf{T}$, then

$$\mathbf{T} \leq_{lo} (\geq_{uo})\tilde{\mathbf{T}} \Longrightarrow \mathbf{R}(\mathbf{T}, \mathbf{Z}) \geq_{uo} (\leq_{lo})\mathbf{R}(\tilde{\mathbf{T}}, \mathbf{Z})$$
$$\text{and} \quad R_i(T_i, Z_i) =_{st} R_i(\tilde{T}_i, Z_i), \ i \in E. \qquad (11.60)$$

Note that neither $\mathbf{R}(\mathbf{T}, \tilde{\mathbf{Z}})$ nor $\mathbf{R}(\tilde{\mathbf{T}}, \mathbf{Z})$ are independent vectors, due to the dependencies injected from $\mathbf{T}$ or $\mathbf{Z}$. However, putting (11.58) and

(11.59) together, we obtain the sufficient conditions under which $\mathbf{R}$ is PUOD (PLOD):

$$\mathbf{T} \leq_{lo} (\geq_{uo}) \tilde{\mathbf{T}} \quad \text{and} \quad \Lambda \geq_{g_{\downarrow}} (\geq_{g_{\uparrow}}) \tilde{\Lambda}$$
$$\implies \mathbf{R}(\mathbf{T}, \mathbf{Z}) \geq_{uo} (\leq_{lo}) \mathbf{R}(\tilde{\mathbf{T}}, \tilde{\mathbf{Z}})$$
$$\text{and} \quad R_i(T_i, Z_i) =_{st} R_i(\tilde{T}_i, \tilde{Z}_i), \ i \in E. \quad (11.61)$$

The above expression implies that, if $\mathbf{T}$ is PLOD (PUOD) and $\mathbf{Z}$ is PUOD (PLOD), then $\mathbf{R}$ is also PUOD (PLOD). Therefore,

$$\Lambda \geq_{g_{\downarrow}} \tilde{\Lambda}(\text{i.e., } \mathbf{Z} \text{ is PUOD}) \text{ and } \mathbf{T} \leq_{lo} \tilde{\mathbf{T}}(\text{i.e., } \mathbf{T} \text{ is PLOD})$$
$$\implies P(\mathbf{R}^K \geq \mathbf{r}^k) \geq \prod_{i \in K} P(R_i \geq r_i), \quad (11.62)$$

$$\Lambda \geq_{g_{\uparrow}} \tilde{\Lambda}(\text{i.e., } \mathbf{Z} \text{ is PLOD}) \text{ and } \mathbf{T} \geq_{uo} \tilde{\mathbf{T}}(\text{i.e., } \mathbf{T} \text{ is PUOD})$$
$$\implies P(\mathbf{R}^K \leq \mathbf{r}^k) \geq \prod_{i \in K} P(R_i \leq r_i). \quad (11.63)$$

As noted in Remark 11.10, it is not true that a more positively lower orthant dependent $\mathbf{R}$ will give a higher due-date fill rate of the *system*. However, the PUOD or PLOD property of $\mathbf{R}$ implies that its system-based performance measure still admits a product-form bound:

$$\Lambda \geq_{g_{\downarrow}} \tilde{\Lambda} \ (\text{i.e., } \mathbf{Z} \text{ is PUOD}) \text{ and } \mathbf{T} \leq_{lo} \tilde{\mathbf{T}}(\text{i.e., } \mathbf{T} \text{ is PLOD})$$
$$\implies \sum_{K \subseteq E} \frac{\lambda^K}{\lambda} P(\mathbf{R}^K \geq \mathbf{r}^K) \geq \sum_{K \subseteq E} \frac{\lambda^K}{\lambda} \prod_{i \in K} P(R_i \geq r_i) \quad (11.64)$$

$$\Lambda \geq_{g_{\uparrow}} \tilde{\Lambda} \ (\text{i.e., } \mathbf{Z} \text{ is PLOD}) \text{ and } \mathbf{T} \geq_{uo} \tilde{\mathbf{T}}(\text{i.e., } \mathbf{T} \text{ is PUOD})$$
$$\implies \sum_{K \subseteq E} \frac{\lambda^K}{\lambda} P(\mathbf{R}^K \leq \mathbf{r}^K) \geq \sum_{K \subseteq E} \frac{\lambda^K}{\lambda} \prod_{i \in K} P(R_i \leq r_i) \quad (11.65)$$

In a typical ATO system, the interarrival times are synchronized, $\mathbf{T} = (T, T \ldots, T)$. The following lemma, known as the *Lorentz inequality*, states that $\mathbf{T}$ is more supermodular dependent than its independent copy $\tilde{\mathbf{T}}$, and hence is PULD and PLOD.

Lemma 11.15 *(Lorentz inequality)* Let $X_1, \ldots, X_m$ be i.i.d. random variables with $X_i =_{st} X$. Then,

$$(X, X, \ldots, X) \geq_{sm} (X_1, X_2, \ldots, X_m).$$

See Tchen [30] Theorem 5A and Rolski [20] Lemma 5 for details.

Example 11.3 (d) Let $\mathbf{Z}$, $\tilde{\mathbf{Z}}$ and $\bar{\mathbf{Z}}$ be defined in Example 11.3 (a). We have shown that $\bar{\mathbf{Z}}$, in which each demand contains a pair of items with rate $\frac{1}{4}$, is NUOD and PLOD. Let $\bar{\mathbf{T}} = (T, T, T)$, where T follows a general distribution. Then the Lorentz inequality states that $\bar{\mathbf{T}}$ is PUOD and PLOD. This implies that $\mathbf{R}$ is PLOD. From (11.62), for any pair (i, j),

$$P(\bar{R}_i \leq r_i, \bar{R}_j \leq r_j) \geq P(\bar{R}_i \leq r_i)P(\bar{R}_j \leq r_j) \quad i, j \in 1, 2, 3, i \neq j.$$

However, we cannot claim that $\mathbf{R}$ is a PUOD. Thus it is generally not true that the survival function of $\bar{\mathbf{R}}^{ij}$ is bounded below by the product of the survival functions of $\bar{R}_i$ and $\bar{R}_j$.

On the other hand, we have also shown that $\mathbf{Z}$, under which the arrival rates of products 1,2,3 and 123 are $\frac{1}{4}$, respectively, is PUOD and NLOD. Thus if $\mathbf{T} = (T, T, T)$, where T follows a general distribution, then $\mathbf{R}$ is PUOD. By (11.61),

$$P(R_i \geq r_i, i = 1, 2, 3) \geq \prod_{i=1}^{3} P(R_i \geq r_i)$$

Unfortunately, we cannot claim that $\mathbf{R}$ is PLOD. Thus the due-date fill rate of product-123 does not have a product-form lower bound.

Remark 11.16 The above example illustrates that the orthant property of $\mathbf{T}$ or $\mathbf{Z}$ alone cannot guarantee the orthant property of $\mathbf{R}$. However, there are several exceptions.

1 If $\mathbf{T}$ is *deterministic*, then the only dependent vector is $\mathbf{Z}$ and its dependence property will be inherited by $\mathbf{R}$. In other words, with interarrival time a constant,

$$\mathbf{Z} \text{ is PUOD (PLOD, NUOD, NLOD)}$$
$$\Longrightarrow \mathbf{R} \text{ is PUOD(PLOD, NUOD, NLOD)}. \qquad (11.66)$$

For instance, in the above example, if interarrival time T is deterministic, then because $\mathbf{Z}$ is PUOD and NOLD, we obtain the product-form *upper* bound of the distribution function of $\bar{\mathbf{R}}^{123}$:

$$P(R_i \leq r_i, \ i = 1, 2, 3) \leq \prod_{i=1}^{3} P(\bar{R}_i \leq r_i).$$

Similarly, if interarrival time T is deterministic, because $\bar{\mathbf{Z}}$ is NUOD and POLD, we obtain the product-form *upper* bound of the survival function of $\mathbf{R}^{i,j}$:

$$P(\bar{R}_i \geq r_i, \bar{R}_j \geq r_j) \leq P(\bar{R}_i \geq r_i)P(\bar{R}_j \geq r_j) \quad i, j \in 1, 2, 3, i \neq j.$$

This example illustrates that simultaneous demands do not automatically lead to a positively dependent performance vector, and the joint distribution (survival) function of a performance vector is not always bounded below by the product of its individual marginal distribution (survival) functions.

2 If an ATO system only serves a single product-type, $\mathbf{Z} = (1, 1, \dots, 1)$ with probability 1, then the only dependent vector is $\mathbf{T}$ and the dependence property of $-\mathbf{T}$ will be preserved by $\mathbf{R}$. In particular, if $\mathbf{T} = (T, T, \dots, T)$, then $\mathbf{R}$ is *always* PUOD and PLOD, independent of distributions of T, $\mathbf{B}$ and $\mathbf{S}$.

3 If $\mathbf{T} = (T, T, \dots, T)$ and T is exponentially distributed, then regardless of the distribution of $\mathbf{Z}$, $\mathbf{R}(\mathbf{T}, \mathbf{Z})$ is always PUOD and PLOD. This can be argued as follows: Construct a tree, $\bar{\Lambda}$, with the weight of node i being the marginal weight of node i in tree Λ, $\bar{\lambda}^i = \lambda_i$, $i \in E$, and the weight of any other node being zero. It is not difficult to see that any tree Λ can be transformed to $\bar{\Lambda}$ via a sequence of $g_\downarrow$ ($g_\uparrow$) transforms. In other words, $\bar{\mathbf{Z}}$ is less PUOD and PLOD (or more NUOD and NLOD) than any other multivariate indicator function of equal marginals:

$$\Lambda \geq_{g_\downarrow} \bar{\Lambda} \text{ and } \Lambda \geq_{g_\uparrow} \bar{\Lambda}.$$

Consequently, by (11.58),

$$\mathbf{R}(\mathbf{T}, \mathbf{Z}) \geq_{uo} (\leq_{lo}) \mathbf{R}(\mathbf{T}, \bar{\mathbf{Z}}).$$

However, the arrival process specified by $\mathbf{T}$ (this vector is more PUOD and PLOD than any other vector of equal marginals) and $\bar{\mathbf{Z}}$ (this vector is less PUOD and PLOD than any other vector of equal marginals) are m *independent* Poisson processes. This implies that $\mathbf{R}(\mathbf{T}, \bar{\mathbf{Z}})$ comprises of independent components, and it further implies that $\mathbf{R}(\mathbf{T}, \mathbf{Z})$ is PUOD and PLOD. Therefore, with simultaneous Poisson demands, the joint distribution (survival) function of $\mathbf{R}$ is always bounded below by the product of its individual marginal distribution (survival) functions. ■

The bounds given by (11.61)–(11.64) are called the *first-order setwise bounds* because they only require the knowledge of univariate marginal distributions. The setwise bounds of higher orders can be similarly derived. We first motivate the idea by an example.

Example 11.17 Consider a four-item ATO system with synchronized interarrival time $\mathbf{T} = (T, T, T, T)$ and the product-type vector $\mathbf{Z}$ distributed as

$$\Lambda = (P^{1234}, P^{12}, P^{23}, P^0) = \left(\frac{1}{4}, \frac{1}{8}, \frac{3}{8}, \frac{1}{4}\right).$$

We wish to find a second-order setwise lower bound for the due-date fill rate of product 1234. To this end, note that (Z_1, Z_2) has the distribution $P(Z_1 = 1, Z_2 = 1) = 3/8$ and $P(Z_1 = 0, Z_2 = 0) = 5/8$. Similarly, (Z_3, Z_4) has the distribution $P(Z_3 = 1, Z_4 = 1) = 5/8$ and $P(Z_3 = 0, Z_4 = 0) = 3/8$. Now let $\mathbf{Z} = (\bar{Z}_1, \bar{Z}_2, \bar{Z}_3, \bar{Z}_4)$ such that $(\bar{Z}_1, \bar{Z}_2) =_{st} (Z_1, Z_2)$, $(\bar{Z}_3, \bar{Z}_4) =_{st} (Z_3, Z_4)$, and $(\bar{Z}_1, \bar{Z}_2)$ and $(\bar{Z}_3, \bar{Z}_4)$ be independent. It is easily verified that the tree associated with $\bar{Z}$ is

$$\bar{\Lambda} = (\bar{P}^{1234}, \bar{P}^{12}, \bar{P}^{34}, \bar{P}^{\emptyset}) = \left(\frac{3}{8} \times \frac{5}{8}, \frac{3}{8} \times \frac{3}{8}, \frac{5}{8} \times \frac{5}{8}, \frac{5}{8} \times \frac{3}{8}\right).$$

Clearly, from (11.19), $g^{\frac{1}{64}, 12, 34}(\Lambda) = \bar{\Lambda}$. Then (11.19) implies $\mathbf{Z} \geq_{sm} \bar{\mathbf{Z}}$. Also, by the Lorantz inequality, $\mathbf{T} \geq_{sm} \bar{\mathbf{T}} := (T, T, T', T')$, where $T =_{st} T'$ and T and T' are independent. Therefore,

$$\mathbf{R}(\mathbf{T}, \mathbf{Z}) \geq_{sm} \mathbf{R}(\bar{\mathbf{T}}, \bar{\mathbf{Z}}).$$

However, due to our construction of $\bar{\mathbf{Z}}$ and $\bar{\mathbf{T}}$, the delivery leadtimes of items 1 and 2 are independent of that of items 3 and 4. Hence, the due-date fill rate of product 1234 admits the second order setwise lower bound:

$$P(\mathbf{R}^{1234} \leq \mathbf{r}^{1234}) \geq P(\mathbf{R}^{12} \leq \mathbf{r}^{12})P(\mathbf{R}^{34} \leq \mathbf{r}^{34}).$$

Note that the above result holds true for arbitrarily distributed interarrival time $\mathbf{T}$. ■

In general, a setwise lower bound of high-order can be constructed as follows. Let $\mathbf{K} = \{K_1, \ldots, K_\nu\}$ be a partition of index set E. Let $\bar{\mathbf{T}}^{\mathbf{K}} = (\bar{\mathbf{T}}^{K_i}, i = 1, \ldots, \nu)$, such that $\mathbf{T}^{K_i} =_{st} \bar{\mathbf{T}}^{K_i}$ and $\bar{\mathbf{T}}^{K_i}$ are independent, for $i = 1, \ldots, \nu$. Similarly, Let $\bar{\mathbf{Z}}^{\mathbf{K}} = (\bar{\mathbf{Z}}^{K_i}, i = 1 \ldots, \nu)$, such that $\bar{\mathbf{Z}}^{K_i} =_{st} \mathbf{Z}^{K_i}$ and $\bar{\mathbf{Z}}^{K_i}$ are independent, for $i = 1, \ldots, \nu$. Let $\bar{\Lambda}^{\mathbf{K}}$ be the tree of $\bar{\mathbf{Z}}^{\mathbf{K}}$. Then, if $\Lambda \geq_{g_\downarrow(g_\uparrow)} \bar{\Lambda}^{\mathbf{K}}$, the following higher order setwise bounds hold for any product L:

$$P(\mathbf{R}^L \geq (\leq) \mathbf{r}^L) \geq \prod_{i=1}^{\nu} P(\mathbf{R}^{K_i \cap L} \geq (\leq) \mathbf{r}^{K_i \cap L}). \qquad (11.67)$$

Evidently, a higher-order setwise bound is tighter than the first-order setwise bound. The tradeoff is that one must evaluate several higher-dimensional joint distribution functions. Song et al. [26] obtain a matrix-geometric solution of an $\{M/M/1/c, i \in E\}$ system. Their solution procedure is especially efficient for small-sized systems, say $|K_i| = 2$. Equation (11.67), coupled with the matrix geometric solutions of the subsystems, provides a high-quality bound for the due-date fill rate (Dayanik et al. [8]).

Next we consider the bounds for the expected product delivery lead-time, $E[\mathbf{R}_{\max}^K]$, assuming that the product is filled when each batch constituting the product is filled. Since

$$E[\mathbf{R}_{\max}^K] = \int_0^\infty P(R_{\max}^K \geq x)\, dx = \int_0^\infty \left[1 - P(R_{\max}^K \leq x)\right] dx,$$

from (11.61) and (11.62), we obtain the sufficient condition under which $E[\mathbf{R}_{\max}^K]$ admits the first-order setwise *upper* bound:

$$\Lambda \geq_{g_\uparrow} \tilde{\Lambda} \text{ (i.e., } \mathbf{Z} \text{ is PLOD) and } \mathbf{T} \geq_{uo} \tilde{\mathbf{T}} \text{(i.e., } \mathbf{T} \text{ is PUOD)}$$

$$\Longrightarrow E[\mathbf{R}_{\max}^K] \leq \int_0^\infty \left[1 - \prod_{i \in K} P(R_i \leq x)\right] dx. \tag{11.68}$$

Similarly, we obtain the sufficient condition under which $E[\mathbf{R}_{\min}^K]$, the expected time until the first batch delivery, admits the first-order setwise *lower bound*:

$$\Lambda \geq_{g_\downarrow} \tilde{\Lambda} \text{ and } \mathbf{T} \leq_{lo} \tilde{\mathbf{T}} \Longrightarrow E[\mathbf{R}_{\min}^K] \geq \int_0^\infty \prod_{i \in K} P(R_i \geq x)\, dx. \tag{11.69}$$

Hence the expected delivery leadtime range has the upper bound:

$$\Lambda \geq_{g_\downarrow} (\leq_{g_\uparrow}) \tilde{\Lambda} \text{ and } \mathbf{T} \leq_{lo} (\geq_{uo}) \tilde{\mathbf{T}} \Longrightarrow E[\mathbf{R}_{\max}^K - \mathbf{R}_{\min}^K]$$

$$\leq \int_0^\infty \left[1 - \prod_{i \in K} P(R_i \leq x) - \prod_{i \in K} P(R_i \geq x)\right] dx. \tag{11.70}$$

Example 11.6 (c) We wish to compute the upper bound of $E[R_{\max}^{12}]$ and the upper bound of $E[R_{\max}^{12}] - E[R_{\min}^{12}]$, using the data given in Example 11.6 (b). Note that for a two-item system, $E[R_{\max}^{12}]$ satisfies

$$E[R_{\max}^{12}] = E[R_1] + E[R_2] - E[R_{\min}^{12}].$$

From (11.69),

$$E[R_{\min}^{12}] = \int_0^\infty P(R_1 \geq x, R_2 \geq x)\, dx$$

$$\geq \int_0^\infty P(R_1 \geq x)P(R_2 \geq x)\, dx$$

$$= \rho_1^{s_1} \rho_2^{s_2} \int_0^\infty e^{-(\mu_1+\mu_2-\lambda_1-\lambda_2)x}\, dx = \frac{\rho_1^{s_1}\rho_2^{s_2}}{\mu_1+\mu_2-\lambda_1-\lambda_2}$$

Hence,

$$E[R_{\max}^{12}] \leq \frac{\rho_1^{s_1}}{\mu_1-\lambda_1} + \frac{\rho_1^{s_1}}{\mu_2-\lambda_2} - \frac{\rho_1^{s_1}\rho_2^{s_2}}{\mu_1+\mu_2-\lambda_1-\lambda_2}. \qquad (11.71)$$

The expected range of the delivery time of product 12 is bounded above by

$$E[R_{\max}^{12}] - E[R_{\min}^{12}] = E[R_1] + E[R_2] - 2E[R_{\min}^{12}]$$

$$\leq \frac{\rho_1^{s_1}}{\mu_1-\lambda_1} + \frac{\rho_1^{s_1}}{\mu_2-\lambda_2} - \frac{2\rho_1^{s_1}\rho_2^{s_2}}{\mu_1+\mu_2-\lambda_1-\lambda_2}.$$

In Section 7.2 we evaluate the upper bounds via numerical examples. ∎

7.2.   Numerical examples

In this section we illustrate, via numerical examples, how to apply the techniques introduced in this chapter to derive performance bounds for various systems. We shall focus on the $\{M/M/1/c_i, i \in E\}$ system, because the exact solution of the system can be derived via matrix geometric solution and will be used to benchmark the quality of setwise bounds. We first study a simple two-item system and obtain first-order setwise lower bounds for the product-based and system-based due-date fill rates. We also obtain first-order setwise upper bounds for the expected product-based and system-based delivery leadtimes. We then examine a three-item system and develop the first-order and second-order setwise bounds for instantaneous fill rates. Finally, We report the numerical result for a six-item system. Some numerical results are taken from Dayanik et al. [8].

7.2.1   A two-item ATO system.   We assume that $\mathbf{T} = (T, T)$ and T is exponentially distributed with rate $\lambda = 9$. Other parameter sets that determine system performance include:

  1 The distribution of product-type indicator vector: We choose two distributions for the product-type indicator vector:

	P^{12}	P^1	P^2	$P^\emptyset$
Λ_1 (HC)	$\frac{8}{18}$	$\frac{1}{18}$	$\frac{1}{18}$	$\frac{8}{18}$
Λ_2 (LC)	$\frac{2}{18}$	$\frac{7}{18}$	$\frac{7}{18}$	$\frac{2}{18}$

From (11.19), pairwise transform $g^{\frac{6}{18},1,2}$ transforms tree Λ_1 to tree Λ_2. An immediate consequence of this result is that, with other things being equal, the due-date fill rate of product 12 with tree Λ_1 is greater than that with tree Λ_2. In addition, product 12 admits the first order setwise lower bound.

2 Traffic intensity ρ: We assume that the production time in facility i is exponentially distributed with rate μ_i, $i = 1, 2$. Let $\rho_i = \lambda_i/\mu_i$. We choose production rates (μ_1, μ_2) such that $\rho = (\rho_1, \rho_2)$ take on values $(0.5, 0.5)$, $(0.9, 0.5)$ and $(0.9, 0.9)$, corresponding to the systems with symmetric and moderate workload, asymmetric workload, and symmetric and heavy workload, respectively.

3 Base-stock vector **s** and buffer capacities **c**: We select $(\mathbf{s}, \mathbf{c}) = (4, 4, 8, 8)$. For $\rho = (0.5, 0.5)$, the probability of lost sales is negligible and the system behaves like an $\{M/M/1, i = 1, 2\}$ system.

Figure 11.1 (a)–(c) depict the due-date file rate distributions of product 12, $P(R_1 \leq r, R_2 < r)$, and their first-order setwise lower-bounds. Product-based lower bounds are computed using (11.62). Figure 11.1 (d)–(e) depict the system-based due-date file rates and their first-order setwise lower-bounds. System-based lower bounds are computed using (11.64).

In Figure 11.1, LC means low demand correlation and HC means high demand correlation. As expected, the bounds are very tight with low demand correlation (with tree Λ_2). The bounds also perform remarkably well with moderate traffic intensity ($\rho = (0.5, 0.5)$), regardless of the degree of demand correlation. As seen in Figure 11.1 (c) and (f), the bounds deteriorate with high demand correlation, high traffic intensity and low due-date fill rates. However, notice that the tightness of the bounds improves as the due-date fill rate increase, and, when due-date fill rates are at least 0.9, the bounds are very close to their exact values under all circumstances.

We also computed the delivery leadtime of product 12, $E[R_{\max}^{12}]$, and its upper bound, under the following parameter settings: Besides the two distributions of the product-type indicator given above, we add another distribution with moderate demand correlation,

$$(P^{12}, P^1, P^2, P^\emptyset) = (0.25, 0.25, 0.5, 0).$$

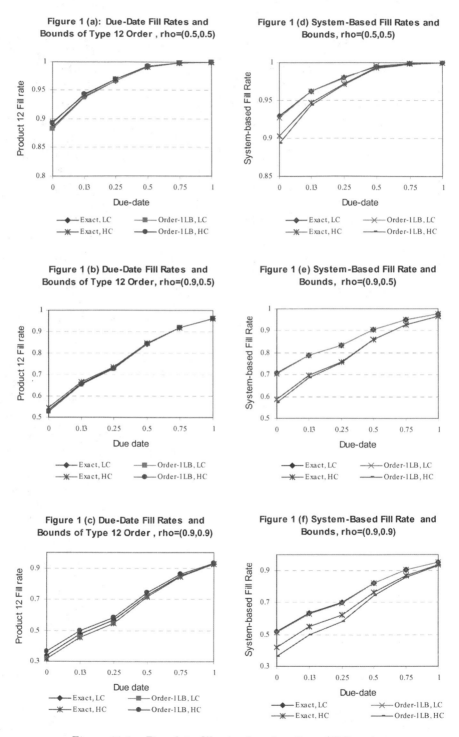

Figure 11.1. Due-date fill rates in a two-item ATO system

The traffic intensity is set to $\rho = (0.5, 0.5)$ to ensure high instantaneous fill rates. Two configurations for base-stock levels and buffer capacities are chosen: $(\mathbf{s}, \mathbf{c}) = (4, 4, 8, 8)$ and $(5, 3, 10, 6)$. Numerical results are reported in Figure 11.2. Figures 1.2(a) and 1.2(b) show $E[R_{\max}^{12}]$ and its setwise upper bound, under $(\mathbf{s}, \mathbf{c}) = (4, 4, 8, 8)$ and $(5, 3, 10, 6)$, respectively. The value of $E[R_{\max}^{12}]$ is evaluated using the matrix geometric solution and its upper bound is computed using (11.71). Observe that as P^{12} increases, $E[R_{\max}^{12}]$ decreases, confirming our analytical result given by (11.42). Observe also that the changes of $E[R_{\max}^{12}]$ corresponding to different levels of demand correlation are very moderate, again confirming our belief that, with moderate traffic intensity, demand correlation

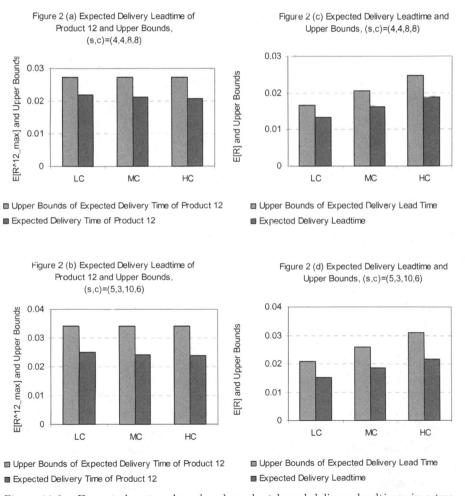

Figure 11.2. Expected system-based and product-based delivery leadtimes in a two-item ATO system

does not have the pronounced impact on a product-based performance measure. Figures 11.2 (c) and 11.2 (d) provide the expected delivery time of an arbitrary product, denoted by $E[R]$, and its upper bound. Note that as P^{12} increases, the system-based delivery leadtime and its bound also increase. This means that demand correlation degrades the system-based performance. Overall, it appears that these upper bounds are of good quality.

7.2.2   A three-item ATO system.   The purpose of this example is to illustrate how to compute the second-order setwise lower bound. We again assume $\mathbf{T} = (T, T, T)$, where T is exponentially distributed with rate $\lambda = 8.1$. Other parameter sets are chosen as follows:

1 Distributions of product-type vectors: The following distributions are selected, for $i = 0, 1, 2, \ldots, 5$,

$$\Lambda_i = \left(P_i^{123}, P_i^1, P_i^2, P_i^3, P_i^\emptyset\right) = \left(\frac{5-i}{15}, \frac{i}{15}, \frac{i}{15}, \frac{i}{15}, \frac{10-2i}{15}\right).$$

Notice that the above distributions have identical marginals. The distributions are chosen in order to examine the impact of demand correlation on system performances.

2 Traffic intensity ρ: We again assume that the production time in facility-i is exponentially distributed with rate μ_i, $i = 1, 2, 3$. The production rates (μ_1, μ_2, μ_3) are selected such that $\rho = (\rho_1, \rho_2, \rho_3) = (0.75, 0.75, 0.75)$ for the symmetric case and $\rho = (0.75, 0.8, 0.9)$ for the asymmetric case, respectively.

3 Base-stock levels: We select the base-stock and buffer capacity vectors as $(\mathbf{s}, \mathbf{c}) = (4, 4, 4, 7, 7, 7)$.

For convenience, we call the system with tree Λ_i System-i. Observe that Λ_{i+1} can be reached from Λ_i by applying sequentially the pairwise transforms $g^{\frac{1}{15}, 12, 3}$ and $g^{\frac{1}{15}, 1, 2}$. This implies $Z_i \geq_{sm} Z_{i+1}$, $i = 0, \ldots, 4$, where $\mathbf{Z}_i$ is the product-type vector corresponding to Λ_i. It further implies that, with other things being equal, the due-date fill rate of product 123 in System-i is higher than its counterpart in System-$(i+1)$, $i = 0, \ldots, 4$. Since the arrival process is Poisson and in System-5 each product requests a single item, the product of its marginal fill rates serves as the first-order setwise lower bound of product 123 for Systems-i, $i = 0, \ldots, 4$.

The second-order setwise lower bounds for product 123 can be computed as

$$P\left(\mathbf{R}_i^{123} \leq \mathbf{r}^{123}\right) \geq P\left(\mathbf{R}_i^{12} \leq \mathbf{r}^{12}\right) P(R_3 \leq r_3)$$

where $P(\mathbf{R}_i^{12} \le \mathbf{r}^{12})$ is evaluated using the matrix geometric solution, with product-type vector distributed as

$$\left(\bar{P}_i^{12}, \bar{P}_i^1, \bar{P}_i^2, \bar{P}_i^{\emptyset}\right) = \left(\frac{5-i}{15}, \frac{i}{15}, \frac{i}{15}, \frac{10-i}{15}\right).$$

Finally, the system-based due-date fill rate can be derived using the expression, for $i = 0, \dots, 4$,

Due-date fill rate of an arbitrary product in System-i

$$= \sum_{j=1}^{3} \frac{i}{15} P(R_i \le r_i) + \frac{5-i}{15} P(\mathbf{R}_i^{12} \le \mathbf{r}^{12}) P(R_3 \le r_3).$$

For simplicity, we shall only report the results of our experiments for system-based instantaneous fill rates ($r_i = 0$, $i = 1, 2, 3$). Figures 11.3(a) and 1.3(b) report system-based instantaneous fill rates and their first and second order setwise bounds for the systems with symmetric and asymmetric traffic intensities, respectively. In the graphs, Order-i LB, $i = 1, 2$, represents the first-order and second-order setwise lower bounds, respectively. We observe that demand correlation has pronounced impact on the system-based due-date fill rate. We also observe that as demand correlation reduces, the quality of the bound improves. From the graphs, it appears that the second-order setwise lower bound is desirable when both demand correlation and traffic intensity are high.

7.2.3   A six-item ATO system.   This example reports the average percentage errors of setwise bounds for a medium-sized problem. The study is based on the PC-Example appeared in Song [24]. It is assumed that there are six items that play a key role in differentiating

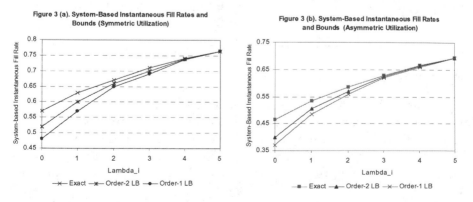

Figure 11.3. System-based instantaneous due-date fill rates in a three-item ATO system

major demand types: (1) interior tape backup; (2) standard hard drive; (3) high profile hard drive; (4) video memory card; (5) Pentium processor; and (6) Pentium-processor. There are six major demand types, 25,35,125, 136,1345 and 1346, resulting from different choices and combinations of those items. We assume that interarrival times of overall order types are i.i.d. exponential random variable with rate $\lambda = 2$. Other parameter sets are chosen as follows:

1 Distribution of product-type vector **Z**: We assume

$$\Lambda = (P^{1346}, P^{1345}, P^{136}, P^{125}, P^{35}, P^{25})$$
$$= (0.05, 0.2, 0.1, 0.15, 0.4, 0.1).$$

2 Traffic intensity ρ: The production time in facility i is exponentially distributed with rate μ_i, $i = 1, 2, \ldots, 6$. We take runs under two levels of traffic intensity: High traffic intensity with $\rho_i = 0.9$, $i = 1, \ldots, 6$ and low traffic intensity $\rho_i = 0.5$, $i = 1, 2, \ldots, 6$.

3 Base-stock and buffer capacity vectors (**s**, **c**): We consider two cases: The partial backlog case with base stock levels $s_i = 4$ and buffer capacities $N_i = 7$, $i = 1, 2, \ldots, 6$; and the lost sales case with base-stock levels $s_i = 7$ and buffer capacities $N_i = 7$, $i = 1, \ldots, 6$.

The following table reports the average percentage errors of setwise bounds for the system-based instantaneous fill rates, under high and low traffic intensities. Overall, both the first and second order setwise lower bounds perform satisfactorily. With high traffic intensity, the second-order lower bound noticeably improves the first-order setwise lower bound.

Table 11.1. Average percentage errors of system-based instantaneous fill rates

	Low Traffic Intensity	High Traffic Intensity
Order-1 Lower Bound	1.02%	10.00%
Order-2 Lower Bound	0.49%	3.25%

8.   Summary

The assemble-to-order strategy has been adopted widely in practice as a competitive weapon to increase customer order responsiveness, delay the point of product-differentiation and prevent product obsolescence. However, performance evaluation of ATO systems is analytically challenging. In this chapter we show that dependence analysis is an indispensable tool in the study of ATO systems.

We present a unified framework to study dependence properties of several prototypical assemble-to-order systems, based on the notions of majorizations of weighted trees and separable Markov chains with independent generating sequences. We show that certain dependence concepts, such as supermodular and orthant orders, are very effective in examining dependence structure of a wide variety of ATO systems. We demonstrate how dependence structure of a performance vector varies in response to changes of input parameters and illustrate that both positive and negative dependencies can favorably and adversely affect system performance. We also derive several easily computable bounds for key product-based and system-based performance measures. Numerical examples show that these bounds are of good quality. Our approach can also be used to analyze other types of ATO systems under base-stock control.

We believe that dependence analysis can also be used to address other issues in ATO systems. One interesting problem is to obtain conditions so that two system-based due-date fill rates can be compared. Such a comparison is interesting because, for a large ATO system with many products, it becomes inevitable to use a single proxy, such as the system-based due-date fill-rate (as oppose to product-based measures), to gauge the performance of the system as a whole. In addition, this line of study will reveal when it is safe to ignore demand correlation and what is the value to identify product types.

The ATO system under other control policies, such as the (s, S) or (r, nQ) policy, defies the exact solution and efficient computational procedures even more so than the ATO system under base-stock control. We believe that dependence analysis can also be used to gain qualitative and quantitative insights of these problems.

9. Appendices

9.1. Appendix A: closure properties of orthant and supermodular order

This appendix summarizes some closure properties of orthant and supermodular orders. The proofs of the closure properties of orthant orders can be found in Shaked and Shanthikumar [22]. The proofs of the closure properties of the supermodular order can be found in Li and Xu [16].

Proposition 11.18 Let $\mathbf{X}$ and $\bar{\mathbf{X}}$ be two m-dimensional random vectors.

1 The supermodular and orthant orders are closed under marginalization. That is, if $\mathbf{X} \geq_{sm} (\geq_{uo}, \leq_{lo}) \bar{\mathbf{X}}$, then for any $K \subseteq E$,

$$\mathbf{X}^K \geq_{sm} (\geq_{uo}, \leq_{lo}) \bar{\mathbf{X}}^K \quad K \subseteq E. \tag{11.72}$$

2 The supermodular order (orthant orders) is (are) closed under monotone (increasing) transformation. That is, if $\mathbf{X} \geq_{sm} (\geq_{uo}, \leq_{lo}) \bar{\mathbf{X}}$, and if $f_i : \mathcal{R} \to \mathcal{R}$, $i \in E$, are monotone in the same direction (increasing), then

$$(f_i(X_i), i \in E) \geq_{sm} (\geq_{uo}, \leq_{lo}) (f_i(\bar{X}_i), i \in E). \tag{11.73}$$

3 Let $\mathbf{Y}$ and $\bar{\mathbf{Y}}$ be another two m-dimensional random vectors such that $\mathbf{X}$ and $\mathbf{Y}$ are independent and $\bar{\mathbf{X}}$ and $\bar{\mathbf{Y}}$ are independent. If $\mathbf{X} \geq_{sm} (\geq_{uo}, \leq_{lo}) \bar{\mathbf{X}}$, $\mathbf{Y} \geq_{sm} (\geq_{uo}, \leq_{lo}) \bar{\mathbf{Y}}$ and if $f_i : \mathcal{R}^2 \to \mathcal{R}$, $i \in E$, are monotone in the same direction (increasing), then

$$(f_i(X_i, Y_i), i \in E) \geq_{sm} (\geq_{uo}, \leq_{lo}) (f_i(\bar{X}_i, \bar{Y}_i), i \in E). \tag{11.74}$$

9.2.   Appendix B: Proof of Theorem 11.9

We prove (11.32) first. Because $Z_{i,n}D_{i,n}$ is increasing in $Z_{i,n}$ for fixed $D_{i,n}$. By (11.19), (11.23) and Proposition 11.18 (2),

$$\Lambda \geq_{g^2} (\geq_{g_\downarrow}, \geq_{g_\uparrow}) \bar{\Lambda} \implies \mathbf{Z}_n \mathbf{D}_n \geq_{sm} (\geq_{uo}, \leq_{lo}) \bar{\mathbf{Z}}_n \mathbf{D}_n,$$
$$Z_{i,n}D_{i,n} =_{st} \bar{Z}_{i,n}D_{i,n}, \quad i \in E. \tag{11.75}$$

Now, $Y_{i,n}$ given by (11.30) is increasing in $T_{i,n+1}$ for fixed $S_{i,j}$, $i \in E$, $j \geq 1$. Again from Proposition 11.18 (2),

$$\mathbf{T}_n \geq_{sm} (\leq_{lo}, \geq_{uo}) \bar{\mathbf{T}}_n,$$
$$\implies \mathbf{Y}_n \geq_{sm} (\leq_{lo}, \geq_{uo}) \bar{\mathbf{Y}}_n, \quad Y_{i,n} =_{st} \bar{Y}_{i,n},$$
$$\implies -\mathbf{Y}_n \geq_{sm} (\geq_{uo}, \leq_{lo}) -\bar{\mathbf{Y}}_n, \quad -Y_{i,n} =_{st} -\bar{Y}_{i,n}. \tag{11.76}$$

Equations (11.74) and (11.75), in conjunction with (11.31), imply that $\mathbf{N}_n^a$ and $\bar{\mathbf{N}}_n^a$ have identical univariate marginals. Finally, because the function defined by (11.31) is increasing in $Z_{i,n}D_{i,n}$ and $-Y_{i,n}$, (11.32) follows from (11.74), (11.75) and Proposition 11.8.

Under stability conditions $\lambda_i < \mu_i$, $i \in E$, Markov chains $\{\mathbf{N}_n^a, n \geq 1\}$ and $\{\bar{\mathbf{N}}_n^a, n \geq 1\}$ converge in distribution. Then, (11.33) holds because supermodular and orthant orders are preserved under weak convergence (Muëller and Scarsini [18]).

To prove (11.34), note that for each $i \in E$, R_i given by (11.29) is increasing in N_i^a, with the values of other components considered fixed. That is, $\mathbf{R}$ is an increasing transformation of $\mathbf{N}^a$. It is also easily seen that $\mathbf{R}$ and $\bar{\mathbf{R}}$ have identical univariate marginals. Finally, (11.34) follows from Proposition 11.18 (3).

9.3. Appendix C: Proof of Theorem 11.12

Let $V_{i,n}$ be the workload found by the nth item-i demand. It is well-known that $V_{i,n}$ satisfies the Lindley recursion:

$$V_{i,n+1} = [V_{i,n} + Z_{i,n}S_{i,n} - T_{i,n+1}]^+, \quad i \in E, \ n \geq 1. \tag{11.77}$$

Then $\{\mathbf{V}_n, n \geq 0\}$ is a separable Markov chain generated by the independent sequences $\mathcal{ZS}$ and $-\mathcal{T}$. Thus,

$$\mathbf{Z}_n\mathbf{S}_n \geq_{sm} (\geq_{uo}, \leq_{lo}) \ \bar{\mathbf{Z}}_n\mathbf{S}_n \quad \text{and} \quad -\mathbf{T}_{n+1} \geq_{sm} (\geq_{uo}, \leq_{lo}) \ -\bar{\mathbf{T}}_{n+1}.$$

By (11.77), $\mathbf{V}_{n+1}$ is an increasing transformation of $\mathbf{Z}_n\mathbf{S}_n$ and $-\mathbf{T}_{n+1}$. From Proposition 11.8,

$$\{\mathbf{V}_n, n \geq 1\} \geq_{sm} (\geq_{uo}, \leq_{lo})\{\bar{\mathbf{V}}_n, n \geq 1\}, \quad \text{and} \quad V_{i,n} =_{st} \bar{V}_{i,n}, \ i \in E, n \leq 1. \tag{11.78}$$

The manufacturing leadtime of an item-i demand thus satisfies

$$L_{i,n} = V_{i,n} + S_{i,n}, \quad i \in E, n \geq 1,$$

where $\mathbf{V}_n$ and $\mathbf{S}_n$ are independent. Then the above equation and (11.77) imply (11.47). Expression (11.48) holds because supermodular and orthant orders are closed under weak convergence. To prove (11.49), we use (11.45) and (11.46). Let $\mathbf{J} = (J_1, \ldots, J_m)$. For any fixed $\mathbf{j} = (j_1, \ldots, j_m)$,

$$P(\mathbf{J} \geq (\leq) \mathbf{j}) = P\left(\sum_{n=1}^{j_i} Z_{i,n} \leq (\geq)s_i, \ i \in E\right). \tag{11.79}$$

Li and Xu [16] have showed that if $\Lambda \geq_{g_\downarrow} (\geq_{g_\uparrow})\bar{\Lambda}$, then

$$\left\{\sum_{n=1}^{j_i} Z_{i,n}, i \in E\right\} \geq_{uo} (\leq_{lo})\left\{\sum_{n=1}^{j_i} \bar{Z}_{i,n}, i \in E\right\}, \ \mathbf{j} \in \mathcal{R}_+^m. \tag{11.80}$$

From (11.45), $\mathbf{J}$ and $\bar{\mathbf{J}}$ must have identical marginals. This, together with (11.79) and (11.80), gives us

$$\mathbf{J} \leq_{lo} (\geq_{uo})\bar{\mathbf{J}} \quad \text{and} \quad J_i = \bar{J}_i, \ i \in E. \tag{11.81}$$

Because $\mathbf{T} \leq_{lo} (\geq_{uo})\bar{\mathbf{T}}$,

$$-\left(\sum_{n=1}^{J_i} T_n, i \in E\right) \geq_{uo} (\leq_{lo}) - \left(\sum_{n=1}^{\bar{J}_i} T_n, i \in E\right). \tag{11.82}$$

Now, (11.46) means that $\mathbf{R}$ is an increasing transformation of two independent random vectors $\mathbf{L}$ and $-(\sum_{n=1}^{J_i} T_n, i \in E)$. Hence we conclude (11.49) from Proposition 11.18 (3).

Acknowledgments

Some results presented in this chapter were developed by the author in the collaboration with Professor Haijun Li of Washington State University. The materials of setwise bounds and some numerical examples are extracted from the author's joint work with Mr. Savas Dayanik of Columbia University and Professor Jing-Sheng Song of University of California at Irvine. The author thanks the anonymous referees for their very constructive comments. The author is supported in part by the NSF Grant DMI9812994 for the project, "A Study of Synchronized Stochastic System with Applications to Manufacturing, Communication and Reliability Systems."

References

[1] ANUPINDI, R. AND S. TAYUR. (1998) Managing Stochastic Multi-Product Systems: Model, Measures, and Analysis. *Oper. Res.*, Vol. 46, S98–S111.

[2] ASXÄTER, S. (2000) *Inventory Control*. Kaluwer Academic Publishers.

[3] BACCELLI, F. AND A.M. MAKOWSKI. (1989) Multidimensional Stochastic Ordering and Associated Random Variables. *Oper. Res.*, Vol. 37, 478–487.

[4] BACCELLI, F., A.M. MAKOWSKI, AND A. SCHWARTZ. (1989) The Fork-Join Queue and Related Systems with Synchronization Constraints: Stochastic Ordering and Computable Bounds, *Adv. Appl. Prob.*, Vol. 21, 629–660.

[5] CHANG, C.S. (1992) A New Ordering for Stochastic Majorization: Theory and Applications. *Adv. Appl. Prob.*, Vol. 24, 604–634.

[6] CHANG, C.S. AND D.D. YAO. (1993) Rearrangement, Majorization and Stochastic Scheduling. *Math. Oper. Res.*, 658–684.

[7] CHEUNG, C.S. AND W. HAUSMAND. (1995) Multiple Failures in a Multi-Item Spare Inventory Model, *IIE Transactions*, Vol. 27, 171–180.

[8] DAYANIK, S., J.S. SONG, AND S.H. XU. (2000) Stochastic Properties and Performance Bounds for an Assemble-to-Order System, in preparation.

[9] DISNEY, R.L., D.C. MCNICKLE, S. HUR, P.R. DE MORAIS, AND R. SZEKLI. (1997). A Queue with Correlated Arrivals, preprint.

[10] ESARY, J.D., F. PROSCHAN, AND D.W. WALKUP. (1967) Association of Random Variables, with Applications, *Ann. Math. Statist.*, Vol. 38, 1466–1474.

[11] GALLIEN, J. AND L.M. WEIN. (1998) A Simple and Effective Procurement Policy for Stochastic Assembly Systems. Working paper, MIT Sloan School.

[12] GLASSERMAN, P. AND Y. WANG. (1998) Leadtime-Inventory Trade-Offs in Assemble-to-Order Systems. *Oper. Res.*, Vol. 46, 858–871.

[13] HAUSMAN, W.H., H.L. LEE, AND A.X. ZHANG. (1998) Order Response Time Reliability in a Multi-item Inventory System. *Euro. J Oper. Res.*, Vol. 109, 646–659.

[14] JOE, H. (1997) *Multivariate Models and Dependence Concepts.* Chapman & Hall.

[15] LEE, H.L. AND C.S. TANG. (1997) Modelling the Costs and Benefits of Delayed Product Differentiation. *Mgmt Sci.*, Vol. 43, 40–53.

[16] LI, H. AND S.H. XU. (2000) On the Dependence Structure and Bounds of Corrrelated Parallel Queues and Their Applications to Synchronized Stochastic Systems. *J. of Appl. Prob.*, Vol. 37, No. 4, 1020–1043.

[17] LI, H. AND S.H. XU. (2001) Stochastic Bounds and Dependence Properties of Survival Times in a Multicomponent Shock Model. *J. of Multivariate Anal.*, Vol. 76, 63–89.

[18] MUËLLER, A. AND M. SCARSINI. (2000) Some Remarks on the Supermodular Order. *J. of Multivariate Anal.*, Vol. 73, 107–119.

[19] NELSON, R. AND A.N. TANTAWI. (1988) Approximate Analysis of Fork/Join Synchronization in Parallel Systems. *IEEE Trans. Comput*, Vol. 37, 739–743.

[20] ROLSKI, T. (1983) Upper Bounds for Single Server Queues with Doubly Stochastic Poisson Arrivals. *Math. of Oper. Res.*, Vol 11, 442–450.

[21] RÜSCHENDORF, L. (1981) Inequalities for the Expectations of Δ-Monotone Functions. *Zeit. Wshr. und Ver. Geb.*, Vol. 57, 349–354.

[22] SHAKED, M. AND J.G. SHANTHIKUMAR. (1994) *Stochastic Orders and Their Applications.* Academic Press, New York, NY.

[23] SHAKED, M. AND J.G. SHANTHIKUMAR. (1997) Supermodular Stochastic Order and Positive Dependence of Random Vectors, *J. Multivariate Analysis*, Vol. 61, 86–101.

[24] SONG, J.S. (1998) On the Order Fill Rate in a Multi-Item, Base-Stock Inventory System. *Oper. Res.*, Vol. 46, 831–845.

[25] SONG, J.S. (1999) Evaluation of Order-Based Backsliders, preprint.

[26] SONG, J.S., S. XU AND B. LIU. (1999) Order-Fulfillment Performance Measures in an Assemble-to-Order System with Stochastic Leadtimes. *Oper. Res.*, Vol. 47, 131–149.

[27] SONG, J.S. AND D.D. YAO. (2000) Performance Analysis and Optimization of Assemble-to-Order Systems with Random Leadtimes. *Oper. Res.* (to appear).

[28] STOYAN, D. (1983) *Comparison Methods for Queues and Other Stochastic Models.* Wiley & Sons.

[29] SZEKLI, R. (1995) *Stochastic Ordering and Dependence in Applied Probability.* Springer-Verlag.

[30] TCHEN, A.H. (1980) Inequalities for distributions with given marginals, *Ann. Probab.*, Vol. 8, 814–827.

[31] TONG, Y.L. (1980) *Probability Inequalities in Multivariate Distributions.* Academic Press, New York.

[32] WANG, Y. (1999) Near-Optimal Base-Stock Policies in Assemble-to-Order Systems under Service Levels Requirements. Working paper, MIT Sloan School.

[33] WOLFF, R.W. (1989) *Stochastic Modelling and the Theory of Queues.* Prentice Hall, Englewood Cliffs.

[34] XU, S.H. (1999) Structural Analysis of a Queueing System with Multi-Calsses of Correlated Arrivals and Blocking. *Oper. Res.*, Vol. 47, 263–276.

[35] XU, S.H. AND H. LI. (2000) Majorization of Weighted Trees — A New Tool to Study Correlated Stochastic Systems. *Math of Oper. Res.*, Vol. 25, 298–323.

[36] ZHANG, R. (1998) Expected Time Delay in a Multi-Item Production-Inventory System with Correlated Demand. Working paper, Department of Industrial and Operations Engineering, University of Michigan, Ann Arbor, MI 48109.

[37] ZIPKIN, P.H. (2000) *Foundations of Inventory Management.* Irwin, McGraw Hill.

Chapter 12

INVENTORY POLICIES FOR SEQUENCES OF MULTI-ITEM DEMANDS WITH NO BACKORDERS PERMITTED

John W. Mamer
Anderson Graduate School of Management
University of California, Los Angeles
Los Angeles, CA 90095
jmamer@anderson.ucla.edu

Stephen A. Smith
Department of Decision and Information Sciences
Leavey School of Business, Santa Clara University
Santa Clara, CA 95053
ssmith@scu.edu

1.   Introduction

We consider a multi-item inventory system of the following sort. Demands arrive randomly in the form of requests for subsets of the multiple items. A demand is met only when its entire subset of items is available. The key difference between this system and many typical inventory models is that it is the number of demands met (or not met) that is important, rather than the number of items supplied (or short).

There are different inventory models for job completion systems, each of which is appropriate for a different class of applications. The classic tool kit problem is concerned with the expected fraction of the arriving repair jobs for which it is adequate or the expected number of jobs that can be completed before an unstocked tool or part is required. The decision variables in the tool kit problem specify only whether or not an item is stocked, and do not consider inventory levels. This is based on the assumption that tools are not consumed and the other items that are in the kit will be stocked with a very high availability. Examples of the tool kit problem arise in the repair of machines in the field, e.g., the repair of computer terminals, photocopiers or other equipment at customer sites.

When the item inventory levels change over time based on the sequence of arriving jobs, there are again different inventory models to consider, depending on whether backorders are permitted. If backorders are permitted, the performance of the system can be based on the expected number of backorders in steady state or the faction of jobs that are completed on time. This is appropriate for purchase order filling systems and assemble-to-order systems which can put incomplete jobs aside for later completion.

If no backorders are permitted, the inventory system has a "mission completion" objective in which it is the expected time until stockout or, alternatively, the probability that there is no stockout until after a specified time that is important. Examples are voyages at sea, space flights or military missions. Assemble-to-order systems can also fall into this category if completing jobs out of their original sequence is prohibitively expensive. This may occur, for example, if the job sequence is matched to a particular shop floor schedule and material requirements plan and revising the sequence is very expensive. This chapter will consider this last version of the job completion inventory problem.

Job completion based inventory systems have been studied in the management science literature in a variety of contexts. Early papers all assumed that the demands for a set of parts arrived simultaneously, but were independent of each other, i.e., the parts for each arriving job were randomly and independently selected. Sherbrooke [14] and Silver [15] considered an inventory system in which a failed line replaceable unit (LRU) could be replaced only if all failed subassemblies were available as spares. Smith, Chambers and Shlifer [18] obtained a closed form optimal stocking condition for this type of problem in the one period case, if the item stock levels are restricted to be $0, 1$. Graves [5] reformulated this as a $0, 1$ knapsack problem, which allows for the inclusion of linear constraints, such as inventory space requirements. Hausman [6] developed an exchange curve analysis for this inventory stocking problem and Schaefer [13] integrated the independent demand job completion formulation into an overall inventory management system. Nahmias and Schmidt [12] developed an efficient heuristic procedure for a multi-item newsboy problem with a linear constraint.

The part independence assumption is likely to be violated in many repair kit, purchase order and assemble-to-order applications, because the arriving jobs typically correspond to predefined subsets of parts.

Subsequent papers on job completion inventory systems have worked on relaxing this assumption. Assuming that the set of parts required for each job type is prespecified, Mamer and Smith [8] developed an optimization algorithm, using the maximum flow/minimum cut algorithm, to solve the 0, 1 stock level, multi-item job fill case and in [9] extended this approach to include the use of spare replacement machines. March and Scudder [11] developed an algorithm for determining optimal 0, 1 stocking policies for this problem, subject to budget constraints. Brumelle and Granot [3] obtain a collection of very general monotone selection results for the 0, 1 job-fill stocking model. These results yield important qualitative sensitivity characterizations as well as enhanced algorithms for finding optimal policies.

Examples of multi-item inventories have also been studied in the context of production processes with parts commonality among products by Baker [2] and Baker, Magazine and Nuttle [1]. In Cohen, Kleindorfer and Lee [4], a dual method and greedy heuristic are developed for the special case in which exactly one of each part is required per job. These authors also sketch the extension of their formulation to the problem considered here and note the importance of finding simple heuristics for the general parts commonality problem.

The case in which backorders are permitted has been formulated and analyzed by Song ([16] and [17]). She considers the problem of calculating the order fill rate and number of back orders in a multi-item base stock inventory system, with multivariate compound Poisson demand and fixed lead times. She obtains computational procedures, suitable for problems of modest size, for calculating the order fill rate based on a decomposition of the Poisson arrival process. Our work is closest in spirit to the work of Song; however, we concentrate on different performance measures: the expected number of jobs and time until stock out. We do not assume a Poisson arrival process (which is critical for the computational procedures proposed by Song).

We consider a formulation in which item demands are dependent in an arbitrary manner, as in the more recent models discussed above. Stock levels are not restricted to 0 or 1 and multiple items of a given type are permitted on a particular job. We concentrate on the performance criteria relevant to the repair kit problem: the probability of completing a fixed number of jobs, the expected number of job encountered up to first stock out, the expected time until stock out and the probability of serving all jobs arriving within a fixed period of time. The formulation assumes a single echelon, one period inventory system. This structure is appropriate for field repair and mission completion problems and has been used as a simplifying approximation in previous parts commonality production inventory analyses. For the performance criteria involving a fixed time interval, this assumption is equivalent to a periodic resupply in which order levels can be readjusted up to the time of delivery.

We derive and compare a variety of upper and lower bounds for system performance. The tightness of the bounds varies by application. Thus, a set of upper and lower bounds is advantageous in that the maximum lower bound and minimum upper bound can be chosen to obtain the tightest bounds for any specific example.

Because of the complexity of the models considered in this chapter, closed form solutions for optimal inventory policies cannot be obtained. In fact, the system performance for a given inventory level can be evaluated exactly only for problems having a small number of different jobs and parts. System performance for a specific stocking policy can always be estimated by Monte Carlo methods, however. When job arrivals form a renewal process, upper and lower bounds on system performance are expressible as simple functions of the item stock levels. The simple forms of the bounds allow optimization, subject to budget constraints and other linear constraints, such as inventory space. Optimization using bounds is important for this inventory system because the exact performance criteria are difficult to compute. We investigate two examples to illustrate the relationships between the bounds and more precise calculations.

2.   The Model

2.1.   Specifications

The key demand data for the problem we consider are contained in the "job matrix" J, which is defined as follows:

$$J_{ij} = \text{the number of items of type } i \text{ required}$$
$$\text{to satisfy a request (complete a job) of type } j.$$

In production applications, the concept of a repair job is replaced by a final assembled product whose bill of materials specifies the items or sub-assemblies. This definition implies that J is made up of non negative integers, and every row and column contains at least one nonzero element, to rule out degenerate jobs or parts.

Assumption 1:

Jobs or requests for items are assumed to have random independent inter arrival times with mean $1/\lambda$. The marginal probabilities for job types are defined as follows:

$$p_j = P\{\text{an arriving job or request is of type } j\}.$$

The type of each job is assumed to be independent of the types of past and future jobs, and independent of the arrival process.

For any fixed time t, we define the random variables

$N_j(t)$ = number of requests (jobs) of type j arriving by time t, $j = 1, \ldots, m$

$n_j(k)$ = number of jobs of type j out of first k jobs

$X_i(t)$ = number of items of type i demanded by time t, $i = 1, \ldots, n$.

$x_i(k)$ = number of items of type i demanded in first k jobs

Letting

$$N(t) = (N_1(t), \ldots, N_m(t)), \qquad (12.1)$$
$$n(k) = (n_1(k), \ldots, n_m(k)),$$
$$X(t) = (X_1(t), \ldots, X_n(t)),$$
$$x(k) = (x_1(k), \ldots, x_n(k))$$

we have the matrix relationships

$$X(t) = JN(t) \quad \text{and} \quad x(k) = Jn(k). \qquad (12.2)$$

Let $\sum_j N_j(t) = M(t)$, $t \geq 0$ denote the number of jobs which have arrived by time t. Throughout this chapter we will assume that $M(t)$ is a renewal process. Since the job types are assigned in a stationary and independent fashion, the processes $N_j(t)$ $j = 1, \ldots, m$ are also renewal processes. The $N_j(t)$ processes are not, however, independent. Even when each job uses a single distinct part, the components of $X(t)$ may be correlated. The dependence among the components of $N(t)$, summarized by the covariance matrix, is given as follows:
For any $t \geq 0$, $1 \leq j_1 \leq m$, $1 \leq j_2 \leq m$,

$$Cov(N_{j_1}(t), N_{j_2}(t)) = p_{j_1} p_{j_2}(Var(M(t)) - E(M(t))) + p_{j_1} \mathbf{1}_{j_1 = j_2} E(M(t)). \qquad (12.3)$$

where $\mathbf{1}_{j_1 = j_2} = 1$ if $j_1 = j_2$ and $= 0$ otherwise. Denoting the covariance matrix of $N(t)$ by $cov(N(t))$ and the covariance matrix of $X(t)$ by $cov(X(t))$, it follows from (12.2) that $cov(X(t)) = J cov(N(t)) J^T$. This gives us a limited characterization of the dependence among the components of $X(t)$:
If $Var(M(t)) - E(M(t)) \geq 0$ then for all $1 \leq i_1 \leq n$, $1 \leq i_2 \leq n$.

$$Cov(X_{i_1}(t), X_{i_2}(t)) \geq 0.$$

2.2.   Defining performance measures

In this subsection, we derive expressions for performance measures as a function of the initial stock level. These are defined in terms of two random variables:

$$\tau = \text{time of the first stock-out}$$
$$\eta = \text{number of jobs arriving up to and including}$$
$$\text{stock-out.}$$

A "stock-out" occurs when a job arrives which cannot be repaired from the set of parts on hand.

The performance measures we shall analyze are:

- probability that stock-out occurs after time t $P\{\tau > t\}$

- expected time to stock-out $E(\tau)$

- probability that at least k jobs arrive until stock-out $P\{\eta > k\}$

- expected number of jobs arriving until stock-out $E(\eta)$

The most appropriate measure depends upon the application. For mission completion problems, $P\{\tau > t\}$ would be appropriate for a mission of given duration t, or $E(\tau)$ if the duration is uncertain. On the other hand, in field repair applications where each repair person is assigned a number of repair calls per shift, $E(\eta)$ or $P\{\eta > k\}$ would be appropriate. In the field repair setting, the occurrence of a stock-out for a single part forces the repairman to return to the warehouse for the needed part and to restock his kit at the same time. For the problem of repairing ships at sea, a stock-out in a single critical part entails a return to port or some form of emergency delivery.

For a given vector $s = s_1, \ldots, s_n$ of stock levels, where s_i is the initial stock level for item type i, the time to first stock-out is defined by

$$\tau = \inf\{t : X_i(t) \geq s_i + 1 \text{ for some } i, \ i = 1, \ldots, n\}.$$

That is, τ is the first time that the demand for some part or parts strictly exceeds supply. Thus

$$P\{\text{All requests filled up to time } t\} = P\{X(t) \leq s\} = P\{\tau > t\}.$$

When the precise time horizon is not specified, a reasonable performance criterion is the expected time to first stock-out, $E(\tau)$,

$$E(\tau) = \int_0^\infty P\{\tau > t\} \, dt.$$

Define η by

$$\eta = \min\{k : x_i(k) \geq s_i + 1, \text{ for some } 1 \leq i \leq n\}.$$

Let Y_i, $i = 1, \ldots$, denote the (random) times between arrivals of jobs, where $E(Y_i) = 1/\lambda$. The Y_i are independent, identically distributed, non-negative random variables with finite mean. Moreover, $\tau = \sum_{i=1}^{\eta} Y_i$ and, by assumption, η is independent of the Y_i. Hence

$$E(\tau) = E(Y_i)E(\eta) = E(\eta)/\lambda. \tag{12.4}$$

For fixed k, the joint probability distribution of $(n_1(k), \ldots, n_m(k))$ is multinomial with parameters k and $p_1, \ldots, p_m$. This follows from the assumption that the types of the arriving jobs are independent. The marginal probability distribution for $n_j(k)$ is binomial for each k

$$P\{n_j(k) = u\} = \binom{k}{u} p_j^u (1 - p_j)^{k-u}. \tag{12.5}$$

Thus,

$$P\{\eta > k\} = \sum_{z \in I_k \cap H} \frac{k!}{z_1! \ldots z_m!} p_1^{z_1} \cdots p_m^{z_m}, \tag{12.6}$$

where $I_k = \{z \mid \sum_i z_i = k, z_i \geq 0\}$ and $H = \{z \mid z \geq 0, Jz \leq s\}$. Now,

$$E(\eta) = \sum_{k=0}^{\infty} \sum_{z \in H \cap I_k} \frac{k!}{z_1! \ldots z_m!} p_1^{z_1} \cdots p_m^{z_m} = \sum_{z \in H} \frac{(\sum_{i=1}^{m} z_i)!}{z_1! \ldots z_m!} p_1^{z_1} \cdots p_m^{z_m}. \tag{12.7}$$

Equation (12.6) can be used to obtain $P\{\tau > t\}$ via the identity,

$$P\{\tau > t\} = \sum_{\ell=0}^{\infty} P\{\eta > \ell \mid M(t) = \ell\} P\{M(t) = \ell\}$$

$$= \sum_{\ell=0}^{\infty} P\{\eta > \ell\} P\{M(t) = \ell\}$$

where the last equality follows since the job types are independent of the job arrival process.

Equations (12.6) and (12.7) can, in principle, be used to calculate the performance criteria for a given stock level. The calculations, however, are quite time consuming even for fairly small numbers of items and stock levels.

2.3.   Bounds

In this section, we develop upper and lower bounds for the performance measures. These bounds are important for two reasons: (1) the exact performance measures are tedious to evaluate and for large problems the computation becomes infeasible (2) our bounds give some insight into the structure of this inventory problem and offer surrogate objective functions that are easily optimized.

We begin with upper and lower bounds on expected time to stock-out.

Proposition 12.1 *With only Assumption 1, we have the following upper bound on expected time to stock-out*

$$E(\tau) \le (1/\lambda)U \tag{12.8}$$

where

$$U = \min_{i} \left\{ \frac{s_i + \max\limits_{j}\{J_{ij}\}}{\sum_j J_{ij}p_j} \right\}. \tag{12.9}$$

Proof: Define

$$\eta_i = \inf\{k : x_i(k) \ge s_i + 1\},$$

corresponding to the job k on which the ith part "stocks out". Clearly,

$$E\eta = E\min\{\eta_1, \ldots, \eta_n\}.$$

It follows from Jensen's inequality that

$$E\eta \le \min\{E\eta_1, \ldots, E\eta_n\}.$$

The process $x_i(k)$ experiences an increment of J_{ij} with probability p_j. If we let $Z_1^{(i)}, Z_2^{(i)}, \ldots$ denote the sequence of increments, then we have

$$s_i + 1 \le \sum_{l=1}^{\eta_i} Z_l^{(i)} \le s_i + \max_{j}\{J_{ij}\},$$

where the term on the far right is the maximum amount by which $x_i(\eta)$ will exceed s_i. Taking expectations on both sides and applying Wald's equation yields

$$\frac{s_i + 1}{E(Z^{(i)})} \le E(\eta_i) \le \frac{s_i + \max_j(J_{ij})}{E(Z^{(i)})}.$$

where $E(Z_i) = \sum_{j=1}^m p_j J_{ij}$. This yields

$$E(\eta) \leq \min_i \left\{ \frac{s_i + \max_j \{J_{ij}\}}{\sum_j J_{ij} p_j} \right\} = U$$

Equation (12.8) now follows by an application of (12.4).

Q.E.D.

The minimum in (12.9) is simple to compute. Thus the right hand side of (12.8) is easily determined.

It is often more important to obtain a lower bound on time to stock-out. The lower bound provides a "pessimistic" estimate of performance. This is useful because, by purchasing sufficient inventory, the pessimistic bound can be brought to any specified level of system performance. This guarantees that acceptable performance will be achieved, albeit with some excess investment in inventory.

Proposition 12.2 *With probability 1, η, the number of jobs until stock-out, satisfies*

$$\eta_* \leq \eta,$$

where

$$\eta_* = \inf\left\{ k \left| \sum_{j=1}^m \frac{n_j(k)}{R_j} \geq 1 \right. \right\}$$

and R_j denotes the maximum number of consecutive jobs of type j that leads to stock-out on the last job

$$R_j = \min_i \left\{ \frac{s_i + 1}{J_{ij}} \right\}.$$

Proof: Note that our nondegeneracy assumption ensures that $J_{ij} > 0$ for some i, hence R_j is always finite. It is sufficient to show that the event $\{JN(k) < s + 1\}$ contains the event $\{\sum_{j=1}^m \frac{n_j(k)}{R_j} < 1\}$. Suppose that, $\sum_{j=1}^m \frac{n_j(k)}{R_j} < 1$, the definition of R_j then gives

$$\sum_{j=1}^m \frac{1}{\min_i\{\frac{s_i+1}{J_{ij}}\}} n_j(k) < 1$$

Now

$$\left(\min_i \left\{ \frac{s_i + 1}{J_{ij}} \right\} \right)^{-1} = \max_i \left\{ \frac{J_{ij}}{s_i + 1} \right\}$$

Thus

$$\sum_{j=1}^{n} \max_{i} \left\{ \frac{J_{ij}}{s_i + 1} \right\} n_j(k) < 1$$

or

$$\sum_{j=1}^{n} J_{ij} n_j(k) < s_i + 1. \quad i = 1, \ldots, n,$$

Q.E.D.

Proposition 12.2 produces a major simplification because the vector relationship $Jn(k) < s + 1$ is reduced to a one dimensional relationship expressed as a weighted sum of multinomial random variables with marginal distributions given by (12.5). Proposition 2 obtains a pessimistic bound for the probability distribution of the number of jobs that are encountered up to and including stock-out. It is reasonable to expect that η_* will accurately approximate η when there is a high level of part commonality among jobs. Indeed, when each job uses exactly one of every part $\eta_* = \eta$.

In some applications, the probability distribution of the time until stock-out is a more appropriate performance measure. It can be noted in the proof of Proposition 12.2, that all the implications are equally valid if $n_j(k)$ is replaced with $N_j(t)$, the number of jobs of type j arriving by time t. Thus, using arguments analogous to those in Proposition 12.2, we can establish the following lower bound on the probability distribution of the time to stock-out.

Corollary 12.3

$$P\{X(t) \le s\} = P\{\tau > t\} \ge P\{\tau_* > t\}$$

where

$$\tau_* = \inf \left\{ t \,\middle|\, \sum_{j=1}^{m} N_j(t)/R_j \ge 1 \right\}$$

and

$$R_j = \min_{i}\{(s_i + 1)/J_{ij}\}.$$

The situation in Proposition 12.2 and Corollary 12.3 is depicted graphically in Figure 12.1. Suppose there are two possible jobs and two parts. The first type of job involves part 1 only and the second parts 1 and 2. Suppose we stock three of part 1 and two of part 2. The axis correspond to the number of jobs of type 1 and 2 observed. The solid lines depict the constraints $Jx \le s+1$. Thus the lattice points in the interior of this region indicate combinations of jobs which can be repaired from stock

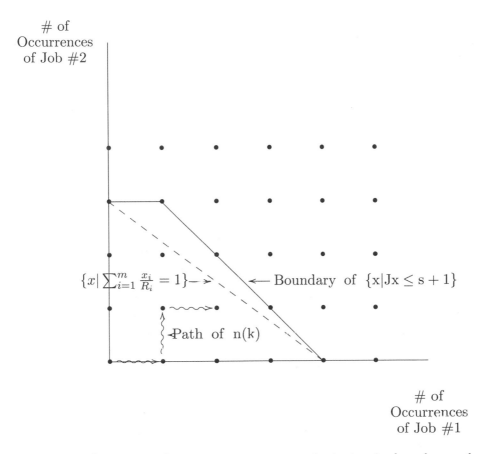

Figure 12.1. Cumulative job occurence process, approximate stopping boundary, and precise stopping boundary

on hand. The dashed line is the set of points such that $\sum_{i=1}^{m} \frac{x_i}{R_i} = 1$. The process $n(k)$ moves from lattice point to lattice point in a random fashion, but always to the northeast. (A typical realization is given by the wiggly line marked with arrows). The proposition states that the $n(k)$ process always crosses the dashed lines before crossing the solid boundary. Proposition 12.2 thus implies that the complementary cumulative distribution of η_* lies to the left of that of η in the sense that $P\{\eta_* > k\} \le P\{\eta > k\}$ all k. Thus the distribution of η_* provides a bound on the distribution of η.

2.4.   Computing the lower bound

Probability of stock-out. The lower bound is of particular interest since it gives a minimum "guaranteed" performance level. Unfortunately the formulae for the lower bounds given by Proposition 12.2 and Corollary 12.3 depend on the distribution of η_*. The distribution of η_* is tedious to calculate exactly, but it is considerably easier to compute than the distribution of η. In this section, we give precise approximations to the lower bounds on stock-out probability and expected time to stock-out. Letting Z_+ denote the non-negative integers, define

$$H_k = \left\{ (\ell \in Z_+^m) : \sum_{j=1}^{m} \ell_j = k, \sum_{j=1}^{m} \frac{\ell_j}{R_j} \leq 1 \right\}.$$

Then

$$P\{\eta_* > k\} = \sum_{\ell \in H_k} \frac{k!}{\ell_1! \ldots \ell_m!} p_1^{\ell_1} \ldots p_m^{\ell_m}. \tag{12.10}$$

In comparing equations (12.10) and (12.6) we note that while the summands are similar, the structure of the set over which the summation takes place is much simpler in equation (12.10).

For large values of k, $P\{\eta_* > k\}$ can be approximated by the tail of a Normal distribution. Since $(n_1(k), \ldots, n_m(k))$ is multinomially distributed, we let

$$\mu(k) = E\left(\sum_{j=1}^{m} \frac{n_j(k)}{R_j} \right) = \sum_{j=1}^{m} \frac{kp_j}{R_j}$$

$$\sigma^2(k) = Var\left(\sum_{j=1}^{m} \frac{n_j(k)}{R_j} \right) = k \sum_{j=1}^{m} (1/R_j)^2 p_j(1 - p_j)$$

$$- 2k \sum_{i=1}^{m-1} \sum_{j=i+1}^{m} (1/R_i)(1/R_j) p_i p_j.$$

Approximating the multinomial by a multivariate Normal with the same mean and covariance matrix yields

$$P\{\eta_* > k\} = P\left\{ \sum_{j=1}^{m} \frac{n_j(k)}{R_j} < 1 \right\} \approx \Phi\left(\frac{1 - \mu(k)}{\sigma(k)} \right),$$

where Φ represents the standard Normal cumulative distribution.

By conditioning on the number of jobs arriving by time t the normal approximation can also be used to compute $P\{\tau_* > t\}$ from Corollary 12.3. These approximations are best when t is such that the $N_j(t)$ are fairly large. This in turn implies that the s_i should be large enough to serve a substantial number of jobs. Thus, the Normal approximations are appropriate when both the stock levels and the number of jobs to be completed before stock-out are large. Exact computations are even less feasible when there are large numbers of jobs to be completed before stock out.

Expected time to stock-out. The expected time to stock-out $E(\tau)$ can be bounded below by using the probability distribution $P\{\tau_* > t\}$. That is,

$$E(\tau_*) = \int_0^\infty P\{\tau_* > t\}dt < \int_0^\infty P\{\tau > t\}dt = E(\tau). \qquad (12.11)$$

Thus $E(\tau_*)$ provides a lower bound on $E(\tau)$. Given the complexity of the analytical expression for the distribution of η_* (and hence τ_*), the integral in (11) may be difficult to calculate. We can, however, develop a simple, precise approximation for $E(\eta_*)$ and hence for $E(\tau_*)$ (since $E(\tau) = E(\eta_*)/\lambda$).

Proposition 12.4 *The pessimistic bound $E(\eta_*)$ on number of jobs until stock-out satisfies*

$$1 \bigg/ \left(\sum_{j=1}^m p_j/R_j \right) \le E(\eta_*) \le [1 + \max_j \{1/R_j\}] \bigg/ \left(\sum_{j=1}^m p_j/R_j \right). \qquad (12.12)$$

Proof: Define W_k by

$$W_k = \sum_{j=1}^m \left(\frac{n_j(k)}{R_j} - \frac{p_j k}{R_j} \right) \qquad k = 0, 1, 2, \ldots$$

W_k is obtained by taking the inner product of the R^m valued process $n(k) - pk$ (define $p = (p_1, \ldots, p_m)^T$) with the m-vector $(1/R_1, \ldots, 1/R_m)$. Note that

$$E(W_{k+1} \mid n(0), \ldots, n(k))$$
$$= E\left(\sum_{j=1}^m \left(\frac{n_j(k+1)}{R_j} - \frac{p_j(k+1)}{R_j} \right) \bigg| n(0), \ldots, n(k) \right).$$

Rewriting $n_j(k+1)$ as $n_j(k+1) - n_j(k) + n_j(k)$, separating terms and using the linearity properties of conditional expectation yields

$$= E\left(\sum_j \frac{n_j(k+1) - n_j(k)}{R_j} - \frac{p_j}{R_j}\Bigg| n(0), \ldots, n(k)\right)$$

$$+ E\left(\sum_j \left(\frac{n_j(k)}{R_j} - \frac{p_j k}{R_j}\right)\Bigg| n(0), \ldots, n(k)\right)$$

$$= \sum_j \left(\frac{E(n_j(k+1) - n_j(k) \mid n(0), \ldots, n(k))}{R_j} - \frac{p_j}{R_j}\right) + W_k$$

$$= E\left(\sum_j \frac{n_j(k+1) - n_j(k)}{R_j} - \frac{p_j}{R_j}\right) + W_k$$

$$= \sum_j \left(\frac{p_j}{R_j} - \frac{p_j}{R_j}\right) + W_k = W_k. \tag{12.13}$$

The second to last inequality follows because the $n_j(k)$ process has stationary, independent increments. In addition, it is easily verified that $E[\|W_k\|]<\infty$. Hence W_k is a martingale with respect to $n(0), \ldots, n(k) \ldots$. Clearly η_* is a stopping time with respect to $n(0), n(1) \ldots$ and since the $n(k)$ process moves only to the "northeast" (increases according to the usual vector partial order) $\eta_* \leq \sum_{i=1}^n (s_i + 1)$. The optional sampling theorem applied to W_k yields

$$0 = E(W_{\eta_*}) = \sum_{j=1}^m \frac{1}{R_j} E(n_j(\eta_*)) - \sum_{j=1}^m E(\eta_*)\frac{p_j}{R_j}. \tag{12.14}$$

By construction

$$1 \leq \sum_{j=1}^m \frac{n_j(\eta_*)}{R_j} \leq 1 + \max_j\{1/R_j\}, \tag{12.15}$$

hence

$$1 \leq \sum_{j=1}^m \frac{p_j}{R_j} E(\eta_*) \leq 1 + \max_j\{1/R_j\},$$

thus

$$L = \left(\sum_{j=1}^m \frac{p_j}{R_j}\right)^{-1} \leq E(\eta_*) \leq \left(\sum_{j=1}^m \frac{p_j}{R_j}\right)^{-1} (1 + \max_j\{1/R_j\}).$$

Q.E.D.

Table 12.1. Summary of bounds obtained

$E(\eta)$	expected number of jobs completed
LOWER	$1/(\sum_{j=1}^{m} p_j/R_j)$
UPPER	$U = \min_i \left\{ \frac{s_i + \max_j \{J_{ij}\}}{\sum_j J_{ij} p_j} \right\}$

$P\{X(t) \le s\} = P\{\text{no stock-out by } t\} = P\{\tau > t\}$	
LOWER	$P\{\sum_{j=1}^{m} N_j(t)/R_j \le 1\}$

$P\{\eta > k\} = P\{\text{at least } k \text{ jobs served}\}$		
LOWER	$P\{\sum_{j=1}^{m} N_j(k)/R_j \le 1\}$	
	$\doteq \Phi([1 - \mu(k)]/\sigma(k))$	for large s, k

---

where:

$R_j = \min_i\{(s_i + 1)/J_{ij}\},$

$\mu(k) = \sum_{j=1}^{m} \frac{kp_j}{R_j}$

$\sigma^2(k) = k \sum_{j=1}^{m}(1/R_j)^2 p_j(1 - p_j) - 2k \sum_{i=1}^{m-1} \sum_{j=i+1}^{m}(1/R_i)(1/R_j)p_i p_j$

Note that in the extreme case in which each job uses exactly one of every part, the bound L is tight. In view of equation (12.4), L/λ is a lower bound for $E(\tau)$. Table 12.1 summarizes the bounds and approximations we have obtained for the performance measures.

3. Two Examples

The rationale behind our upper bound on expected time to stock-out is clear. Ignoring the dependency brought about by the fact that stock-out occurs when the first part stocks out yields an upper bound on expected time to stock-out. Yet it is exactly this sort of reasoning that lies behind most practical ordering decisions. Indeed, few inventory management systems are equipped to detect dependencies among part demands, and concentrate instead on marginal demand rates. The lower bound, on the other hand operates more subtly; it makes a pessimistic estimate of the number of demands of each type that can be satisfied. This approach works best when there is a high degree of part commonality.

We illustrate the importance of lower bound with a specifically chosen, but in our experience typical, repair kit-type example: the part usage

Table 12.2. Part usage matrix for the first example

parts					jobs					
	1	2	3	4	5	6	7	8	9	10
1	1	1	1	0	1	0	1	0	1	1
2	0	1	0	1	0	1	0	1	0	1
3	1	0	1	0	1	0	1	0	1	0
4	0	1	0	1	0	1	0	1	0	1
5	1	0	1	0	1	0	1	0	1	1
6	0	1	0	1	0	1	0	1	0	0
7	0	0	0	0	0	0	1	0	0	0
8	0	0	0	0	0	0	0	1	0	0
9	0	0	0	0	0	0	0	0	1	0
10	0	0	0	0	0	0	0	0	0	1
11	0	0	0	0	1	0	0	0	0	0
12	0	0	0	1	0	0	0	0	0	0
13	1	0	0	0	0	0	0	0	0	0
14	0	1	0	0	0	0	0	0	0	0
15	0	0	0	0	1	0	0	0	0	0
16	0	0	1	0	0	0	0	0	0	0
17	0	0	0	1	0	0	0	0	0	0
18	0	0	0	0	0	1	0	0	0	0
19	0	0	0	0	0	0	1	0	0	0
20	0	0	0	0	0	0	0	0	0	1

matrix given in Table 12.2, the job probabilities are each 0.1 and the job arrival rate is 1.0. If we look only at the marginal part demand rates then the demand rates are 0.7, 0.5, 0.5, 0.5, 0.6, and 0.4 for Parts 1 through 6 and 0.1 for Parts 7 through 20 (items per period). Concentrating only on marginal demand rates would indicate that Parts 1 through 6 should be stocked at a higher level than Parts 7 through 10. To make matters more concrete, suppose that our criterion for system performance is the expected time to stock-out, and we wish to stock so as to achieve an expected time of stock-out of 25 time units. From equation (12.4) this is seen to be equivalent to finding a policy for which the expected number of jobs up to and including stock-out is 25. The expected number of units of item i that will be demanded by a particular arrival to the inventory system is $\xi_i = \sum_{j=1}^{m} J_{ij} p_j$. Were we to consider part i in isolation, the expected number of demands arriving up to the time of stock-out would be $(s_i + 1)/\xi_i$ and the expected time to stock-out $(s_i + 1)/\lambda \xi_i$. For the first item a stock level of 17 units gives an expected time to stock-out just over 25. Reasoning in a similar fashion gives stock levels of 17, 12, 12, 12, 14, and 9 for parts 1 through 6 and a stock level of 2 for parts 7 through 20 (rounding to the nearest integer). Based on marginal reasoning (and in accord with current practice), this policy should give a stock-out time of at least 25 time units[1]. Of course, it falls far short. The lower bound on expected time to stock-out for this policy is 3.0 time units. Monte Carlo simulation[2] reveals an estimated average time to stock-out of approximately 9.7 units. In this case, the lower bound is closer to the true expected time to stock-out than the estimate based on part usage rates. A closer examination of the item usage matrix in Table 12.2 reveals why calculations based on individual part usage rates are so optimistic. Each demand consists of a set of parts common to many demands and a set of parts specific to that demand. The marginal arrival rate of demands for the common parts was is high; however, *no single demand could be satisfied without the necessary special items.* Any stocking policy based purely on marginal demand rates tends, therefore to systematically under stock items with apparent low marginal demand rates.

The lower bound offers an avenue for finding conservative stocking polices. One can easily find a policy that stocks nearly the same total number of items as the policy based on marginal demand rates, but improves the lower bound. This policy will have the effect of giving guaranteed performance. For the example at hand, the stocking policy so obtained stocks 5 of each part and has a lower bound expected time to stock-out of 6.0 time units, and Monte Carlo simulation reveals an average time to stock-out of 7.9 time units. While this performance is not as good as the policy based on marginal usage rates, it has the advantage of predictability: the expected time to stock out is more accurately predicted by the lower bound than for the policy based on the individual part demands.

The problem of deciding how to stock parts is even more difficult when the item stocking decisions must be made subject to resource constraints. For example, it may be required that the total value of parts stocked in a mobile repairman's car fall below some fixed amount, or that the weight or volume of items not exceed some threshold. A single side constrained model for the one job case was posed and solved in Mamer and Shogan [7]. There is no particular reason to stop with a single side constraint. Some situations may be well modeled using a collection of side constraints describing limitations on the weight, value and volume of spare parts allocated to a single repairperson's car, for example. Letting A denote the constraint matrix, our requirement is that for any stocking policy s, $As \leq b$, where b is a vector of resource availabilities.

To restate the problem, we seek a stocking policy which maximizes the expected time to stock-out, subject to constraints on the quantity of each item stocked. A natural extension of the marginal-part-usage-rate reasoning is to find a policy that maximizes the minimum expected time to stock-out for each part considered separately (where part dependencies are ignored) subject to the budget constraint. Written mathematically,

$$\max_{(s_1,\ldots,s_n)} \left\{ \min_i \left\{ \frac{s_i + 1}{\sum_{j=1}^m J_{ij} p_j} \right\} \right\} (1/\lambda)$$

s.t.

$$As \leq b,$$
$$s_i \geq 0 \quad i = 1, \ldots, n.$$

When the entries in the matrix J_{ij} are all either 0 or 1, this optimization amounts to maximizing the upper bound given in (12.8). Rearranged to produce a linear program, this problem becomes

$$\hat{V}_p = \max_{t, s_1, \ldots, s_n} \quad t/\lambda \tag{12.16}$$

$$\text{s.t.} \quad t \leq (s_i + 1) \left(\sum_{j=1}^m J_{ij} p_j \right)^{-1}, \quad i = 1, \ldots, n$$

$$As \leq b$$
$$t \geq 0, \ s_i \geq 0, \quad i = 1, \ldots, n.$$

The difficulty with the marginal usage rate approach, of course, is that it ignores inter-item dependencies. This difficulty is exacerbated when item stocking decisions must be made in the face of resource constraints. An alternative, and considerably more conservative, approach is to find the policy that maximizes the value of the lower bound on the expected time to stock-out. This approach, while potentially far from optimal, yields a policy and a lower bound on the expected time to stock-out. Rendering the problem of maximizing the lower bound into an easily optimized form is straightforward. Let V_L denote the maximum value of the lower bound.

$$\hat{V}_L = \max_{(s_1,\ldots,s_n,R_1,\ldots,R_m)} \left(\sum_{j=1}^m \frac{p_j}{R_j} \right)^{-1} (1/\lambda) \tag{12.17}$$

$$\text{s.t.} \quad As \leq b,$$
$$R_j \leq (s_i + 1)/J_{ij} \quad \text{each } i \text{ and } j \text{ such that } J_{ij} > 0,$$
$$s_i \geq 0, \quad i = 1, \ldots, n,$$

which is equivalent to

$$\left[\min_{(s_1,\ldots,s_n;R_1,\ldots,R_m)} \sum_{j=1}^{m} p_j/R_j \right]^{-1} (1/\lambda) \qquad (12.18)$$

$$\text{s.t. } J_{ij}R_j - s_i \le 1, \quad \text{each } i \text{ and } j \text{ such that } J_{ij} > 0,$$

$$As \le b,$$
$$s_i \ge 0, \quad i = 1,\ldots,n.$$

The optimization in (12.18) has a separable convex objective and linear constraints, and thus may be solved efficiently using a piece-wise linear approximation.

To illustrate this problem we add to the previous expected time to stock-out example a single side constraint on the set of parts that can be placed in inventory. Suppose that each item has a cost h_i, and we have a budget b. Define $h_1 = \cdots = h_6 = 4.0$ and $h_7 - \ldots h_{20} = 0.1$ and $b = 500$. The stock level optimization problems as stated in (12.16) and (12.18) are mixed integer programming problems. However, it will suffice for our purposes to use the solution to the corresponding linear programming relaxations of the problem and "round" the solutions to the nearest integer[3]. Maximizing the minimum of the marginal expected times to stock-out (12.16) and then rounding yields a policy that stocks 27 units of part 1, 19 units of parts 2 through 4, 23 of part 5, 15 of part 6 and 3 of parts 7 through 20. The objective function value associated with this policy is 40.00, the lower bound applied to this policy yields 4.00. A Monte Carlo simulation[4] of the problem yields an estimated expected time to stock-out of 15.5 time units. Optimizing the lower bound yields a policy that stocks 20 of each item, and a lower bound of 21, Monte Carlo simulation using this stocking policy yields an estimated average time to stock-out of 29.6 time units. We see that in this case the policy which maximizes the minimal marginal expected time to stock-out over-invests in the common items at the expense of the specialized items.

Table 12.3 illustrates a situation in which there is little part commonality between jobs, as depicted in the job matrix in Table 12.3, one expects that the lower bound optimization will do no better than the the optimization of the minimum marginal time to stock-out. Note that, in this second example, each part is used in but a single job. The holding costs for the parts are $h_1 = h_2 = 1$, $h_3 = h_4 = 2$, $h_5 = h_6 = 3$, $h_7 = h_8 = 4$, $h_9 = h_{10} = 5,\ldots,h_{19} = h_{20} = 10$. The budget remains 500, as

Table 12.3.   Part usage matrix for the second example

parts					jobs					
	1	2	3	4	5	6	7	8	9	10
1	1	0	0	0	0	0	0	0	0	0
2	0	1	0	0	0	0	0	0	0	0
3	1	0	0	0	0	0	0	0	0	0
4	0	1	0	0	0	0	0	0	0	0
5	0	0	1	0	0	0	0	0	0	0
6	0	0	0	1	0	0	0	0	0	0
7	0	0	1	0	0	0	0	0	0	0
8	0	0	0	1	0	0	0	0	0	0
9	0	0	0	0	1	0	0	0	0	0
10	0	0	0	0	0	1	0	0	0	0
11	0	0	0	0	1	0	0	0	0	0
12	0	0	0	0	0	1	0	0	0	0
13	0	0	0	0	0	0	1	0	0	0
14	0	0	0	0	0	0	0	1	0	0
15	0	0	0	0	0	0	1	0	0	0
16	0	0	0	0	0	0	0	1	0	0
17	0	0	0	0	0	0	0	0	1	0
18	0	0	0	0	0	0	0	0	0	1
19	0	0	0	0	0	0	0	0	1	0
20	0	0	0	0	0	0	0	0	0	1

in the previous example. The marginal usage rate based optimization yields a policy of stocking exactly 5 units of each part and gives an objective function value of 60.0 time units. Monte Carlo simulation reveals an estimated average time to stock-out for this policy of 28.5 time units. The lower bound for this policy is 6.0 time units. The optimization that maximizes the value of the lower bound yields a policy which stocks higher levels of the cheaper parts instead of choosing equal stock levels. This policy stocks 10 of items 1–4, 6 of items 5–8, 5 of items 9–12, 4 of items 13–16 and 3 of items 17–20. The lower bound time to stock-out for this policy is 5.9 time units, a Monte Carlo simulation reveals an average time to stock-out of 23.4 time units. In this case the policy identified by optimizing the lower bound is inferior to the policy obtained from (12.16).

4.   Conclusion

One purpose of this chapter has been to call attention to the importance of the dependence among item demands in a multi-item inventory system. The type of dependence considered in this chapter arises naturally in many multi-item settings, and is characteristic of systems in which demands for different items arrive in groups (as in the repair problems discussed in the introduction). When part or item demands show significant interdependence, the potential error arising from ignoring it can be large. While ignoring the dependence between item demands may lead to serious errors in assessing system performance, exact stock level optimization with item demand dependence seems fraught with difficulty. Evaluation of the exact formulas for system performance for a fixed stocking policy is computationally demanding, even for systems with very modest size. Monte Carlo is effective for the evaluation of individual stocking policies, but is not well suited for optimization because the objective functions are neither concave nor convex and have no simple representation.

Our solution to this problem is to develop easily computable bounds on system performance. In Section 3, these bounds were used as surrogate objective functions to find policies which maximized the performance of the bound, subject to budget constraints. The bounds give rise to relatively simple surrogate optimization problems. Our linear programming optimization of the bounds scales well to larger number of parts, jobs and constraints. Monte Carlo is then used to test the actual performance of the stock levels determined by the optimization.

Our experience with these bounds leaves us cautiously optimistic. Clearly, many interesting and unsolved problems remain in analyzing stocking policies for interdependent multi-item inventory systems. Given the increased interest in multi-item inventories in production, as well as in repair kit inventory problems, we hope that the results obtained in this chapter will stimulate further progress in this research area.

Notes

1. Since we round the stock level to the nearest integer, the expected times to stock out for each of the parts is not exactly the same. In this case the minimal expected time to stock out is 25.0 time units (part 5 and 6).

2. Our estimate is based on the average number of jobs up to an including stock out for a sample of 20,000 simulated stockout cycles. In view of (12.4) and the assumption that the job arrival rate is 1.0 per time unit, this is also an estimate of the expected time to stock out.

3. This procedure does not guarantee that the solutions obtained are either feasible or optimal, however the solutions so obtained will suffice to demonstrate the qualitative properties of the two optimization strategies.

4. Again, we make use of the fact that the job arrival rate is 1.0 and hence by equation (12.4) the expected time to stock out is equal to the expected number of jobs up to and including stockout. We estimate the expected number of jobs up to and including stockout by simulating 20,000 stockout cycles and averaging the total number of jobs per cycle.

References

[1] BAKER, K.R., M.J. MAGAZINE, AND H.L. NUTTLE. (1986) The Effect of Commonality on Safety Stock in a Simple Inventory System. *Management Science*, Vol. 32, No. 8, 982–988.

[2] BAKER, K.R. (1985) Safety Stocks and Component Commonality. *Journal of Operations Management*, Vol. 6, No. 1, 13–21.

[3] BRUMELLE, S. AND D. GRANOT. (1993) The Repair Kit Problem Revisited. *Operations Research*, Vol. 41, No. 5, 994–1006.

[4] COHEN, M.A., P. KLEINDORFER, AND H. LEE. (1989) Near Optimal Service Constrained Policies for Spare Parts. *Operations Research*, Vol. 37, No. 1, 104–117.

[5] GRAVES, S.C. (1982) A Multiple-Item Inventory Model with a Job Completion Criterion. *Management Science*, Vol. 28, No. 11, 1334–1337.

[6] HAUSMAN, W.H. (1982) On Optimal Repair Kits Under a Job Completion Criterion. *Management Science*, Vol. 28, No. 11, 1350–1351.

[7] MAMER, J.W. AND A.W. SHOGAN. (1987) A Constrained Capital Budgeting Problem with Applications fo Repair Kit Selection. *Management Science*, Vol. 33, No. 7, 800–806.

[8] MAMER, J.W. AND S.A. SMITH. (1982) Optimal Field Repair Kits Based on Job Completion Rate. *Management Science*, Vol. 28, No. 11, 1325–1333.

[9] MAMER, J.W. AND S.A. SMITH. (1985) Job Completion Based Inventory Systems: Optimal Policies for Repair Kits and Space Machines. *Management Science*, Vol. 31, No. 6, 703–718.

[10] MAMER, J.W. AND S.A. SMITH. (1988) A Lower Bound Heuristic for Selecting Multi-Item Inventories for a General Job Completion Objective. Manuscript, Santa Clara University.

[11] MARCH, S.T. AND G.D. SCUDDER. (1984) On Optimizing Field Repair Kits Based on Job Completion Rate. *Management Science*, Vol. 30, No. 8, 1025–1028.

[12] NAHMIAS, S. AND C.P. SCHMIDT. (1984) An Efficient Heuristic for the Multi-Item Newsboy Problem with a Single Constraint. *Naval Research Logistics Quarterly*, Vol. 31, 463–474.

[13] SCHAEFER, M.K. (1983) A Multi-Item Maintenance Center Model for Low Demand Repairable Items. *Management Science*, Vol. 29, No. 7, 1062–1068.

[14] SHERBROOKE, C.C. (1971) An Evaluator for the Number of Operationally Ready Aircraft in a Multi Level System. *Operations Research*, Vol. 19, No. 3, 618–635.

[15] SILVER, E.A. (1972) Inventory Allocation Among an Assembly and Its Repairable Sub-Assemblies. *Naval Research Logistics Quarterly*, Vol. 19, No. 2, 261–280.

[16] SONG, J.-S. (1998) On the Order Fill Rate in a Multi-Item, Base-Stock Inventory System. *Operations Research*, Vol. 46, No. 2, 831–845.

[17] SONG, J.-S. (2001) Order-Based Backorders and their Implications in Multi-Item Inventory Systems. Manuscript, p. 25.

[18] SMITH, S.A., J. CHAMBERS, AND E. SHLIFER. (1980) Optimal Inventories Based on Job Completion Rate for Repairs Requiring Multiple Items. *Management Science*, Vol. 26, No. 8, 849–852.

Index